Electronic Instrumentation with MATLAB®

Electronic instrumentation systems are electronic systems for implementing specific desired functions such as measurement, control, and monitoring. This book discusses sensors and conditioning circuits for electronic instrumentation. This book is unique. It is the first time MATLAB® computational powers are being used to describe and explore various aspects of electronic instrumentation, including electronic signal conditioning circuits and instrumentation sensors. In addition, it is the first time an electronic instrumentation system is written to introduce the Internet of Things (IoT) and popular sensors for IoT applications.

- MATLAB is used for computations, plotting, curve fitting, statistical analysis, and parameter identification of instrumentation systems.
- An in-depth presentation of signal-conditioning circuits for instrumentation sensors, such as amplifiers, filters, and analog-to-digital conversion circuits is included.
- There are broad descriptions of the underlying principles of temperature, displacement, pressure, flow, and electro-optical sensors. In addition, this book discusses popular sensors used in IoT applications.
- Several worked examples for solving electronic instrumentation problems by using MATLAB are provided. The examples help the reader understand the operations of signal-conditioning circuits and explore the characteristics of various instrumentation sensors.
- The book includes end-of-chapter problems to test and help solidify the reader's understanding of the material.

This book can be used by students, professional engineers, and technicians who seek knowledge of signal-conditioning circuits used in instrumentation systems, along with a detailed description of instrumentation sensors.

Electronic Instrumentation with MATLAB®

John Okyere Attia

CRC Press
Taylor & Francis Group
Boca Raton London New York

CRC Press is an imprint of the
Taylor & Francis Group, an **informa** business

Designed cover image: Shutterstock

First edition published 2027
by CRC Press
2385 NW Executive Center Drive, Suite 320, Boca Raton FL 33431

and by CRC Press
4 Park Square, Milton Park, Abingdon, Oxon, OX14 4RN

CRC Press is an imprint of Taylor & Francis Group, LLC

ISBN: 9781041286165 (hbk)
ISBN: 9781041286172 (pbk)
ISBN: 9781003765554 (ebk)

DOI: 10.1201/9781003765554

Typeset in Times
by codeMantra

Access the Instructor and Student Resources: www.routledge.com/9781041286165

To my immediate family members (Christine, John II and Angela) for their ceaseless love.

Contents

Preface

Electronic instrumentation systems are electronic systems for implementing specific desired functions such as measurement, control, and monitoring. This instrumentation system will normally have a measured system, an input transducer, signal-conditioning circuits, and an output transducer. The measured system is the physical quantity that is measured by the instrumentation system. It can be the temperature inside a room, the force on a structure, and the flow of a liquid in a pipe. A transducer converts one form of energy to another. Examples of transducers are displacement, force, pressure, temperature, and flow velocity. The electrical output from a transducer is desirable because of the relative ease with which electrical signals can be conditioned, displayed, or stored. Signal-conditioning circuits are used to modify the output signal of the transducer to allow the signal to be in a form acceptable for the output stage. The signal conditioning stage may include circuits for amplification, filtering, and analog-to-digital conversions. Normally, the output of an instrumentation system should be in a form that can easily be perceived by an observer. In this book, a sensor is termed as a device that senses a physical quantity, while a transducer converts energy from one form to another. This book discusses sensors and conditioning circuits for electronic instrumentation.

MATLAB® is primarily a tool for matrix computations. It has numerous functions for data processing and analysis. In addition, MATLAB has a rich set of plotting capabilities, which are integrated into the MATLAB package. Furthermore, since MATLAB is also a programming environment, a user can extend MATLAB's functional capabilities by writing new modules. In this book, MATLAB is used to (i) perform computations associated with instrumentation systems, (ii) perform curve fitting of data obtained from instrumentation sensors, (iii) perform statistical analysis of data obtained from instrumentation systems, (iv) plot relationships between variables associated with electronic instrumentation systems, and (v) identify the parameters that characterize various sensors.

The goals of writing this book are to (i) provide the reader with the understanding of conditioning circuits for electronic instrumentation, (ii) describe the operation of popular sensors of instrumentation systems, (iii) use MATLAB for computation, plotting, obtaining statistical information from instrumentation systems, and identifying parameters of sensors, (iv) provide examples of electronic instrumentation systems, (v) introduce popular sensors used in IoT to the reader, and (vi) empower engineering students and practitioners to explore the characteristics of sensors and electronic conditioning circuits through end-of-chapter problems.

This book is unique. It is the *first* time that MATLAB's computational powers are being used to describe and explore various aspects of electronic instrumentation (electronic signal-conditioning circuits and instrumentation sensors). MATLAB is integrated throughout this book for computations, plotting, curve fitting, statistical analysis, and parameter identification. In addition, it is the *first* time that an electronic instrumentation book is written that discusses the popular sensors of the IoT applications.

Figure slides are available on the Routledge website (www.routledge.com/9781041286165).

AUDIENCE

This book can be used by students, professional engineers, and technicians who want knowledge of signal-conditioning circuits used in instrumentation systems and the underlying principles of instrumentation sensors. Chapter 1 covers an introduction to instrumentation systems, and it will be useful to all engineering students and practitioners. Chapters 2–5 discuss signal-conditioning circuits that are applicable to instrumentation systems. The electronic circuits form the backbone of instrumentation systems, and they might be of interest to those who want to understand the electronic circuits for instrumentation. Chapters 6–11 provide broad descriptions of the underlying principles of popular instrumentation sensors. The chapters may be studied by students and practicing engineers who want to study instrumentation sensors.

ORGANIZATION

In Chapter 1, system approaches to instrumentation are presented. Common errors in instrumentation systems are discussed, and methods of reducing those common errors are provided. The differences between accuracy, precision, and resolution are discussed.

Chapters 2–5 cover the signal-conditioning circuits. Chapter 2 presents the various amplifiers that are used in instrumentation systems. These include inverting, non-inverting, differential, and instrumentation amplifiers. The features, advantages, and limitations of the amplifiers are provided. The design of both passive and active filters is provided in Chapter 3. The advantages and limitations of active filters are discussed. Chapter 4 describes the circuits that can be used to perform various mathematical operations on signals from transducers. The circuits include integrators, differentiators, weighted summer, limiters, Schmidt trigger, and peak detector. Chapter 5 discusses analog-to-digital and digital-to-analog conversion circuits. For digital-to-analog conversion, the following techniques are described: (i) weighted-resistance D/A converter and (ii) R-2R Ladder D/A converter. For analog-to-digital conversion, the following methods are discussed: (i) successive approximation converter, (ii) flash A/D converter, (iii) dual-slope A/D converter, and (iv) sigma–delta converter.

Chapters 6–10 provide broad descriptions of the underlying principles of temperature, displacement, pressure, flow, and electro-optical sensors. The temperature sensors that are described in Chapter 6 include the following: (i) pn junction thermometer, (ii) integrated-circuit temperature sensor, (iii) resistance temperature detector, (iv) thermistors, and (v) thermocouples. Chapter 7 discusses displacement sensors realized through resistive, capacitive, inductive, and piezoelectric methods. Potentiometers and strain gages are described for displacement measurement. Various forms of capacitive transducers for displacement measurement are shown. In addition, the linear-variable differential transformer is described as an example of an inductive displacement transducer. Furthermore, piezoelectric transducers are discussed for measuring displacement. Pressure and force sensors are discussed in Chapter 8. The differences between absolute, gage, and differential pressures are presented. The following pressure sensors are discussed: (i) capacitive, (ii) strain gage, and (iii) displacement pressure sensors. The chapter ends with a discussion of load cells for force measurement. Chapter 9 describes the differences between mass flow rate, volume flow rate, and average flow rate. In addition, the following sensors are described: (i) thermal flow meters, (ii) electromagnetic flowmeter, (iii) ultrasonic flow sensors (transit-time and Doppler flow meters), and (iv) differential pressure sensors using Venturi Tube, Orifice Plate, and Pitot Tube). Chapter 10 discusses photon sensors with detectors such as the following: (i) photoconductive cells, (ii) photovoltaic cells, (iii) photo-emissive cells, (iv) photodiodes, and (v) phototransistors. In addition, Chapter 10 discusses thermal detectors such as the following: (i) bolometers, (ii) thermopiles, and (iii) pyroelectric devices.

Chapter 11 discusses sensors for IoTs. A block diagram of a typical IoT application is provided. The following popular sensors for IoT applications are discussed: (i) accelerometer, (ii) barometric pressure, (iii) force, (iv) gas, (v) Hall effect, (vi) light, (vii) magnetic field, (viii) PIR motion, (ix) rainfall, (x) capacitive soil moisture, (xi) sound, (xii) temperature, (xiii) touch, (xiv) ultrasonic, (xv) vibration, and (xvi) water flow.

The end-of-chapter problems are available to give the reader additional exercises for the exploration of sensors and signal-conditioning circuits of instrumentation systems. A significant number of problems require the use of MATLAB for solving. In addition, each chapter has its own extensive bibliography to allow the reader to explore further the topics covered in this book.

Acknowledgments

I thank my professors who introduced me to biomedical and industrial instrumentation systems. I also thank my students who offered constructive criticism of the content of this book when they took my course in electronic instrumentation systems.

I especially thank Nora Konopka, editorial director at CRC Press/Taylor & Francis Group, and Sifat Kaur Keer at the CRC Press for their support in getting this book to the market.

Author

John Okyere Attia is Professor Emeritus of Electrical and Computer Engineering at Prairie View A&M University. He was the department head of the Electrical and Computer Engineering Department from 1997 to 2013. He earned a Ph.D. in electrical engineering from the University of Houston, an M.S. from the University of Toronto, and a B.S. from the Kwame Nkrumah University of Science and Technology.

He has taught graduate and undergraduate courses in electrical and computer engineering in the fields of electronics, circuit analysis, electronic instrumentation systems, digital signal processing, and VLSI & ULSI design. His research interests include hands-on learning, power electronics for microgrids, innovative electronic circuit designs for radiation environments, and radiation testing methodologies of electronic devices and circuits.

He has worked on projects funded by the National Science Foundation, NASA, Texas Higher Education Coordinating Board, the Texas Workforce Commission, and Apple Inc. Attia has more than 100 technical publications. He has published refereed journal papers and refereed conference papers. In addition, he is the author of six books published by the CRC Press. His most recent books are *Electronics and Circuit Analysis Using MATLAB®*, 2nd Edition, *PSPICE and MATLAB® for Electronics: An Integrated Approach,* 2nd Edition, and *Circuits and Electronics: Hands-on Learning with Analog Discovery.*

He has received several honors, including the Most Outstanding Senior in the Graduating Class, the Outstanding Teacher Award, Exemplary University Achiever, and the University Leader Award. He is a member of the following honor societies: Sigma Xi, Tau Beta Pi, Kappa Alpha Kappa, and Eta Kappa Nu. In addition, he is an ABET program evaluator for electrical and computer engineering programs. Furthermore, he is a senior member of the Institute of Electrical and Electronics Engineers and a registered professional engineer in the State of Texas.

1 Introduction

1.1 GENERAL INSTRUMENTATION SYSTEMS

Instrumentation systems are, in general, electrical, mechanical, acoustical, and electronic systems for implementing specific desired functions, including measurement, control, and monitoring. Figure 1.1 shows a block diagram of a generalized instrumentation system. Every instrumentation system has at least some of the functional components shown in the figure. The transducer converts energy or information from the measured system into another form of energy. The output signal from the transducer is processed and displayed for viewing.

The measurand or the measured system is the physical quantity that is measured by the instrumentation system. It can be the temperature inside a room, the force on a structure, or the flow velocity of a river.

The transducer is a device that converts one form of energy to another. Examples of transducers include displacement, force, pressure, temperature, and flow velocity sensors. Electrical output from the transducer is desirable because of the relative ease at which electrical signals can be conditioned, displayed, or stored. In addition, electrical signals can be filtered, amplified, modulated, demodulated, transmitted, recorded, and analyzed.

Signal conditioning circuits are used to modify the output signal of the transducer to allow the signal to be in a form acceptable for the output stage. The signal conditioning stage may include circuits for amplification, filtering, impedance matching, and analog-to-digital and digital-to-analog conversions.

The output of an instrumentation system should be in a form that can easily be perceived by an observer and may be numerical, graphical, discrete, and continuous. Although most visual displays are most common, some outputs may be perceived by other senses. For example, Doppler ultrasound signal is best heard by our auditory senses.

Feedback and control signals may be required to monitor and adjust the output of transducer or to display output signals through signal conditioning circuits. To examine the data that precede some critical conditions, data may be stored for offline processing. Communication system techniques may be employed to transmit data to remote locations.

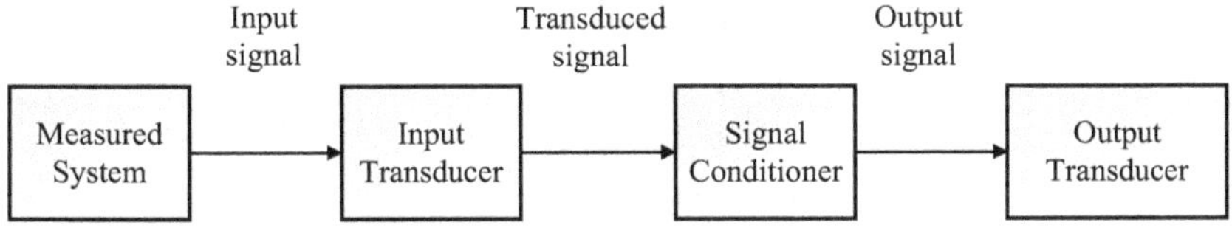

FIGURE 1.1 Block diagram of a generalized instrumentation system.

DOI: 10.1201/9781003765554-1

1.2 SIGNAL CHARACTERIZATION

There are at least four types of signals that need to be processed in instrumentation systems:

(i) periodic signals, (ii) transient signals, (iii) stochastic signals, and (iv) sampled signals

In this section, we will discuss some of the features of the above mentioned signals.

1.2.1 Periodic Signals

Periodic signals occur in nature, and they can also be generated by electronic circuits. Examples of periodic signals are square waveform, triangular waveform, the oscillation of a pendulum, and frequency of the power supply voltage supplied to households. Figure 1.2 shows a periodic signal.

Periodic signals can be expanded into infinite series of harmonics through the Fourier series expansion. If a function $g(t)$ is periodic with period T_p, i.e.

$$g(t) = g(t \pm T_p) \tag{1.1}$$

and in any finite interval $g(t)$ has at the most a finite number of discontinuities and a finite number of maxima and minima (Dirichlet's conditions), then the integral of $g(t)$ over one period will converge. That is,

$$\int_0^{T_p} g(t)\,dt < \infty \tag{1.2}$$

The function $g(t)$ can be expressed with a series of sinusoids. That is,

$$g(t) = \frac{a_0}{2} + \sum_{n=1}^{\infty} a_n \cos(nw_0 t) + b_n \sin(nw_0 t) \tag{1.3}$$

where

$$w_0 = \frac{2\pi}{T_p} \tag{1.4}$$

and the Fourier coefficients a_n and b_n are determined by the following equations.

FIGURE 1.2 A periodic rectangular wave signal.

$$a_n = \frac{2}{T_p}\int_{t_o}^{t_o+T_p} g(t)\cos(nw_0 t)dt \qquad n = 0, 1, 2, \ldots \tag{1.5}$$

$$b_n = \frac{2}{T_p}\int_{t_o}^{t_o+T_p} g(t)\sin(nw_0 t)dt \qquad n = 0, 1, 2, \ldots \tag{1.6}$$

Equation (1.3) can be rewritten as

$$g(t) = \frac{a_0}{2} + \sum_{n=1}^{\infty} A_n \cos(nw_0 t + \Theta_n) \tag{1.7}$$

where

$$A_n = \sqrt{a_n^2 + b_n^2} \tag{1.8}$$

and

$$\Theta_n = -\tan^{-1}\left(\frac{b_n}{a_n}\right) \tag{1.9}$$

Although the Fourier series with an infinite number of terms can exactly represent a periodic signal, in practice, only a finite number of terms have to be used to approximate a given waveform. The number of terms included in the Fourier series will depend on the nature of the waveform and the accuracy desired. In general, the more abrupt and discontinuous a waveform is, the greater the number of terms that must be used to obtain a reasonable approximation to the periodic waveform.

Example 1.1: Fourier Series Expansion of a Square Wave

Using Fourier series expansion, a square wave with a period of 1 ms, a peak-to-peak value of 3 V, and an average value of 0 V can be expressed as

$$g(t) = \frac{6}{\pi}\sum_{n=1}^{\infty} \frac{1}{(2n-1)}\sin[(2n-1)2\pi f_0 t] \tag{1.10}$$

where

$f_o = 1000$ Hz

if $a(t)$ is given as

$$a(t) = \frac{6}{\pi}\sum_{n=1}^{12} \frac{1}{(2n-1)}\sin[(2n-1)2\pi f_0 t] \tag{1.11}$$

Write a MATLAB program to plot $a(t)$ from 0 to 4 ms. at intervals of 0.05 ms. and show that $a(t)$ is a good approximation of $g(t)$.

Solution

MATLAB® CODE

```
% Fourier series expansion
f = 1000; c = 4/pi; dt = 5.0e-05;
tpts = (4.0e-3/5.0e-5) + 1;
for n = 1: 12
for m = 1: tpts
s1(n,m) = (6/pi)*(1/(2*n - 1))*sin((2*n - 1)*2*pi*f*dt*(m-1));
end
end
for m = 1:tpts
    s2 = s1(:,m);
s3(m) = sum(s2);
end
f1 = s3';
t = 0.0:5.0e-5:4.0e-3;
plot(t,f1)
xlabel('Time, s')
ylabel('Amplitude, V')
title('Fourier Series Expansion')
```

Figure 1.3 shows the plot of $a(t)$. It can be seen from Figure 1.3 that $a(t)$ is a good approximation to the square wave $g(t)$.

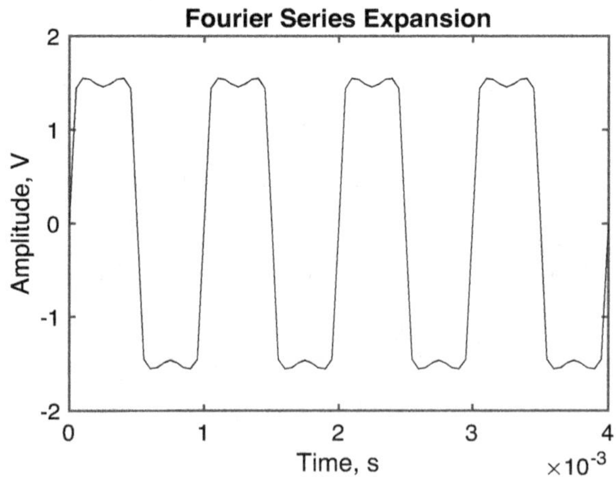

FIGURE 1.3 An approximation to a square wave.

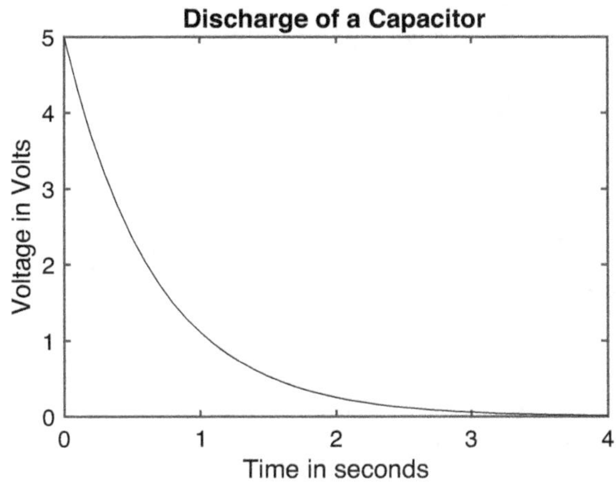

FIGURE 1.4 A discharging of a capacitor.

1.2.2 Transient Signals

A transient signal occurs during a short duration of time and may be described by an explicit mathematical expression. In the time domain, a transient signal may also be characterized by its zero crossings, peak value, average value, and the root-mean-squared (rms) value. An example of a transient signal is the signal picked up by microphones from reflections at various strata of the earth when strong impact excitation (explosion) is directed to specific areas of the earth during well-logging operations. Another example of a transient signal is the voltage across the discharging capacitor of a resistor–capacitor (RC) circuit. For a parallel RC circuit, if V_m is the initial voltage across the capacitor, then the voltage across a discharging capacitor is given as

$$v_0(t) = V_m e^{-\left(\frac{t}{\mathrm{CR}}\right)} \tag{1.12}$$

where

CR is the time constant

Figure 1.4 shows the discharging of a capacitor.

1.2.3 Stochastic Signals

Stochastic or random signals are those signals that cannot be described by an explicit mathematical expression. Stochastic signals may be described statistically by using mean squared value, probability density functions, autocorrelation, and power spectral density functions. Examples of stochastic signals are (i) the electrical activity of the brain measured at the human cortex (electroencephalogram) and (ii) noise signals in electronic systems. Figure 1.5 shows a noisy time-domain signal.

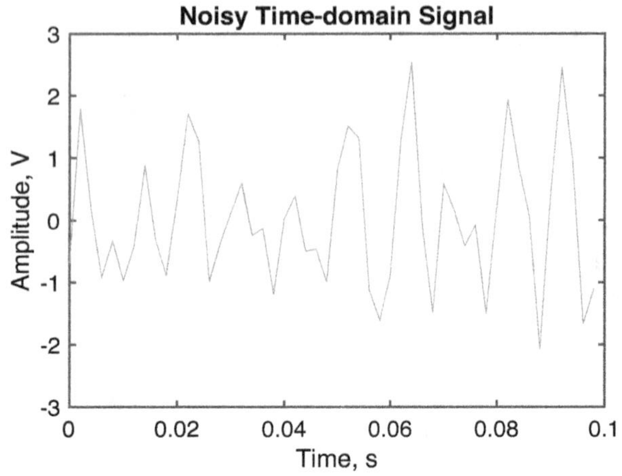

FIGURE 1.5 A noisy time-domain signal.

The mean squared value or the normalized power represents the general intensity of the random signal. For a stochastic signal $g(t)$, the mean squared value is given as

$$\overline{g(t)} = \lim_{T\to\infty} \frac{1}{T} \int_{-T/2}^{T/2} g^2(t)\,dt \tag{1.13}$$

The probability density function of a signal describes the probability that the random signal will take a value within some specified range at any instant of time. The probability density function may be Gaussian, Weibull, or chi-square distributions.

The autocorrelation function describes the similarity of the signal with itself when shifted by time. The autocorrelation function of a signal $g(t)$ is given as

$$R_g(\tau) = \lim_{T\to\infty} \frac{1}{T} \int_{-T/2}^{T/2} g(t)g(t+\tau)\,dt \tag{1.14}$$

The power spectral density describes the frequency decomposition of a signal in terms of the spectral density of its mean squared value. For a random signal $g(t)$, the power spectral density is given as

$$S(f) = \lim_{x\to\infty} \frac{1}{T} \left|G_T(f)\right|^2 \tag{1.15}$$

where

$$G_T(f) = \int_{-\infty}^{\infty} g(t)\exp(-j2\pi ft)\,dt \tag{1.16}$$

and

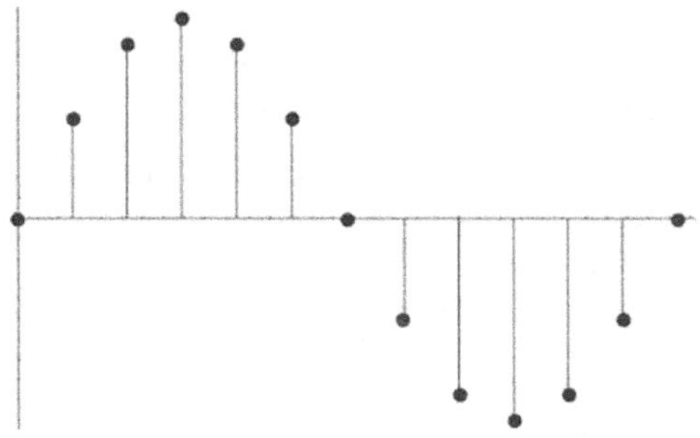

FIGURE 1.6 Sampled sinusoidal signal.

$$g_T(t) = g(t) \quad -T/2 < t < T/2$$
$$= 0 \quad \text{elsewhere} \tag{1.17}$$

The autocorrelation function and the power spectral density are Fourier transform pairs. A stochastic signal has a continuous spectrum. However, periodic signals have a discrete spectrum.

1.2.4 Sampled and Digital Signals

Sampled signals are those signals that are defined only at discrete time points. The amplitude might take any value. For the discrete signal to have the same information as the analog signal, the analog signal has to be sampled at the rate which is more than twice the highest occurring frequency in the analog signal according to the Nyquist sampling theorem, which will be discussed in Chapter 5. Figure 1.6 shows a sampled signal.

If the amplitude of the sampled signal is quantized, then the sampled signal becomes a digital signal. It should be noted that there are errors involved in converting analog signals into digital ones. The amplitude of a sampled signal has to be represented by a finite number of states or levels. The more the levels used in the conversion, the less the error in approximation of the quantized levels with the actual signal. Sampling allows several signals to be transmitted through a channel. This is called time-division multiplexing. Sampled and digital signals will be discussed in detail in Chapter 5. The following section describes some MATLAB functions that may be used to characterize signals.

1.3 MATLAB® STATISTICAL FUNCTIONS

In MATLAB®, data analyses are performed on column-oriented matrices. Different variables are stored in individual column cells, and each row represents different observations of each variable. A data set consisting of ten samples in four variables would be stored in a 10-by-4 matrix. Functions act on the elements in the column. Table 1.1 gives a brief description of a few MATLAB functions for performing statistical analysis.

TABLE 1.1
Statistical Analysis Functions

Function	Description
`corrcoef(x)`	Determines correlation coefficients.
`hist(x)`	Draws the histogram or the bar chart of x.
`max(x)`	Obtains the largest value of x. If x is a matrix, max(x) returns a row vector containing the maximum elements of each column.
`[y, k]=max(x)`	Obtains the maximum value of x and the corresponding locations (indices) of the first maximum value for each column of x.
`mean(x)`	Determines the mean or the average value of the elements in the vector. If x is a matrix, mean(x) returns a row vector that contains the mean value of each column.
`median(x)`	Finds the median value of the elements in the vector x. If x is a matrix, this function returns a row vector containing the median value of each column.
`min(x)`	Finds the smallest value of x. If x is a matrix, min(x) returns a row vector containing the minimum values from each column.
`[y, k]=min(x)`	Obtains the smallest value of x and the corresponding locations (indices) of the first minimum value from each column of x.
`sort(x)`	Sort the rows of a matrix a in ascending order
`std(x)`	Calculates and returns the standard deviation of x if it is a one-dimensional array. If x is a matrix, a row vector containing the standard deviation of each column is computed and returned.

Example 1.2: Statistics of Resistors in a Bin

The measurement of the dc current gain of transistor 2N2222 operating at a particular current were taken in a laboratory. The values were 76, 78, 79, 74, 75, 74, 80, 77, 81, and 75. Determine (a) the average value and (b) the standard deviation of the dc current gain of the transistor.

Solution

MATLAB® CODE

```
%  This program computes the mean,
% and standard deviation of transistor gains
% the data is stored in matrix b
b = [76, 78, 79, 74, 75, 74, 80, 77, 81, 75];
%
% Calculate the mean
mean_b = mean(b);
% Calculate the standard deviation
std_b = std(b);
% Print out the results
```

```
fprintf('Statistics of Transistor Current Gain Measured in
Lab\n\n')
fprintf('Mean of transistor current gain = %7.3e \n', mean_b)
fprintf('Standard Deviation of transistor current gain =
%7.3e\n', std_b)
```

MATLAB results are as follows:

Statistics of Transistor Current Gain Measured in Lab
Mean of transistor current gain = 7.690e+01
Standard Deviation of transistor current gain = 2.514e+00

1.4 SYSTEM CHARACTERIZATION

1.4.1 System Equation

The instrumentation system is an arrangement of circuits or physical components connected in such a manner that will allow a system to perform some desired operations. In general, systems can be classified as linear and nonlinear. For linear systems, superposition can be applied. The following differential equation is a general mathematical model that encompasses a wide range of systems:

$$a_n \frac{d^n y(t)}{dt^n} + a_{n-1} \frac{d^{n-1} y(t)}{dt^{n-1}} + \cdots + a_1 \frac{dy(t)}{dt} + a_0 y(t) = b_m \frac{d^m x(t)}{dt^m} + b_{m-1} \frac{d^{m-1} x(t)}{dt^{m-1}} + \cdots + b_1 \frac{dx(t)}{dt} + b_0 x(t) \tag{1.18}$$

where $a_n, a_{n-1}, \ldots, a_0, b_m, b_{m-1}, \ldots b_0$ are real constants, $x(t)$ is the input, and $y(t)$ is the output.

Equation (1.18) is applicable to physical systems with the extra condition that $x = 0$ for $t < t_0$ and $y(t) = 0$ for all $t < t_0$. The differential Equation (1.18) corresponds to a system of nth order since it is described by an equation of the highest-order derivative of response $\frac{d^n y(t)}{dt^n}$.

Equation (1.18) represents a linear system. A linear system normally results if the characteristics of the components in the system do not change with respect to the magnitude of the excitation applied to the system. For most systems, the characteristics of the component will change if the magnitude of the excitation is large enough. When we speak of a linear system, we are implying that, for normal magnitudes of excitation, the system characteristics do not change significantly with respect to the excitation.

In Equation (1.18), if the coefficients of the differential equation are a function of time, then it represents a time-varying system. A time-varying system results when any of the physical components or configuration of the system changes with time. Except for special cases, a closed-form analytic solution is not possible for differential equations with time-varying coefficients.

If the differential equation has constant coefficients, then it will represent a fixed time-invariant or stationary system. The latter system is obtained when the configuration and the physical components of the system do not change with time.

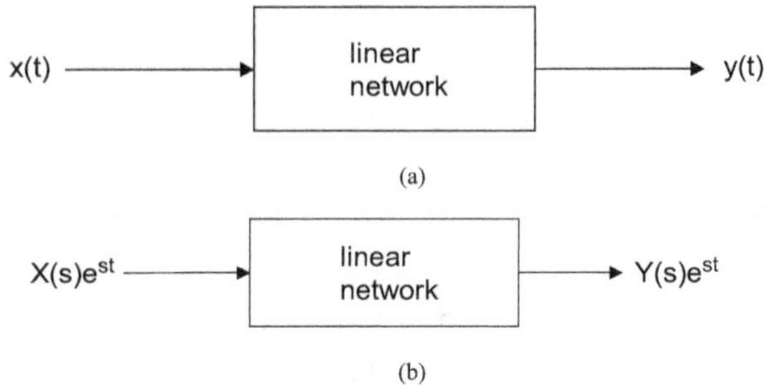

FIGURE 1.7 Linear network representation: (a) time domain; (b) s- domain.

Figure 1.7a shows a linear network with input $x(t)$ and output $y(t)$. Its complex frequency representation is also shown in Figure 1.7b.

If $x(t) = X(s)e^{st}$, then the output must have the form $y(t) = Y(s)e^{st}$, where $X(s)$ and $Y(s)$ are phasor representations of $x(t)$ and $y(t)$, respectively. From Equation (1.18), we have

$$(a_n s^n + a_{n-1}s^{n-1} + \cdots + a_1 s + a_0)Y(s)e^{st} = (b_m s^m + b_{m-1}s^{m-1} + \cdots + b_1 s + b_0)X(s)e^{st} \tag{1.19}$$

and the network function is

$$H(s) = \frac{Y(s)}{X(s)} = \frac{b_m s^m + b_{m-1}s^{m-1} + \cdots b_1 s + b_0}{a_n s^n + a_{n-1}s^{n-1} + \cdots a_1 s + a_0} \tag{1.20}$$

The network function can be rewritten in factored form as

$$H(s) = \frac{k(s-z_1)(s-z_2)\cdots(s-z_m)}{(s-p_1)(s-p_2)\cdots(s-p_n)} \tag{1.21}$$

where

k is a constant
$z_1, z_2, \ldots, z_m$ are zeros of the network function.
$p_1, p_2, \ldots, p_n$ are poles of the network function.

The network function can also be expanded using partial fractions as

$$H(s) = \frac{r_1}{s-p_1} + \frac{r_2}{s-p_2} + \cdots + \frac{r_n}{s-p_n} + k(s) \tag{1.22}$$

Frequency response is the response of a network to a sinusoidal input signal. If we substitute $s = jw$ in the general network function, $H(s)$, we get

$$H(s)\big|_{s=jw} = M(w)\angle\theta(w) \tag{1.23}$$

where

$$M(w) = |H(jw)| \tag{1.24}$$

and

$$\theta(w) = \angle H(jw) \tag{1.25}$$

The plot of $M(\omega)$ versus ω is the magnitude characteristics or response. Also, the plot of $\theta(w)$ versus ω is the phase response.

The poles and zeros of a transfer function can be found by the MATLAB function **roots**. The magnitude and phase characteristics can be obtained using MATLAB function **freqs**. The MATLAB functions **roots** and **freqs** are described in the following sections.

1.4.2 MATLAB® Function roots

The MATLAB function **roots** determines the roots of a polynomial. The general form of the roots function is

$$r = \text{roots}(p) \tag{1.26}$$

where

p is a vector containing the coefficients of the polynomial in descending order
r is a column vector containing the roots of the polynomials

For example, given the polynomial

$$f(x) = x^3 + 9x^2 + 23x + 15,$$

the commands to compute and print out the roots of $f(x)$ are

```
p  = [1  9  23  15]
r  = roots (p)
```

and the values printed are

```
r  =
        -1.0000
        -3.0000
        -5.0000
```

Example 1.3: Roots and Zeros of a Transfer Function

For the system with network function

$$H(s) = \frac{s^2 + 14s + 20}{s^3 + 12s^2 + 41s + 30}$$

find the poles and zeros of $H(s)$.

Solution

MATLAB® CODE

```
% Solution to Example 1.3
% Poles and zeros determination
% num contains the coefficients of numerator polynomial
% den contains the coefficients of denumerator polynomials
num = [1     14    20];
den = [1     12    41       30];
disp('the zeros are')
z = roots(num)
disp('the poles are')
p = roots(den)
The zeros are
z =
  -12.3852
   -1.6148
The poles are
p =
   -6.0000
   -5.0000
   -1.0000
```

1.4.3 MATLAB® Function freqs

MATLAB function **freqs** is used to obtain the frequency response of transfer function $H(s)$. The general form of the frequency function is

$$hs = \text{freqs}\left(\text{num}, \text{den}, \text{range}\right) \tag{1.27}$$

where

$$H(s) = \frac{Y(s)}{X(s)} = \frac{b_m s^m + b_{m-1} s^{m-1} + \cdots b_1 s + b_0}{a_n s^n + a_{n-1} s^{n-1} + \cdots a_1 s + a_0} \tag{1.28}$$

$$\text{num} = \left[b_m \; b_{m-1} \ldots b_1 \; b_0 \right] \tag{1.29}$$

$$\text{den} = \left[a_n \; a_{n-1} \ldots a_1 \; a_0 \right] \tag{1.30}$$

range is the range of frequencies for the case
hs is the frequency response (in complex number form)

Example 1.4: Magnitude Characteristic of a Transfer Function

The transfer function of a system is as follows:

$$H(s) = \frac{s^2 + 1.51x10^6}{s^2 + 3.02x10^6 s + 4.53x10^6} \tag{1.31}$$

Use MATLAB to plot the magnitude response.

Solution

MATLAB® CODE

```
num = [1 0 1.51e6];
den = [1 3.02e6 4.53e6];
w = logspace(-2, 8);
 h = freqs(num, den, w);
 f = w/(2*pi);
mag = 20*log10(abs(h));
semilogx(f, mag)
title('Magnitude Response')
xlabel('Frequency, Hz')
ylabel('Gain, dB')
```

The magnitude response is shown in Figure 1.8.

1.5 SYSTEM SPECIFICATION

In general, a system can be specified in terms of input and output parameters. Some terms used for system specification include resolution, accuracy, precision, sensitivity, linearity, dynamic range, and errors. They will be described in the following sub-sections.

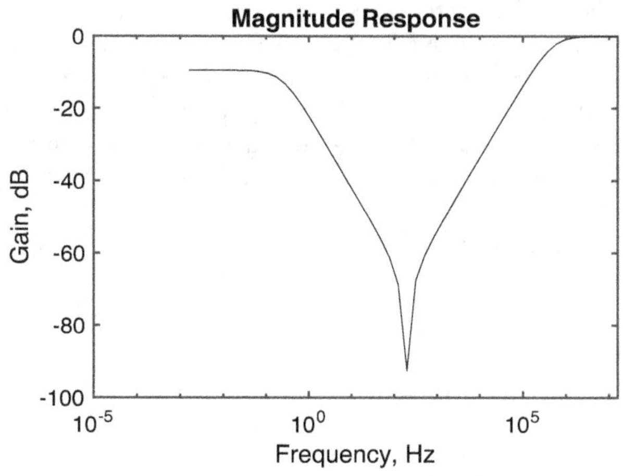

FIGURE 1.8 Frequency response of a transfer function of Example 1.4.

1.5.1 Resolution

Resolution is the smallest change in the input signal that will initiate the smallest distinguishable reading at the output of a system. For example, a sensor may be said to have a resolution of 0.1°C if an increment of 0.1°C produces a readily measurable output.

1.5.2 Accuracy

Accuracy is a measure of how close the measured value of a variable is close to the true value of the variable (quantity) being measured. The accuracy error is the difference between the measured quantity and its true value. Thus,

$$\text{Accuracy Error} = \text{Measured value} - \text{True value} \tag{1.28}$$

Accuracy may be quoted as plus or minus some value of a variable. For example, a resistor with a value of $1000 \pm 5\%\ \Omega$ will have measured values between 950 and 1050 Ω. Another method of representing accuracy is to quote it as a full-scale deflection (f.s.d) of the instrument. For example, a voltmeter may have accuracy quoted as $\pm 1\%\ f.s.d$. This means that the accuracy of the reading of the voltmeter when used for any reading within the range of 0–10 V is ±1% of 10 V, which is equal to ±0.1 V.

1.5.3 Precision

Precision is a measure of a system or an instrument to reproduce its readings of the same input signal and under the same conditions. Precision implies the repeatability of the set of readings. It should be noted that precision does not imply accuracy. The difference between precision and accuracy is illustrated by a series of the targets shown in Figure 1.9. If the arrow on the line represents the "true value," then the *X*'s shown in Figure 1.9a make the data to be neither precise nor accurate. The *X*'s shown in Figure 1.9b make the data to be precise (reproducible) but not accurate (far from arrow on the line). The *X*'s shown in Figure 1.9c make the data have both precision and accuracy.

It should be noted that an instrument that has not been calibrated or is partially defective might give results which are highly repeatable from a series of measurements (it is precise), but the results might be far from the true value (not accurate). However, an instrument that is accurate implies that the readings are repeatable; thus, accuracy implies precision.

Precision is estimated by the deviation of a reading from mean. It can be expressed mathematically as

$$\text{precision} = 1 - \left| \frac{x_i - \overline{x}}{x_i} \right| \tag{1.29}$$

where

x_i is the value of the *i*th measurement
$\overline{x}$ is the mean value of N measurements, where

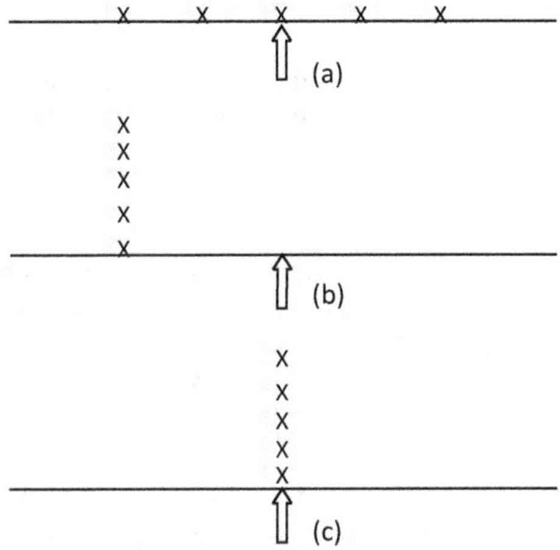

FIGURE 1.9 (a) Low precision and low accuracy, (b) high precision and low accuracy, and (c) high precision and high accuracy.

$$\bar{x} = \frac{1}{N}\sum_{i=1}^{N} x_i \tag{1.30}$$

Example 1.5: Precision of a Measurement

Six voltage measurements taken from a circuit with a voltmeter are 98.2, 98.6, 98.4, 98.3, 98.5, and 98.4 V. If the true value is 100 V, (a) determine the mean value of the measurements, (b) Are the measurements precise? (c) Are the measurements accurate?

Solution

a. Using Equation (1.30), the mean value is given as follows:

$$\bar{x} = \frac{98.2+98.6+98.4+98.3+98.5+98.4}{6} = 98.4 \text{ V}$$

b. Using Equation (1.29), we have

x_i	$\bar{x}$	$\bar{x}-x_i$	$\lvert x_i-\bar{x}\rvert$	$\left\lvert\frac{x_i-\bar{x}}{x_i}\right\rvert$	$1-\left\lvert\frac{x_i-\bar{x}}{x_i}\right\rvert$
98.2	98.4	−0.2	0.2	0.002037	0.99796
98.6	98.4	0.2	0.2	0.002028	0.99797
98.4	98.4	0	0	0.000000	1.00000
98.3	98.4	−0.1	0.1	0.001017	0.99898
98.5	98.4	0.1	0.1	0.001015	0.99898
98.4	98.4	0	0	0.000000	1.00000

Thus, the precision values are from 0.99 to 1.0. The measurements are precise.

c. However, since the mean value of the measurements (98.4 V) is far different from the true value of 100 V, the measurements are not accurate.

1.5.4 Sensitivity

For a system with input variable $x(t)$and the output $y(t)$, the sensitivity is defined as the ratio of the change in $y(t)$ caused by the change in $x(t)$. That is:

$$\text{Sensitivity, } S = \frac{\Delta y(t)}{\Delta x(t)} \tag{1.31}$$

where $\Delta y(t)$ and $\Delta x(t)$ are changes in $y(t)$ and $x(t)$, respectively.

If $x(t)$ and $y(t)$ are linearly related, then the sensitivity is proportional to the slope of the straight-line relation between $x(t)$ and $y(t)$.

1.5.5 Linearity

A system exhibits a linear behavior if the output variable is directly proportional to the input variable. A system would be nonlinear if the sensitivity is a function of $y(t)$ and $x(t)$. The nonlinearity of a system can be defined as the maximum deviation of a specified curve from the least-squares straight line, expressed as a percentage of the full-scale input. Examples of static nonlinearity characteristics are depicted in (i) transistor and diode characteristics and (ii) hysteresis curves. Figure 1.10 shows the current versus voltage of silicon diode.

Diodes and transistors have nonlinear characteristics. If they are operated in a small region of their characteristics, the devices can be considered linear. However, if the region of operation is large, they exhibit nonlinear characteristics.

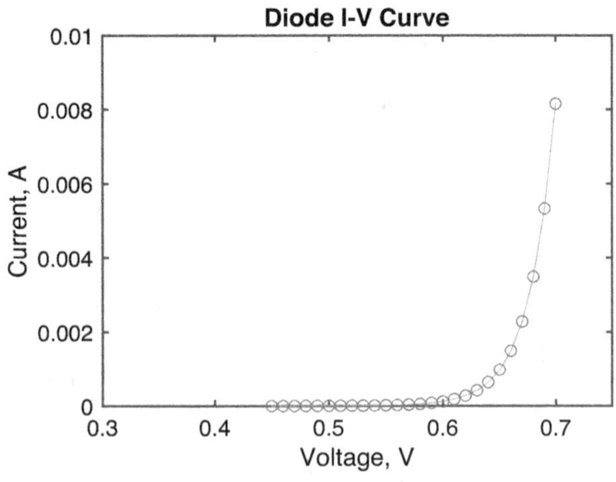

FIGURE 1.10 Diode current–voltage characteristics.

1.5.6 Dynamic Range

Dynamic Range is the ratio of the span of the signal to the minimum signal level that a system can handle while maintaining a specified output signal. Typically, a minimum signal level will be constrained by system noise. The maximum value at the output of a system will be limited by the operational characteristics of devices used to build the system and also the power supply voltage of the system. Usually, the dynamic range is expressed by a single number or in terms of decibel (dB). For voltage-like quantities, the dynamic range (in dB) is given as follows:

$$\text{Dynamic Range} = 20\log\left(\frac{\text{span of quantity in voltage}}{\text{lowest measurable voltage}}\right) \tag{1.32}$$

For power-like quantities, the dynamic range (in dB) is given as follows:

$$\text{Dynamic Range} = 10\log\left(\frac{\text{span of quantity in power}}{\text{lowest measurable power}}\right) \tag{1.33}$$

In digital sensors and analog-to-digital converters, the signal levels change in increments of bits. In general, one can define a dynamic range based on the digital representation. In an N-bit system, the ratio between the highest and lowest quantized level is $2^N/1 = 2^N$. Thus, the dynamic range may be written for digital sensors and converters as follows:

$$\text{Dynamic range} = 20\log(2^N) = 20N\log(2) = 6.026N \text{ dB} \tag{1.34}$$

Example 1.6: Dynamic Range of the Analog-to-Digital Converter

A 12-bit analog-to-digital converter is used to convert analog music to digital format. The amplitude of the analog music varies from −5V to +5V. (a) Calculate the smallest signal increment that may be used and (b) determine the dynamic range of the analog-to-digital conversion.

Solution

a. A 12-bit analog-to-digital converter can represent $2^{12} = 4096$ levels of the signal. The smallest signal increment is as follows:

$$\Delta V = \frac{10}{2^{12}} = 2.441 \times 10^{-3} \text{ V}$$

b. From Equation (1.34), the dynamic range of the analog-to-digital converter is

$$\text{Dynamic range} = 6.026N = 6.026 \times 12 \text{ dB} = 72.24 \text{ dB}$$

1.5.7 Errors

Whenever measurements are taken, the value obtained might deviate from the true value. The error is the difference between the true and the measured value. Errors will always be present irrespective of the accuracy of the measuring instrument or the carefulness employed in taking the measurements. It is thus advisable to specify the size and type of the expected error.

The absolute error is defined as follows:

$$\text{Absolute error} = |x_m - x_t| \tag{1.35}$$

where

x_t is the true or expected value, and
x_m is the measured value.

The percent error is defined as

$$\text{Percent error} = \left|\frac{x_m - x_t}{x_t}\right| \times 100 \tag{1.36}$$

and

$$\text{The standard deviation, } \sigma = \sqrt{\frac{\sum_{i=1}^{N}\left(x_i - \bar{x}\right)^2}{N}} \tag{1.37}$$

where

$\bar{x}$ is the mean value of N measurements, which is given in Equation (1.30).

Errors may come from three major sources: (i) gross, (ii) systematic, and (iii) random.

Gross errors, as the name implies, can be quite large. They are largely human errors and result generally from the carelessness of human operators. Examples of gross errors are failure to correctly record a measurement and not reading the correct scale on a meter. One can minimize gross errors by cross-checking of results and paying more attention to the measurements.

Systematic errors are those errors that remain unchanged with repeated measurements. These errors can arise from inaccuracies in the manufacture of an equipment, lack of calibration of equipment, the use of an equipment in an environment that it was not designed for, and not using the specified loads for instruments. In some cases, systematic errors can be measured and compensated for. In other cases, they can be eliminated by periodic calibration of instruments, using an instrument in the specified environment, and by choosing a suitable instrument for a particular measurement application.

Random errors are those errors that are due to unknown causes. They are the errors that remain when gross and systematic errors have been eliminated or substantially reduced. Some of the sources of random errors are noise and environmental effects. Noise that is generated in the system or picked up by a system may cause the output signal of a system to fluctuate. In addition, changes in the conditions external to an instrument, such as pressure, temperature, and humidity, may also cause random errors. Random errors are difficult to predict or correct, but they can be minimized by increasing the number of readings and by using statistical techniques to obtain the best estimate of the true value of the quantity being measured.

Example 1.7: Absolute and Percent Error Calculations

A voltmeter reads 4.8 V, and the true value of the voltage is 5.0 V. Determine (a) the absolute error and (b) its relative percentage error.

Solution

The true value, $x_t = 5.0$ V. The measured value, $x_m = 4.8$ V.
Using Equations (1.35) and (1.36), we have
Absolute error $= |x_m - x_t| = |4.8 - 5| = 0.2$ V

$$\text{Percent error} = \left|\frac{x_m - x_t}{x_t}\right| \times 100 = \frac{0.2}{5} * 100 = 4.0\%$$

PROBLEMS

Problem 1.1 The triangular waveform, shown in Figure P1.1, can be expressed as

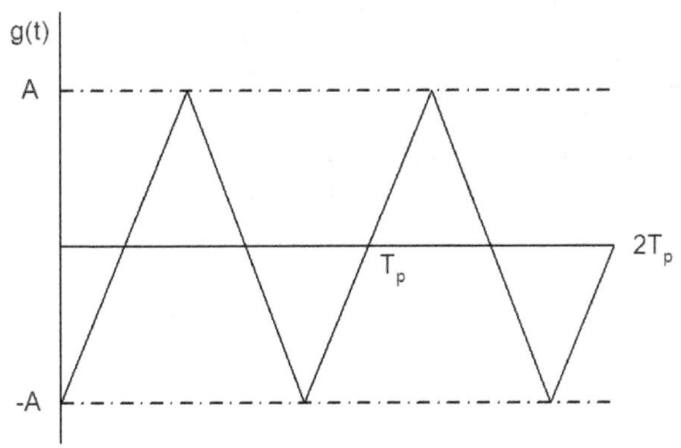

FIGURE P1.1 Triangular waveform.

$$g(t) = \frac{8A}{\pi^2} \sum_{n=1}^{\infty} \frac{(-1)^{n+1}}{4n^2 - 1} \sin((2n-1)w_0 t)$$

where

$$w_0 = \frac{1}{T_p}$$

If $A=2$, $T=8$ ms., and sampling interval is 0.1 ms, write a MATLAB program to re-synthesize $g(t)$ if 20 terms are used.

Problem 1.2 For the system with network function

$$H(s) = \frac{s^3 + 4s^2 + 16s + 4}{s^4 + 20s^3 + 12s^2 + s + 10}$$

find the poles and zeros of $H(s)$.

Problem 1.3 The transfer function of a filter is as follows:

$$H(s) = \frac{5x10^3}{s^2 + 64 \times 10^3 s + 25 \times 10^3}$$

(i) Use MATLAB to plot the magnitude response; (ii) determine the approximate bandwidth of the filter.

Problem 1.4 A transfer function of a system with several poles is given as

$$H(s) = \frac{3.65 \times 10^{21}}{(s + 600\pi)(s + 12\pi \times 10^6)(s + 2.7\pi \times 10^8)}$$

(i) Use MATLAB to plot the magnitude response; (ii) determine the frequency where the magnitude is unity.

Problem 1.5 Two individuals took seven measurements of a voltage across a thermistor. The values are shown in Table P1.1.

If the true value of the voltage is 5 V, (i) determine which measurements are precise and (ii) which measurements are accurate.

Problem 1.6 For a signal that varies from −2.5 V to 2.5 V, if a 10-bit A/D converter is used to convert the analog signal into digital, (i) determine the smallest signal increment and (ii) calculate the dynamic range of the A/D converter.

TABLE P1.1
Voltage Across a Thermistor

Person	Data #1	Data #2	Data #3	Data #4	Data #5	Data #6	Data #7
Person A	4.82	4.84	4.81	4.83	4.82	3.84	4.85
Person B	5.01	4.98	4.97	5.03	5.04	5.02	4.99

Problem 1.7 A loudspeaker is rated at 8 W, and it requires a minimum power of 0.01 W to change the mechanical mechanism of the speaker's cone to allow an output signal to be heard. What is the dynamic range of the loudspeaker?

Problem 1.8 A resistor with a nominal value of 5000 Ω was purchased for a project. What will be the possible range of values of the 5000-Ω resistor, if the tolerance is (i) 5%, (ii) 10%, and (iii) 20%.

Problem 1.9 A voltmeter reads 9.7 V, and the true value of the voltage is 10 V. Determine the absolute error and the relative percentage error.

Problem 1.10 The transistor gain β of eight 2N3904 transistors, measured in the laboratory, are 212, 207, 223, 210, 217, 219, 225, and 215. Use MATLAB to (i) determine the mean value of β, (ii) minimum and maximum value of β, and (iii) the standard deviation of β.

BIBLIOGRAPHY

Alkin, Oktay, *Discrete-Time Signals and Systems: A MATLAB® Integrated Approach*, 2nd Edition, CRC Press, 2025.

Attia, John O., *Electronics and Circuit Analysis Using MATLAB®*, CRC Press, 2nd Edition, 2004.

Bolton, William, *Instrumentation and Control Systems*, 3rd Edition, Elsevier, 2021.

Chapman, Stephen J., *MATLAB® Programming for Engineers*, 7th Edition, Cengage Learning, 2024.

Etter, Delores, *Engineering Problem Solving with MATLAB®*, 3rd Edition, Pearson, 2021.

Morris, Alan S. and Reza Langari, *Measurement and Instrumentation: Theory and Applications*, 3rd Edition, Academic Press, 2020.

Northrop, Robert B., *Introduction to Instrumentation and Measurements*, 3rd Edition, CRC Press, 2014.

Sadiku, Matthew N.O., Warsame H. Ali, and Sarhan M. Musa, *Signals and Systems: A Primer with MATLAB®*, 2nd Edition, CRC Press, 2024.

Tripathi, A.K., Gupta Rajeev, and Kumar Basant, *Signals and Systems*, Mechasoft Publishers, 2024.

Ulaby, Fawwaz and Andrew E. Yagle, *Signals & Systems: Theory and Applications*, 2nd Edition, Michigan Publishing Services, 2024.

Valentine, Daniel T. and Brian D. Hahn, *Essential MATLAB® for Engineers and Scientists*, 8th Edition, Academic Press, 2022.

Vector, Maxwell, *MATLAB® Fundamentals for Electrical Engineers*, MATLAB Fundamentals Book Series, 2025.

2 Amplifiers

2.1 USES OF AMPLIFIERS IN INSTRUMENTATION SYSTEMS

2.1.1 Amplification

One of the major uses of amplifiers is to amplify signals. For most instrumentation systems, signals to be conditioned, transmitted, or displayed may be small in amplitude, and amplifiers are, therefore, needed to increase the amplitude. For an AC signal, we need AC amplifiers, but for DC signals, DC amplifiers are needed.

Difference amplifiers can be used to amplify the difference between two signals. Difference amplifiers can also be used to amplify one signal if the other input of these amplifiers is connected to ground. In addition, amplifiers can be used to amplify the sum of several signals.

2.1.2 Input Isolation

Amplifiers can also be used to provide input isolation. Some transducers, specifically ultrasonic transducers, have very high input impedance. These transducers, if connected directly to a voltage amplifier with low input impedance, will have most of the transducers' signals lost inside them and not amplified. An isolation amplifier or a buffer amplifier is normally placed between the transducer and the voltage amplifier. The isolation amplifiers will normally have high input impedance to match high impedances of some transducers.

2.1.3 Reduction on Nonlinear Distortion

Negative feedback amplifiers can be used to reduce the noise and distortion at the output of an amplifier. In addition, negative feedback amplifiers can provide gain stability and increased bandwidth of an amplifier. Thus, the use of negative feedback amplifiers can be used to obtain output signals not affected significantly by noise.

2.1.4 Driver to Output Stages

Amplifiers can also be used to supply power to drive loads. Power amplifiers are used to supply large amounts of power to loads.

2.2 AMPLIFIER PARAMETERS

When designing or choosing an amplifier for instrumentation systems, the following parameters should be considered: input impedance, output impedance, gain, bandwidth, noise, and common-mode rejection. Relationships exist between some of the above parameters. For example, the gain-bandwidth product of an amplifier will

DOI: 10.1201/9781003765554-2

normally be constant. In addition, some of the parameters are related to the input elements and the load connected to the amplifier. The parameters will now be discussed to show their importance in instrumentation systems.

2.2.1 Input Impedance

Input impedance is the impedance seen at the input of an amplifier. As shown in Figure 2.1, if the input impedance of the amplifier is Z_{in}, then the voltage at the input of the amplifier, V_{in} can be expressed as.

$$V_{\text{in}} = \frac{Z_{\text{in}}}{Z_{\text{in}} + Z_s} V_s \tag{2.1}$$

where

V_s is the source voltage (it can be a signal from a sensor) and.
Z_s is the source impedance.

From Equation (2.1), if $Z_{\text{in}} \geq Z_s$, then V_{in} is approximately equal to V_S. It is advisable to have the input impedance of the amplifier be far greater than the source impedance such that most of the sensor signals can be amplified and processed. In practice, it is acceptable to have the input impedance Z_{in} be ten times the source impedance Z_S.

Example 2.1: Effect of Input Impedance of an Amplifier

As shown in Figure 2.1, for $Z_S = 1000\ \Omega$, find the gain $\frac{V_{\text{in}}}{V_S}$ if Z_{in} varies from 100 Ω to 50,000 Ω.

Solution

Using Equation (2.1), $\frac{V_{\text{in}}}{V_S}$ was calculated. Table 2.1 shows the calculated values of $\frac{V_{\text{in}}}{V_S}$, and Figure 2.2 represents the plot of the $\frac{V_{\text{in}}}{V_S}$ versus Z_{in}. It can be seen from Table 2.1 and Figure 2.2 that as the input impedance Z_{in} becomes larger than Z_S, the voltage gain, $\frac{V_{\text{in}}}{V_S}$, approaches unity.

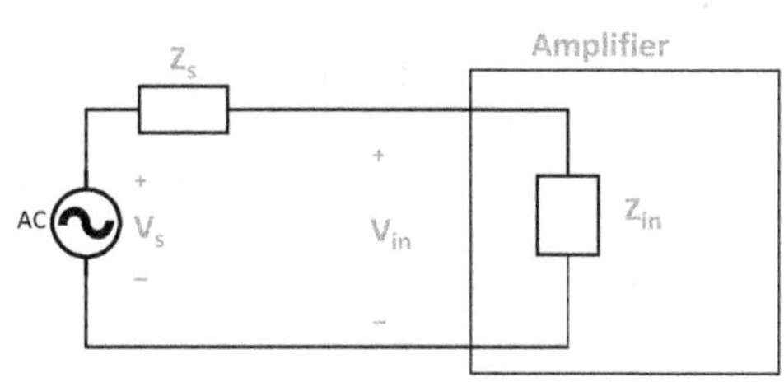

FIGURE 2.1 Amplifier with a source voltage.

TABLE 2.1
Z_{in} versus $\frac{V_{in}}{V_S}$ for $Z_S = 1000\ \Omega$ of Figure 2.1

Input Impedance, Z_{in}	Voltage and Gain $\frac{V_{in}}{V_S}$
100	0.091
250	0.2
500	0.333
750	0.429
1000	0.5
2000	0.667
5000	0.833
10000	0.909
20000	0.952
50000	0.980392

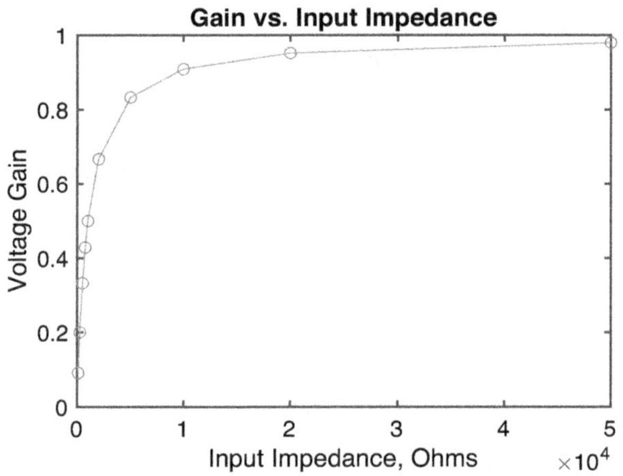

FIGURE 2.2 Relation between amplifier input impedance Z_{in} and gain $\frac{V_{in}}{V_S}$.

2.2.2 Output Impedance

The output impedance of an amplifier is the impedance seen at the output terminals of an amplifier. In Figure 2.3, Z_0 is the output impedance of the amplifier and V_0 is the open-circuit voltage (voltage at output when output is open-circuited) of the amplifier.

The voltage across the load V_L can be expressed as

$$V_L = \frac{Z_L}{Z_o + Z_L} V_o \tag{2.2}$$

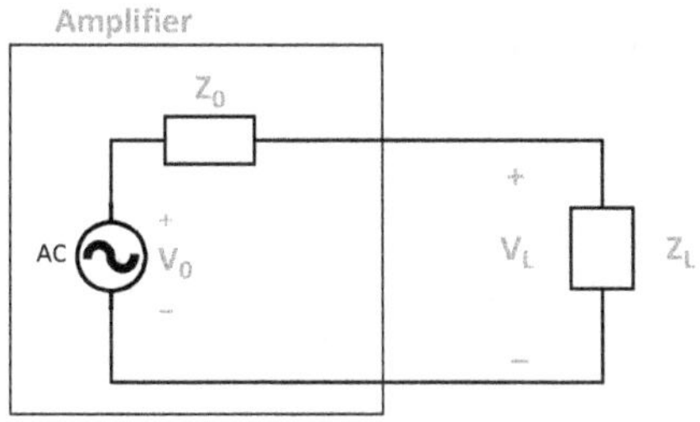

FIGURE 2.3 Amplifier output with load.

Provided the load impedance Z_L is far greater than the output impedance of the amplifier Z_0, the output voltage will be approximately equal to the amplifier open-circuit voltage V_0. For instrumentation systems, where Z_L can be adjusted (for example, the input impedance of a stage next to an amplifier), the load impedance should be changed to be at least ten times greater than the output impedance of the amplifier.

For some instrumentation systems, one might be interested in transferring the maximum power from the amplifier to the load. In that case, we need to make the load impedance equal to the output impedance of the amplifier, as required by the maximum power transfer theorem. In other instrumentation applications, one might be interested in obtaining a voltage across the load which is linearly proportional to the input and free from distortion. In that case, the exact relationship between $Z_L$and Z_o in Equation (2.2) might not be significant.

2.2.3 Gain and Frequency Response

Figure 2.4 shows an amplifier with a source voltage V_S and load impedance Z_L. The voltage, current, and power gains for an amplifier can be written as follows:

$$\text{Voltage Gain, } G_V = \frac{V_L}{V_s} \tag{2.3}$$

$$\text{Current Gain, } G_I = \frac{I_L}{I_s} \tag{2.4}$$

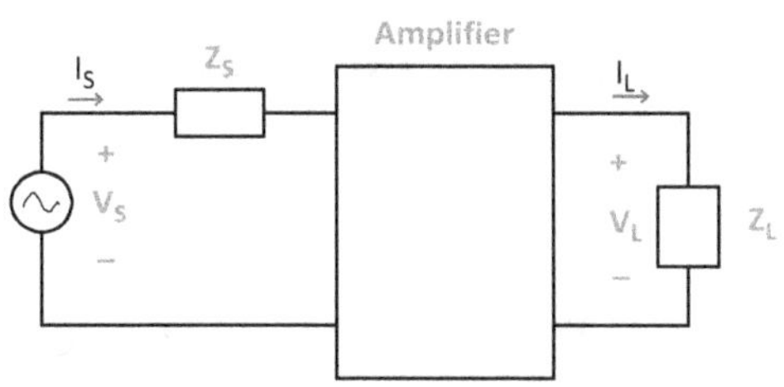

FIGURE 2.4 Amplifier with a voltage source and a load.

$$\text{Power Gain, } G_P = G_V G_I = \frac{V_L I_L}{V_s I_s} \tag{2.5}$$

where

V_S and V_L are the source and load voltages, respectively.
I_S and I_L are the source and load currents, respectively.

The voltage or current gains when plotted with respect to frequency result in frequency responses. If w_L and w_H are the low and high cut-off frequencies, the bandwidth of a capacitively coupled amplifier or a tuned amplifier is given as

$$\text{Bandwidth, BW} = w_H - w_L \tag{2.6}$$

The product of the maximum gain and the bandwidth of an amplifier is termed the gain–bandwidth product. The latter is a constant for a particular amplifier. It is normally possible to trade gain for bandwidth and vice versa. The bandwidth of an amplifier should be chosen to amplify only the frequencies of interest. If practicable, the bandwidth chosen for an amplifier should be small, such that noise generated by an amplifier will also be small.

2.2.4 Noise

Noise is customarily used to designate unwanted signals that tend to mar the processing and transmission of signals in an instrumentation system. Noise is generated by both passive and active electrical elements. Since amplifiers are built from these elements, noise is generated inside amplifiers.

2.2.5 Common-Mode Rejection

For a difference amplifier, when two inputs are tied together and a signal applied to the two inputs, the output will be non-zero. This is illustrated in Figure 2.5a.

The common-mode gain, A_{cm}, is defined as

$$A_{\text{cm}} = \frac{v_o}{v_{i,\text{cm}}} \tag{2.7}$$

The differential-mode gain, A_d, is defined as (see Figure 2.5b)

$$A_d = \frac{v_o}{v_{id}} \tag{2.8}$$

For an amplifier with arbitrary input voltages, V_1 and V_2, the differential input signal, v_{id}, is

$$v_{id} = V_2 - V_1 \tag{2.9}$$

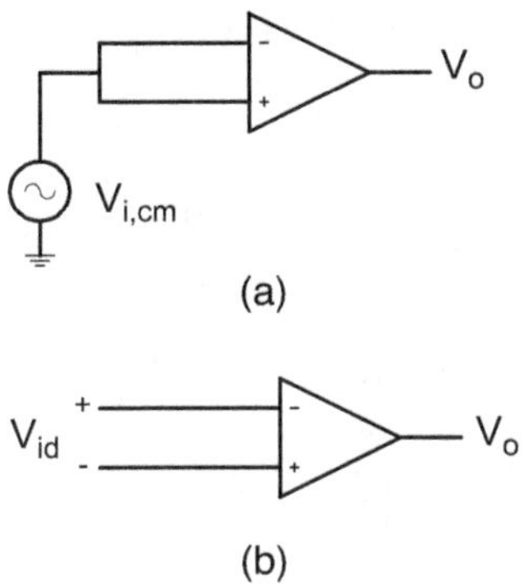

FIGURE 2.5 Circuits showing the definitions of (a) common-mode gain and (b) differential-mode gain.

and the common-mode input voltage is the average of the two input signals

$$V_{i,\mathrm{cm}} = \frac{V_2 + V_1}{2} \tag{2.10}$$

The output of the op amp can be expressed as

$$V_O = A_d v_{id} + A_{\mathrm{cm}} v_{i,\mathrm{cm}} \tag{2.11}$$

The common-mode rejection ratio (CMRR) is defined as

$$\mathrm{CMRR} = \left| \frac{A_d}{A_{\mathrm{cm}}} \right| \tag{2.12}$$

The CMRR represents the op amp's ability to reject signals that are common to the two inputs of an amplifier. A good difference amplifier has a very high CMRR. That means, common-mode gain should be very small and differential gain should be large. Typical values of the CMRR range from 80 to 120 db.

Example 2.2: CMRR of an Amplifier

A differential amplifier has two input signals, V_1 and V_2 whose values are 5 V each. The common-mode rejection ratio of the amplifier is 80 dB, and its differential gain is 120. Find the output voltage of the amplifier

Solution

The CMRR is given in decibels (dB). If it is expressed not in terms of dB, we have

$$\mathrm{CMRR} = 20\log_{10}(\mathrm{CMRR})$$

$$80 = 20\log_{10}(\mathrm{CMRR}), \text{ thus the CMRR} = 10{,}000$$

Knowing the differential gain, A_d, and CMRR, one can find the common-mode gain A_{cm}

From Equation (2.12), we have

$$A_{\mathrm{cm}} = \frac{A_d}{\mathrm{CMRR}} = \frac{120}{10{,}000} = 0.012$$

Using Equations (2.9)–(2.11), we have

$$v_o = A_d(V_1 - V_2) + A_{\mathrm{cm}}\left(\frac{V_1 + V_2}{2}\right) = A_d(0) + A_{\mathrm{cm}}(5) = 0.06 \text{ V}$$

2.3 OPERATIONAL AMPLIFIERS

Most amplifiers and signal conditioning circuits are built using operational amplifiers. The latter amplifiers are very versatile, and they can be used to design several complex instrumentation circuits. In this section, the properties of the ideal operational amplifier will be discussed.

An ideal op amp is a DC amplifier. Its equivalent circuit is shown in Figure 2.6. The properties of the ideal op amp are as follows:

i. The gain, A, normally referred to as open-loop gain, is infinite. If the output signal is finite, then the voltage $V_e = V_1 - V_2$ must approach zero since A is infinite.
ii. The input resistance R_{in} of the amplifier is infinitely large. This implies that the input currents should be zero. That is, the op amp does not take any current at the two inputs.
iii. The output resistance R_O of the amplifier is zero.
iv. The gain A_O is independent of frequency, and the bandwidth of the amplifier is infinitely large.
v. The common-mode rejection and the power-supply rejection are infinitely large. The former implies that the op amp amplifies only the input differential voltage and has no gain for signals common to the two inputs. In addition, the amplifier has no output changes due to the changes in the power supply voltage.

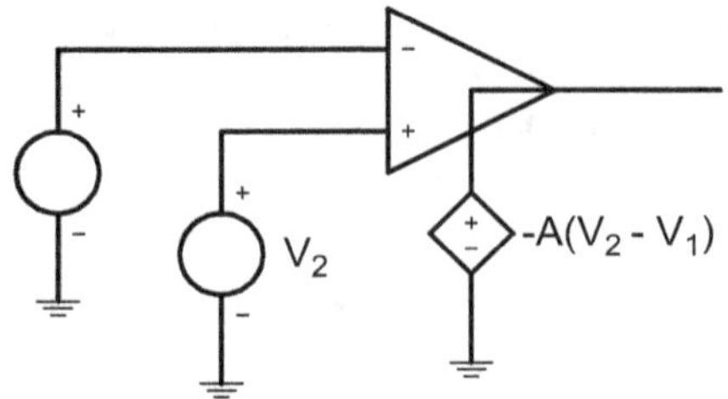

FIGURE 2.6 Equivalent circuit of ideal op amp.

TABLE 2.2
Properties of Real Op Amps

Property	Typical Value for LM741 Op Amp	Typical Value for OP484 Op Amp
Open loop gain	2.0×10^5	1.5×10^5
Offset voltage	1 mV	60 μV
Input bias current	30 nA	80 nA
Unity-gain bandwidth	1 MHz	4.25 MHz
Common-mode rejection ratio	95 dB	90 dB
Slew rate	0.7 V/μV	4.0 V/μV

Integrated circuit op amps have characteristics very close to those of ideal op amps. In addition, the behavior of op amps is close to their predicted theoretical performance. Real op amps have characteristics that are different from those of ideal op amps but are closely related to those of ideal op amps. A typical op amp might have open loop gain in excess of 10^5. The common-mode voltage gain of real op amps is non-zero but small. The CMRR of a typical op amp might be in the range of 60 to 120 dB. Op Amps built with FET inputs may have an input resistance greater than 10^{10} Ω, while an op amp with BJT inputs may have a resistance of about 2 MΩ. Typical values of the output resistance range from 50 to 125 Ω. The offset voltage is small but finite, and the frequency response will deviate considerably from the infinite frequency response. An op amp with internal compensation might have a dominant pole (or 3 dB frequency) in tens of Hertz and a unity gain bandwidth of 10^6. The gain decreases with frequency. Table 2.2 shows the properties of the general-purpose op amps LM741 and OP484.

Whenever there is a connection from the output of the op amp to the inverting input, we have a negative feedback connection. With negative feedback and finite output voltage, Equation (2.14) implies that the two input voltages are also equal. This condition is termed as the concept of the virtual short circuit.

$$V_0 = A(V_2 - V_1) \tag{2.13}$$

Since the open-loop gain is very large,

$$(V_2 - V_1) = \frac{V_0}{A} \cong 0 \tag{2.14}$$

In addition, because of the large input resistance of the op amp, the latter is assumed to take no current at the input terminals for most calculations.

2.4 BASIC AMPLIFIER CONFIGURATIONS

Three amplifier configurations will be discussed in this section: inverting, non-inverting, and adder. These circuits have their special features, and their applications will be discussed.

2.4.1 Inverting Configuration

An op amp circuit connected in an inverted closed-loop configuration is shown in Figure 2.7.

Using nodal analysis at node A, we have

$$\frac{V_a - V_{in}}{Z_1} + \frac{V_a - V_O}{Z_2} + I_1 = 0 \tag{2.15}$$

From the concept of a virtual short circuit (virtual ground at node A),

$$V_a = 0 \tag{2.16}$$

Because of the large input resistance, $I_1 = 0$. Thus, Equation (2.15) simplifies to

$$\frac{V_O}{V_{in}} = -\frac{Z_2}{Z_1} \tag{2.17}$$

The minus sign implies that V_O and V_{in} are out of phase by 180°. The input impedance Z_{in} is given as

$$Z_{in} = \frac{V_{in}}{I_1} = Z_1 \tag{2.18}$$

For $Z_1 = R_1$ and $Z_2 = R_2$, the inverting amplifier is shown in Figure 2.8. The closed-loop gain of the inverting amplifier is given as

$$\frac{V_O}{V_{in}} = -\frac{R_2}{R_1} \tag{2.19}$$

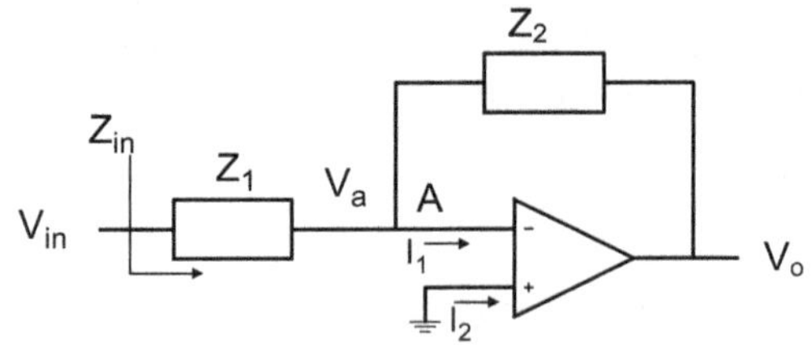

FIGURE 2.7 Inverting configuration of an op amp.

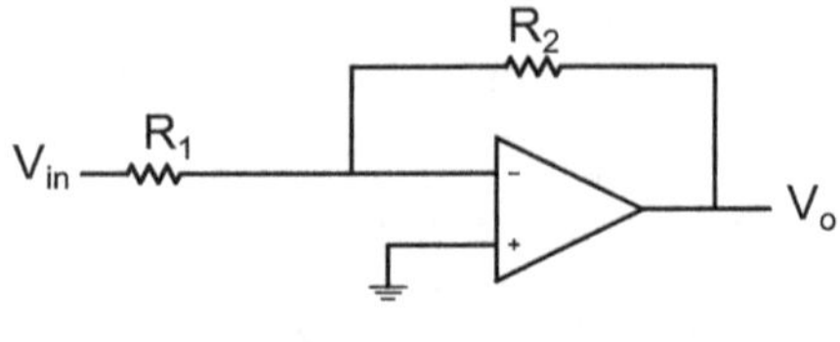

FIGURE 2.8 Inverting amplifier.

From Equation (2.19), it can be seen that the closed-loop gain depends on the external components R_1 and R_2. The gain is independent of the open-loop gain of the op amp. The input resistance of the closed-loop inverting amplifier is

$$R_{in} = R_1 \tag{2.20}$$

To obtain a high gain, we need increase R_2 and decrease R_1. The resistance R_2 cannot be excessively large. The maximum values selected for R_2 are limited by the input resistance of the op amp.

If the open-loop gain of the op amp, A, is not infinite, it can be shown that the closed-loop gain is given as follows:

$$\frac{V_0}{V_{in}} = -\frac{R_2/R_1}{1+\left(1+R_2/R_1\right)/A} \tag{2.21}$$

It should be noted that as the open-loop gain approaches infinity, the closed-loop gain becomes

$$\frac{V_0}{V_{in}} \cong -\frac{R_2}{R_1}$$

The above expression is identical to Equation (2.19). To minimize the dependence of the closed-loop gain on the value of the open-loop gain, A, we should make

$$\left(1+\frac{R_2}{R_1}\right) \ll A \tag{2.22}$$

This is illustrated by the following example.

Example 2.3: Effect of Finite Open-Loop Gain

In Figure 2.8, $R_1 = 1\ \text{k}\Omega$ and $R_2 = 70\ \text{k}\Omega$. Plot the closed-loop gain as the open-loop gain of the op amp increases from 10^2 to 10^8.

Solution

```
% Effect of finite open-loop gain
a = logspace(2,8);
r1 = 1e3; r2 = 70e3; r21 = r2/r1;
g = [];
n = length(a);
for i = 1:n
g(i) = r21/(1+(1+r21)/a(i));
end
semilogx(a,g,'k',a,g,'kx')
xlabel('Open Loop Gain')
ylabel('Closed Loop Gain')
title('Effect of Finite Open Loop Gain')
axis([1.0e2,1.0e8,10,80])
```

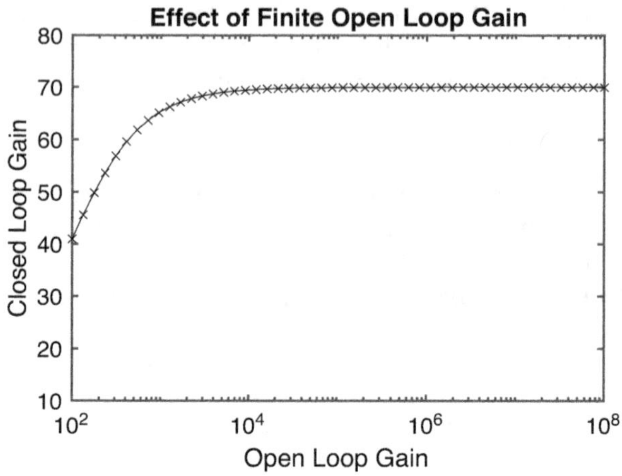

FIGURE 2.9 Closed-loop gain versus open-loop gain.

Figure 2.9 shows the characteristics of the closed-loop gain as a function of the open-loop gain. It can be seen that the closed-loop gain is dependent on the open-loop gain at low values. However, for an open-loop gain greater than 10^5, the closed-loop gain depends only on the external resistors R_1 and R_2.

2.4.2 Non-Inverting Configuration

An op amp connected in a non-inverting configuration is shown in Figure 2.10.

Using nodal analysis at node A

$$\frac{V_a - 0}{Z_1} + \frac{V_a - V_0}{Z_2} + I_1 = 0 \tag{2.23}$$

From the concept of a virtual short circuit,

$$V_{in} = V_a \tag{2.24}$$

Because of the large input resistance, $I_1 = 0$. Thus, Equation (2.23) is simplified to

$$\frac{V_0}{V_{in}} = 1 + \frac{Z_2}{Z_1} \tag{2.25}$$

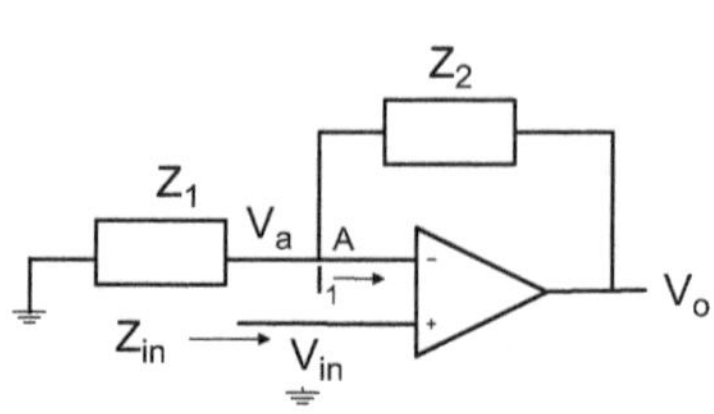

FIGURE 2.10 Non-inverting configuration.

It can be seen from Equation (2.25) that the gain of the non-inverting amplifier is positive. The input impedance of the amplifier Z_{in} approaches infinity since the current that flows into the positive input of the op-amp is almost zero.

If $R_1 = Z_1$ and $R_2 = Z_2$, then the configuration shown in Figure 2.10 is of a voltage follower with gain, as shown in Figure 2.11.

The voltage gain is

$$\frac{V_0}{V_{in}} = 1 + \frac{R_2}{R_1} \tag{2.26}$$

It should be noted the closed-loop gain of a non-inverting amplifier is always positive, (i.e., output and input voltages are in phase). Since no current flows into the input of the op amp, the input impedance is infinite.

If the open-loop gain, A, is not infinite, then the closed-loop gain is given as follows:

$$\frac{V_o}{V_{in}} = \frac{\left(1 + {R_2}/{R_1}\right)}{1 + \left(1 + {R_2}/{R_1}\right)/A} \tag{2.27}$$

If the resistance R_1 approaches infinity (open-circuited) and resistance $R_2 = 0$ the circuit becomes a voltage follower. A voltage follower has high input impedance, low output impedance, and unity gain. Because of these properties, the voltage follower can be used as a buffer amplifier to connect a source with high impedance to a low impedance load.

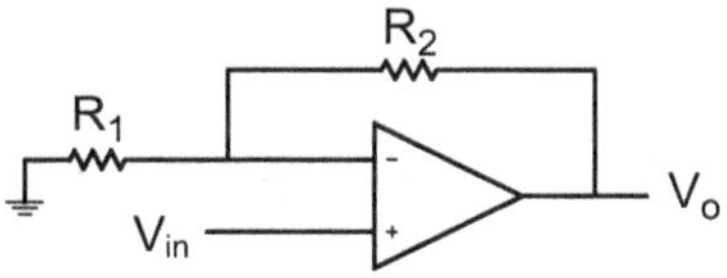

FIGURE 2.11 Voltage follower with gain.

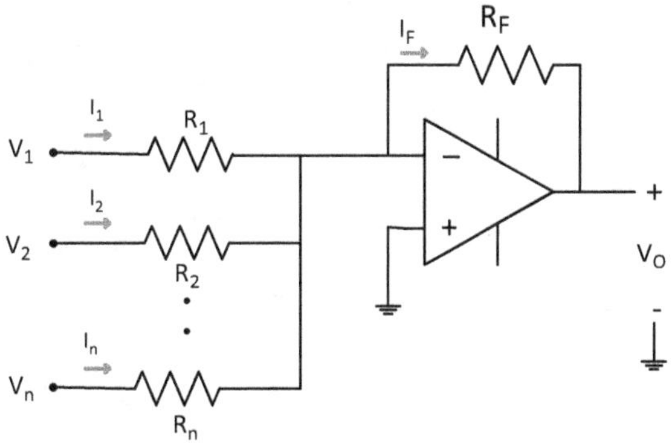

FIGURE 2.12 Weighted summer circuit.

2.4.3 Weighted Summer

A circuit for the weighted summer is shown in Figure 2.12. Using Ohm's Law, we get

$$I_1 = \frac{V_1}{R_1}, \quad I_2 = \frac{V_2}{R_2}, \quad ...., \quad I_n = \frac{V_n}{R_n} \tag{2.28}$$

Using Kirchhoff's Current Law, we get

$$I_F = I_1 + I_2 + ...I_N \tag{2.29}$$

$$V_O = -I_F R_F \tag{2.30}$$

Substituting Equations (2.28) and (2.29) into Equation (2.30), we have

$$V_O = -\left(\frac{R_F}{R_1} V_1 + \frac{R_F}{R_2} V_2 + \cdots + \frac{R_F}{R_n} V_n \right) \tag{2.31}$$

The weighted summer can be used to add DC offset to an alternating current signal as in an LED modulation circuit, the DC component is needed to keep the LED in its linear operating range. It can also be used to sum several signals in audio mixers.

Example 2.4: Design of an Adder Circuit

Design an op amp circuit with inputs V_1 and V_2 whose output V_0 is given as $V_o = -3V_1 - 5V_2$. Use resistors between 100 Ω and 10 kΩ in your design.

Solution

A circuit that can be used to realize the solution is shown in Figure 2.13. From Equation (2.31), we have

$$V_o = -\left(\frac{R_F}{R_1} V_1 + \frac{R_F}{R_2} V_2 \right) = -3V_1 - 5V_2$$

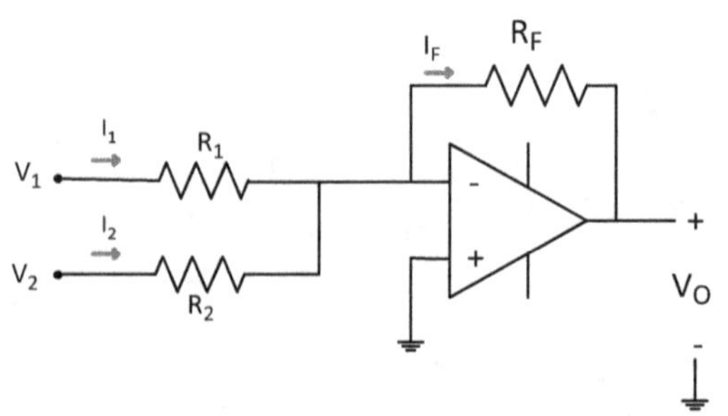

FIGURE 2.13 An adder circuit.

If we choose $R_F = 15\text{k}\Omega$, then $R_1 = 5\text{k}\Omega$ and $R_2 = 3\text{k}\Omega$

It should be noted that there are other values of R_f, R_1, and R_2 that will satisfy the design.

2.5 DIFFERENTIAL AMPLIFIER

A differential amplifier amplifies the difference between two inputs. An op amp circuit that can be used to realize a differential amplifier is shown in Figure 2.14. The circuit can be analyzed using the principle of superposition. The output voltage is given as

$$V_O = -\frac{R_2}{R_1}V_1 + \left(1 + \frac{R_2}{R_1}\right)\left(\frac{R_4}{R_3 + R_4}\right)V_2 \quad (2.32)$$

If we make

$$R_2 = R_4 \text{ and } R_1 = R_3 \quad (2.33)$$

then Equation (2.32) becomes

$$V_0 = \frac{R_2}{R_1}(V_2 - V_1) \quad (2.34)$$

The input resistance of the differential amplifier is

$$R_{\text{in}} = R_1 + R_3 = 2R_1 \quad (2.35)$$

If the op amp is to have a large differential gain, then R_1 will, by necessity, be relatively small and the input resistance will also be small. This is one of the disadvantages of the differential amplifier. The instrumentation amplifier, discussed in Section 2.6, corrects this problem by using two additional op amps to increase the input resistance.

It can be shown that the common mode of the difference amplifier is given as follows:

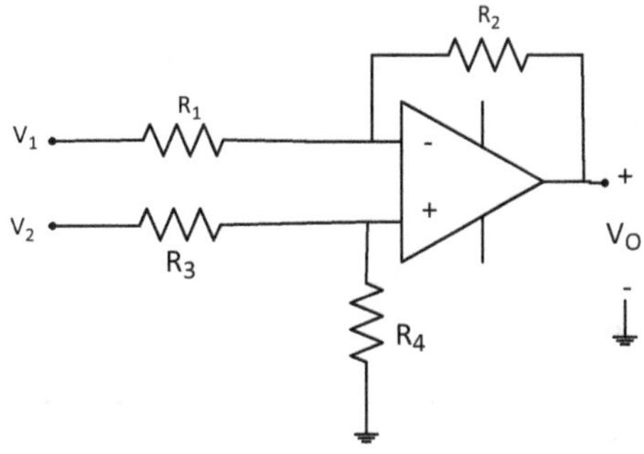

FIGURE 2.14 Differential amplifier.

$$A_{cm} = \frac{V_o}{V_{cm}} = \left(\frac{R_4}{R_4 + R_3}\right)\left(1 - \frac{R_2 R_3}{R_1 R_4}\right) \tag{2.36}$$

If $\frac{R_4}{R_3} = \frac{R_2}{R_1}$, then $A_{cm} = 0$. Example 2.5 will examine the situation when $\frac{R_4}{R_3} \neq \frac{R_2}{R_1}$

Example 2.5: Common-Mode Gain of Differential Amplifier

For the differential amplifier shown in Figure 2.14, $R_1 = 1.0\,\text{k}\Omega$, $R_2 = 50\,\text{k}\Omega$, and $R_3 = 1.0\,\text{k}\Omega$. If the resistor $R_4 = 50\,\text{k}\Omega \pm 10\%$, determine the common-mode gain for the range of resistor values of R_4.

Solution

Equation (2.36) will be used to find the common-mode gain, changing the value of the resistance $R_4 = 50\,\text{k}\Omega \pm 10\% = 45\,\text{k}\Omega$ to $55\,\text{k}\Omega$. The MATLAB program is as follows.

MATLAB® CODE

```
% Effect of Resistance Variation on common-mode gain
r4 = 45.0e3:1.0e3:55.0e3;
r1 = 1.0e3; r2 = 50.0e3; r3 = 1.0e3;
acm = [];
n = length(r4);
for i = 1:n
ac(i)= r4(i)/(r4(i)*r3);
cm(i)= 1-(r2*r3)/(r1*r4(i));
acm(i) = ac(i)*cm(i);
end
plot(r4,acm,'k',r4,acm,'kx')
xlabel('Resistance R4 in Ohms')
ylabel('Common-mode Gain')
title('Common-mode Gain vs. Resistance Change')
```

Figure 2.15 shows the common-mode gain versus change in resistance R_4.

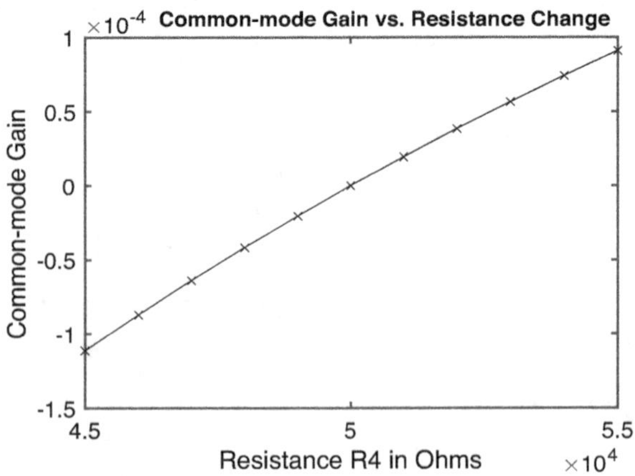

FIGURE 2.15 Common-mode gain versus change in resistance R_4 given in Figure 2.14.

It can be seen from Figure 2.15 that as R_4 changes, the common-mode gain changes. It is zero when $R_4 = 50\ \text{k}\Omega$.

2.6 INSTRUMENTATION AMPLIFIER

The instrumentation amplifier behaves in a way similar to that of the differential amplifier, but it has a high input resistance. A circuit diagram of the instrumentation amplifier is shown in Figure 2.16.

It can be shown that the overall gain of the amplifier is given as

$$V_O = -\frac{R_4}{R_3}\left(1 + \frac{2R_2}{R_1}\right)(V_1 - V_2) \tag{2.37}$$

The instrumentation amplifier has high input impedance, high common-mode rejection ratio, and high voltage gain. The gain can be changed by adjusting the value of resistance R_1. The instrumentation amplifiers can be used to amplify signals from low impedance sensors such as strain gages.

Example 2.6: Gain of an Instrumentation Amplifier

For the instrumentation amplifier shown in Figure 2.17, $V_1 = 2\text{mV}$, $V_2 = 1\text{ mV}$, $R_1 = 1\text{k}\Omega$, $R_2 = 10\text{k}\Omega$, $R_3 = 2\text{k}\Omega$, and $R_4 = 20\text{k}\Omega$. (i) find currents i_1 and i_2 and (ii) determine voltage V_o.

Solution

$$i_1 = 0\ \text{A}\left(\text{op amp has infinite input resistance}\right)$$

$$i_2 = \frac{V_1 - V_2}{R_1} = \frac{(2-1)\ \text{mV}}{1\text{k}\Omega} = 1.0 \times 10^{-6}\ \text{A}$$

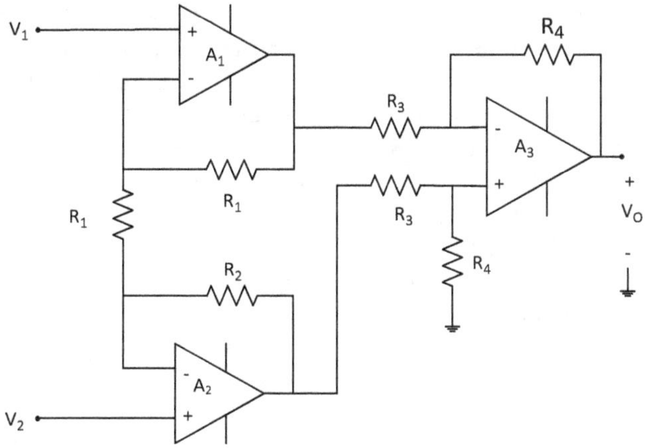

FIGURE 2.16 Instrumentation amplifier.

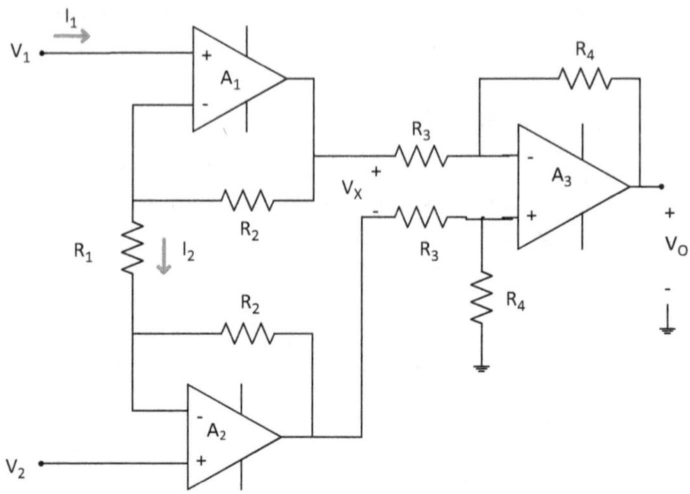

FIGURE 2.17 Instrumentation amplifier circuit for Example 2.6.

$$V_X = (R_2 + R_1 + R_2)i_1 = (2R_2 + R_1)i_1 = 21 \text{ mV}$$

$$V_o = \frac{-R_4}{R_3} V_X = -(10)(21\text{mV}) = -210\text{mV}$$

2.7 CHOPPER-STABILIZED AMPLIFIER

Chopper-stabilized amplifiers are used for applications that demand very low input DC-offset voltage and very small offset voltage drift. For example, for applications where thermocouples are used, it is sometimes difficult to distinguish between the drift in the DC amplifier from the thermocouple-generated signal. In that particular case, the drift in the amplifier is removed using a chopper-stabilized amplifier. A block diagram of the chopper-stabilized amplifier is shown in Figure 2.18.

A chopper circuit converts DC input into an AC signal. The AC signal is then amplified by an AC amplifier. Any DC level at the input of the AC amplifier is removed from reaching the output by the AC amplifier, thus removing DC drift problems. At the output of the amplifier, a rectifying circuit synchronized to the chopper converts the AC signal to an equivalent DC level. The chopped waveform is the product of the input signal and the square wave.

If V_C is a square wave signal and V_S is the input signal, then as given in Figure 2.18, the input to the AC amplifier V_A is given as

$$V_A(t) = V_C(t)V_S(t) \tag{2.38}$$

V_B, the output of the AC amplifier, is given as follows:

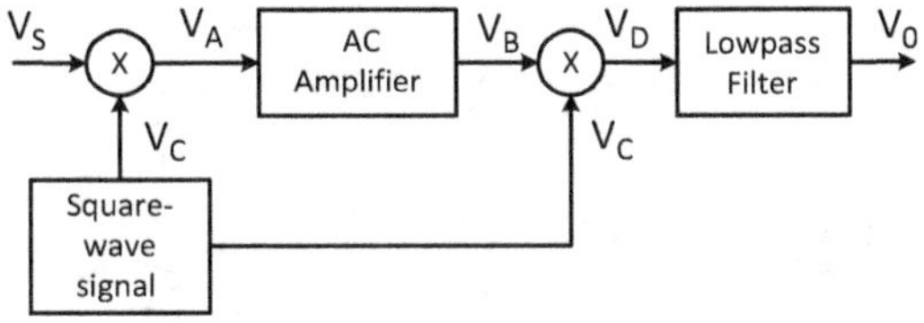

FIGURE 2.18 Chopper-stabilized amplifier.

$$V_B(t) = G_A V_A(t) \tag{2.39}$$

where G_A is the gain of the AC amplifier. The input to the low-pass filter V_D is given as follows:

$$V_D(t) = V_B(t) V_C(t) \tag{2.40}$$

Substituting Equations (2.38) and (2.39) into Equation (2.40), we have

$$\begin{aligned} V_D(t) &= G_A V_A(t) V_C(t) = G_A V_S(t) V_C(t) V_C(t) \\ &= G_A V_S(t) V_C^2(t) \end{aligned} \tag{2.41}$$

$V_C(t)$ is a square wave, and it can be expressed using Fourier series expansion as follows:

$$V_C(t) = C_0 + C_1 \cos(w_1 t) + C_2 \cos(2w_1 t) + C_3 \cos(3w_1 t) + \cdots \tag{2.42}$$

Using a low-pass filter, the high-frequency components are filtered out, and the output voltage $V_0(t)$ is given as follows:

$$V_o(t) = kV_S(t) \tag{2.43}$$

where $k = G_A C_0^2$ and it is a constant.

It can be seen from Equation (2.43) that the output voltage is proportional to the input voltage.

PROBLEMS

Problem 2.1 A force-measuring sensor has an open-circuit output voltage VS of 100 mV and an output impedance R1 of 500 Ω. To amplify the signal, it is connected to an amplifier with a gain of 10. Estimate the input loading error if the amplifier has an input impedance R2 of (a) 5 kΩ or (b) 1 MΩ (Figure P2.1).

Problem 2.2 As given in Figure 2.2, for $Z_0 = 5000$ Ω, find the gain $\frac{V_L}{V_0}$, if Z_L varies from 500 Ω to 100 kΩ.

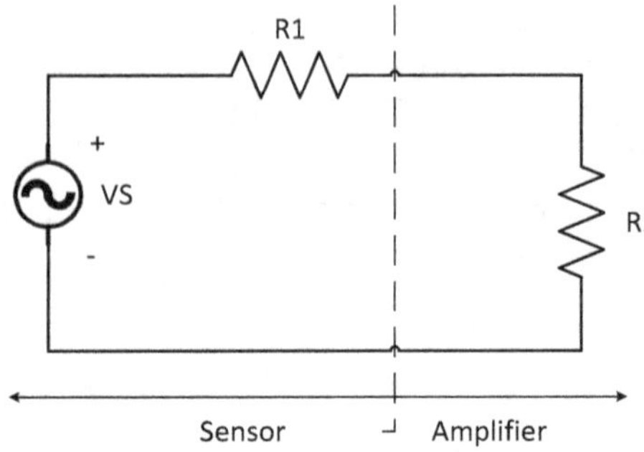

FIGURE P2.1 A force-measuring transducer.

Problem 2.3 A differential amplifier has two input signals, V_1 and V_2, whose values are 3 V each. Its differential gain is 110. If the common-mode rejection ratios of amplifiers range from 60 to 120 dB, use MATLAB to determine the output voltage for each value of common-mode rejection.

Problem 2.3 A differential amplifier has a differential gain of 200 and a common-mode rejection of 75 dB. (a) What is the common-mode gain (b) If the two input signals, V_1 and V_2, of the amplifier are given as $V_1 = 3.09$ V and $V_2 = 3.08$ V, and determine the output voltage of the amplifier.

Problem 2.5 Design an op amp circuit with input V_1 whose output voltage V_0 is given as $V_0 = 20V_1$. Only use resistors between 500 Ω and 20 kΩ in your design.

Problem 2.6 Design an op amp inverter with an input resistance of 10 kΩ and a gain of -20.

Problem 2.7 Design an op amp circuit with inputs V_1, and V_2 whose output voltage V_0 is given as $V_0 = -(4V_1 - 5V_2)$. Only use resistors with resistance between 1000 Ω and 50 kΩ in your design.

Problem 2.8 Design an op amp circuit with inputs V_1, V_2, and V_3, whose output voltage V_0 is given as $V_0 = -(2V_1 + 7V_2 + 4V_3)$. Only use resistors with resistance between 2000 Ω and 100 kΩ in your design.

Problem 2.9 For the differential amplifier shown in Figure 2.14, $R_1 = 2.0\,\text{k}\Omega$, $R_2 = 70\,\text{k}\Omega$, and $R_4 = 70.0\,\text{k}\Omega$, if the resistor $R_3 = 2.0\,\text{k}\Omega \pm 10\%$, determine the common-mode gain for the range of resistor values of R_3.

Problem 2.10 Design an instrumentation amplifier with input voltages of V_1 and V_2 and output voltage V_0 given as $V_0 = -150(V_1 - V_2)$

Problem 2.11 For the instrumentation amplifier shown in Figure 2.17, $R_1 = 1$ kΩ, $R_2 = 16$ kΩ, $R_3 = 3$ kΩ, $R_4 = 70$ kΩ, and the current $I_2 = 10^{-6}$ A, determine the voltages $(V_1 - V_2)$, V_X, and V_0 (assume that the op amps are ideal).

BIBLIOGRAPHY

Alexander, Charles and Matthew N.O. Sadiku, *Fundamentals of Electric Circuit, ISF*, 7th Edition, McGraw-Hill Education, 2020.

Asadi, Farzin, *Essential Circuit Analysis using NI Multisim and MATLAB®*, 1st Edition, Springer, 2022.

Attia, John O., *Electronics and Circuit Analysis Using MATLAB®*, CRC Press, 2nd Edition, 2004.

Chapman, Stephen J., *MATLAB® Programming for Engineers*, 7th Edition, Cengage Learning, 2024.

Etter, Delores, *Engineering Problem Solving with MATLAB®*, 3rd Edition, Pearson, 2021.

Hayes, Thomas C., David Abrams, and Paul Horowitz, *Learning the Art of Electronics: A Hands-On Lab Course*, 2nd Edition, Cambridge University Press, 2025.

Hayt, William, Jack Kemmely, Jamie Phillips, and Steven M. Durbin, *Engineering Circuit Analysis*, 10th Edition, McGraw Hill, 2023.

Horowitz, Paul and Winfield Hill, *The Art of Electronics*, 3rd Edition, Cambridge University Press, 2015.

Irwin, J. David and R. Mark Helms, *Engineering Circuit Analysis*, 12th Edition, Wiley, 2021.

Jaeger, Richard C. and Travis Blalock, *Microelectronic Circuit Design (ISE)*, 6th Edition, McGraw Hill, 2022.

Kay, Art and Marek Lis, "Optimizing Chopper Amplifier Accuracy." Texas Instruments Technical White Paper, SBOA586A, February 2024 and Revised May 2025. https://www.ti.com/lit/wp/sboa586a/sboa586a.pdf?ts=1752659637804, Viewed July 16, 2025.

Kitchin, Charles and Lew Counts, "A Designer's Guide to Instrumentation Amplifiers." Analog Devices, 2006. https://www.analog.com/media/en/training-seminars/design-handbooks/designers-guide-instrument-amps-complete.pdf. Viewed July 16, 2025.

Pandiey, Ivailo Milanov, "Design and Implementation of Difference and Instrumentation Amplifier's Laboratory System for Education in Analog Electronics," *2020 XXIX International Scientific Conference Electronics (ET)*, Sozopol, Bulgaria, pp. 1–4, 2020.

Razavi, Behzad, *Fundamentals of Microelectronics*, 3rd Edition, Wiley, 2021.

Sedra, Adel S. and Kenneth Smith, *Microelectronic Circuits*, 8th International Edition, Oxford University Press, 2020.

Ulaby, Fawwaz, Michael M. Maharbiz, and Cynthia M. Furse, *Circuit Analysis and Design*, Michigan Publishing Services, 2024.

Valentine, Daniel T. and Brian D. Hahn, *Essential MATLAB for Engineers and Scientists*, 8th Edition, Academic Press, 2022.

Vector, Maxwell, *MATLAB Fundamentals for Electrical Engineers*, MATLAB Fundamentals Book Series, 2025.

3 Filters

3.1 FILTER CHARACTERISTICS

From the point of view of amplitude response, most filters can be classified as low pass, high pass, band pass or band-reject. Ideal filters have two bands: passband and stopband. The passband is a range of frequencies that are passed or transmitted through a filter. On the other hand, the stopband is a range of frequencies that are rejected. Figure 3.1 shows the amplitude response of some ideal filters.

For ideal filter responses shown in Figure 3.1a and b, w_C is the low cutoff frequency of low-pass filters and high cutoff frequency of high-pass filters. As shown in Figure 3.1c and d, w_{C1} is the low cutoff frequency of band-pass and band-rejection filters. In addition, w_{C2} is the high cutoff frequency of band-pass and band-reject filters. The bandwidth of the band-pass and band-reject filters is given as B, where

$$B = w_{C2} - w_{C1} \tag{3.1}$$

The ideal filter characteristics can never be reached, but as it is approached, the complexity of the filter circuit increases. Practical filters have an additional band, called the transition band. This band lies between the passband and the stopband. Figure 3.2 shows the transition band for a low-pass filter.

For a practical filter, the number of poles determines the roll-off rate of the filter. For example, a Butterworth response produces –20 dB/decade/pole. This means that:

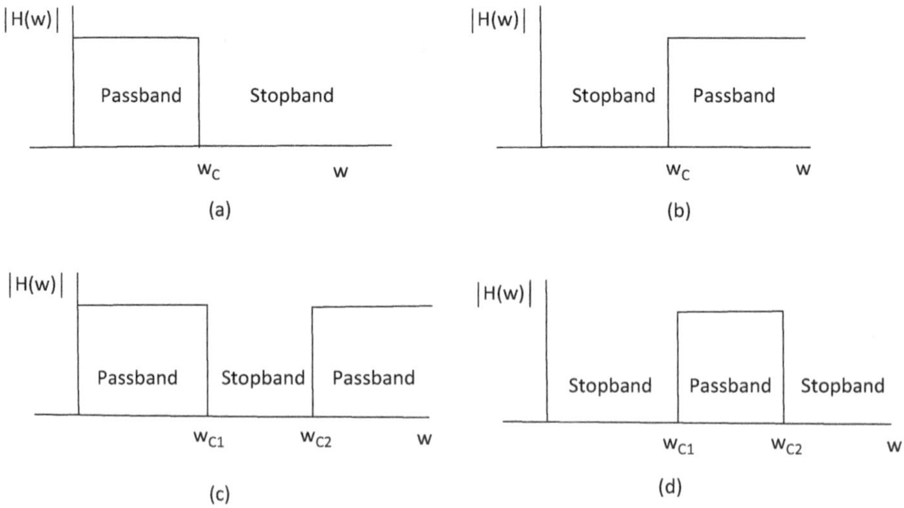

FIGURE 3.1 Amplitude response of ideal: (a) low-pass filter, (b) high-pass filter, (c) band-reject filter, and (d) band-pass filter.

DOI: 10.1201/9781003765554-3

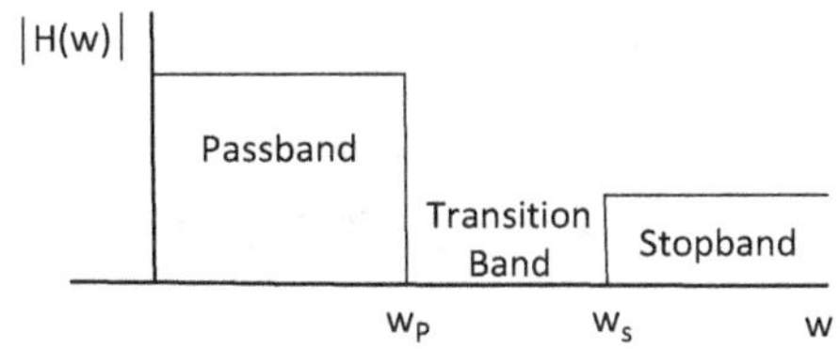

FIGURE 3.2 Characteristics of a practical low-pass filter.

(i) A one-pole (first-order) filter has a roll-off of –20 dB/decade, (ii) a two-pole (second-order) filter has a roll-off of –40 dB/decade, and (iii) a three-pole (third-order) filter has a roll-off of −60 dB/decade.

3.2 FILTER AMPLITUDE RESPONSE

In specifying filter requirements, the concept of attenuation is commonly used. The relative attenuation, measured in decibels (dB), is defined as

$$A_{dB}(f) = 20\log_{10}\left(\frac{A_0}{A(f)}\right) \tag{3.2}$$

where:

A_0 is the maximum passband amplitude response of a particular filter
$A(f)$ is the amplitude response at some frequency f.

There are well-known amplitude response characteristics for filters. The popular filter amplitude response characteristics are as follows: (i) Butterworth, (ii) Chebyshev, and (iii) maximally flat time delay. They are discussed below.

The transfer function of the *n*th-order filter is given as

$$H(s) = \frac{Kb_0}{s^n + b_{n-1}s^{n-1} + b_{n-2}s^{n-2} + \,..\, + b_1 s + b_0} \tag{3.3}$$

where $b_0, b_1, \ldots b_{n-1}$ and K are obtained based on the order of the filter. The Butterworth, Chebyshev, and Bessel filter circuits differ only by the values of the coefficients b_i, which will give different amplitude response characteristics. Tables 3.1–3.4 show the coefficients for Butterworth, Chebyshev, and Bessel filters.

3.2.1 Butterworth Response

The Butterworth response is also called maximum flat approximation. Butterworth response has a flat response at zero frequency. The Butterworth low-pass amplitude squared function is defined as

TABLE 3.1
Coefficient of Normalized Butterworth Filters with Transfer Function given by Equation (3.3)

n	b_0	b_1	b_2	b_3	b_4	b_5	b_6
1	1.000						
2	1.000	1.414					
3	1.000	2.000	2.000				
4	1.000	2.613	3.414	2.613			
5	1.000	3.236	5.236	5.236	3.236		
6	1.000	3.863	7.464	9.142	7.464	3.864	
7	1.000	4.494	10.098	14.592	14.592	10.098	4.494

TABLE 3.2
Coefficient of Normalized Chebyshev Filters for 0.1-dB Passband Ripple with Transfer Function Given by Equation (3.3)

n	b_0	b_1	b_2	b_3	b_4	b_5	b_6
1	6.552						
2	3.313	2.372					
3	1.638	2.630	1.939				
4	0.829	2.026	2.627	1.804			
5	0.410	1.436	2.397	2.771	1.744		
6	0.207	0.902	2.048	2.779	2.996	1.712	
7	0.102	0.562	1.483	2.705	3.169	3.184	1.693

TABLE 3.3
Coefficient of Normalized Chebyshev Filters for 1.0-dB Passband Ripple with Transfer Function Given by Equation (3.3)

n	b_0	b_1	b_2	b_3	b_4	b_5	b_6
1	1.965						
2	1.103	1.098					
3	0.491	1.238	0.988				
4	0.276	0.743	1.454	0.953			
5	0.123	0.581	0.974	1.689	0.937		
6	0.069	0.307	0.939	1.202	1.931	0.928	
7	0.031	0.214	0.549	1.358	1.429	2.176	0.923

TABLE 3.4
Coefficient of Normalized Bessel–Thomson Filters with Transfer Function Given by Equation (3.3)

n	b_0	b_1	b_2	b_3	b_4	b_5	b_6
1	1						
2	3	3					
3	15	15	6				
4	105	105	45	10			
5	945	945	420	105	15		
6	10395	10395	4725	1260	210	21	
7	135135	135135	62370	17325	3150	378	28

$$A(f) = \frac{1}{1+\left(\frac{f}{f_c}\right)^{2n}} \tag{3.4}$$

where

f_c is the cutoff frequency
n is the order of the filter

The attenuation of the Butterworth low-pass filter can be expressed in decibels as follows:

$$A_{\text{dB}}(f) = 10\log_{10}\left[1+\left(\frac{f}{f_c}\right)^{2n}\right] \tag{3.5}$$

where $\left(\frac{f}{f_c}\right)$ is the ratio of the given frequency f to the 3-dB cutoff frequency f_c.

The Butterworth amplitude characteristics are shown in Figure 3.3. The filter response is characterized by a flat amplitude response in the passband. Butterworth filters provide a roll-off rate of −20 dB/decade/pole and are normally used when all frequencies in the passband must have almost the same gain.

Table 3.1 shows the denominator polynomials of n-order Butterworth filters. Butterworth approximation is a monotonically decreasing function of frequency. As the order of the Butterworth filter increases, the response becomes "flatter" in the passband and the attenuation is greater in the stopband.

The Butterworth function is normally used when flat magnitude in the passband is most important (without ripple) and where linearity of phase or freedom from transient overshoot to a step input is of least importance.

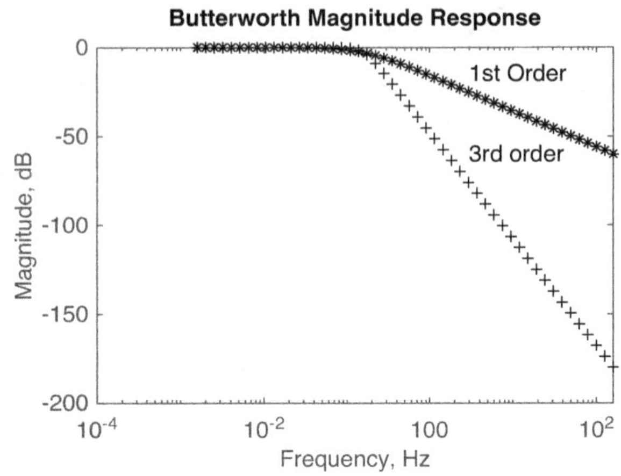

FIGURE 3.3 Amplitude responses of 1st- and 3rd-order Butterworth filters.

Example 3.1: Amplitude Response of a Butterworth Low-pass Filter

Plot the magnitude response for the first- and third-order Butterworth filters by using MATLAB®.

Solution

As from Table 3.1 and Equation (3.3), the transfer function of first-order and third-order Butterworth filters are $H_{B1}(s)=\dfrac{1}{s+1}$ and $H_{B3}(s)=\dfrac{1}{s^3+2s^2+2s+1}$, respectively.

The MATLAB code for the magnitude responses is shown below. The magnitude responses are shown in Figure 3.3.

MATLAB® CODE

```
%
% Magnitude response of a Butterworth Lowpass Filter
num  =  [ 1 ];  %  coefficients of numerator polynomial
den1  =  [1   1];  %  coefficients of 1st order Butterworth
polynomial
w  =  logspace (-2, 3); % range of frequencies
hs1  =  freqs(num, den1, w);
f  =  w/(2*pi);  %  frequency in Hz.
hs_mag1  =  20*log10(abs(hs1));  % Magnitude in decibels
den2  =  [1 2 2 1 ];  %  coefficients of 3rd order Butterworth r
polynomial
hs2  =  freqs(num, den2, w);
hs_mag2  =  20*log10(abs(hs2));  % Magnitude in decibels
% Plot the magnitude response
semilogx(f, hs_mag1,'k*',f,hs_mag2,'k+')
% semilogx(f, hs_mag1,f,hs_mag2)
title ('Butterworth Magnitude Response')
xlabel('Frequency, Hz')
```

```
ylabel('Magnitude, dB')
text(50, -40, '1st Order')
text(3, -60, '3rd order')
```

3.2.2 Chebyshev Response

The Chebyshev amplitude response is defined by

$$A^2(f) = \frac{\alpha}{1 + \varepsilon^2 C_n\left(\frac{f}{f_c}\right)} \tag{3.6}$$

where

n is the order of the Chebyshev polynomial or the order of the corresponding transfer function.
ε is a parameter chosen to provide the upper passband ripple.
α is a constant chosen to determine the proper DC gain.
f_c is the cutoff frequency.

$$\begin{aligned} C_n(x) &= \cos\left(n \cos^{-1} x\right) && \text{for } |x| < 1 \\ C_n(x) &= \cosh\left(n \cosh^{-1} x\right) && \text{for } |x| > 1 \end{aligned} \tag{3.7}$$

The Chebyshev polynomial, $C_n(x)$, includes a set of orthogonal functions possessing the following properties:

i. The Chebyshev polynomials have equi-ripple amplitude characteristics over the range $-1 \le x \le 1$, with the ripple oscillating between −1 and 1.
ii. The Chebyshev polynomial increases more rapidly for $x > 1$ than any other polynomial of order n bounded by the limits mentioned in part (i).

The parameter ε in Equation (3.6) is given as

$$\varepsilon = \sqrt{10^{\left(\frac{A_{\text{dB}}}{10}\right)} - 1} \tag{3.8}$$

where

A_{dB} is the passband ripple in dB, given as

$$A_{\text{dB}} = 10 \log_{10}\left[1 + \varepsilon^2 C_n^2(x)\right] \tag{3.9}$$

and

$C_n(x)$ is a Chebyshev polynomial whose magnitude oscillates between ± 1 for $x \le 1$.

The Chebyshev polynomial up to the 7th order is shown in Tables 3.2 and 3.3. The Chebyshev filter response is characterized by overshoot or ripples in the passband. The filter provides a roll-off rate greater than −20 dB/decade/pole. Filters with the Chebyshev response can be implemented with fewer poles and less complex circuitry for a given roll-off rate.

Chebyshev function is similar to the Butterworth function, in that it is an all-pole function. In the passband, the Chebyshev response has a magnitude that varies with specified tolerance limits. It has equal ripple magnitude within the passband. The Chebyshev polynomials are very efficient, in that they offer a faster transition from the passband to the stopband compared to Butterworth filters. However, Chebyshev filters have a highly nonlinear phase change with frequency. The nonlinearity of the phase worsens as the ripple increases, thereby seriously worsening the transient response.

Example 3.2: Amplitude Response of Chebyshev Low-pass Filters

Plot the magnitude responses for the second-order and third-order Chebyshev filters by using MATLAB.

Solution

As from Table 3.3, the transfer functions of second-order and third-order Chebyshev filters with 1.0 dB passband ripple are $H_{C2}(s) = \dfrac{1.103}{s^2 + 1.098s + 1.103}$ and $H_{C3}(s) = \dfrac{0.491}{s^3 + 0.988s^2 + 1.238s + 0.491}$, respectively.

The MATLAB code for plotting the magnitude responses is given as follows. The magnitude responses are shown in Figure 3.4.

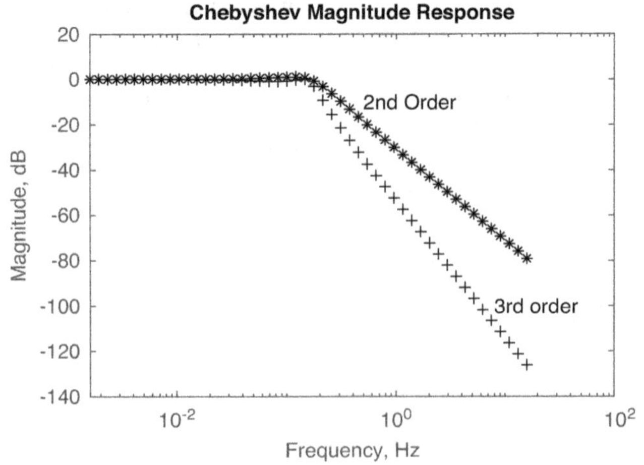

FIGURE 3.4 Amplitude responses of 2nd- and 3rd-order Chebyshev filters.

MATLAB® CODE

```
%
% Magnitude response Chebyshev Lowpass Filter
num  =  [ 1.103 ];  %  coefficients of numerator polynomial
den1  =  [1    1.098   1.103];  %  coeff. of 2nd order Chebyshev
polynomial
w  =  logspace (-2, 2); % range of frequencies
hs1  =  freqs(num, den1, w);
f  =  w/(2*pi);  %  frequency in Hz.
hs_mag1  =  20*log10(abs(hs1));  % Magnitude in decibels
num2 = [0.491];
den2  =  [1 0.988 1.238 0.491 ];  %  coef. of 3rd order Chebyshev
polynomial
hs2  =  freqs(num2, den2, w);
hs_mag2  =  20*log10(abs(hs2));  % Magnitude in decibels
% Plot the magnitude response
semilogx(f, hs_mag1,'k*',f,hs_mag2,'k+')
% semilogx(f, hs_mag1,f,hs_mag2)
title ('Chebyshev Magnitude Response')
xlabel('Frequency, Hz')
ylabel('Magnitude, dB')
text(0.5, -10, '2nd Order')
text(8, -100, '3rd order')
```

3.2.3 Bessel–Thomson Filters

The Bessel–Thomson or the maximally flat time-delay filter has a linear phase. The transfer function has been optimized such that all frequency components have a nearly constant delay. The frequency response is much less selective than those of Butterworth and Chebyshev filters.

The low-pass approximation to a constant time-delay can be expressed by the transfer function.

$$H(s) = \frac{1}{\sinh(s) + \cosh(s)} \tag{3.10}$$

Expanding Equation (3.10) for various lengths and using the continued fraction expansion will result in a Bessel function. The relative attenuation of the Bessel low-pass filter can be approximated by the expression (i.e., the initial stopband attenuation)

$$A_{\text{dB}} = 3\left(\frac{f}{f_c}\right)^2 \tag{3.11}$$

where $\left(\frac{f}{f_c}\right)$ is the ratio of the given frequency f to the 3-dB cutoff frequency f_c.

Equation (3.11) is reasonably accurate for $\left(\frac{f}{f_c}\right)$ ranging from 0 to 2. For values of $\left(\frac{f}{f_c}\right)$ greater than 2, a straight-line approximation of 6 dB per octave per element can be used.

The general applications of the Bessel–Thomson filters are in the design of systems where the linearity of phase is required or in situations where a lack of overshot or oscillation in response to a step input is of importance. The Bessel–Thomson filter polynomial up to the 7th order is shown in Table 3.4.

Example 3.3: Amplitude Response of Bessel–Thomson Low-pass Filters

Plot the magnitude and phase responses of third-order Bessel–Thomson lowpass filters by using MATLAB.

Solution

As from Table 3.4 and Equation (3.3), the transfer function of third-order Bessel–Thomson filters is given as $H_{BT3}(s)=\dfrac{15}{s^3+6s^2+15s+15}$.

The MATLAB code for plotting the magnitude and phase responses is shown below. Figure 3.5 shows the magnitude and phase responses.

MATLAB® CODE

```
% Amplitude and Phase Response of Bessel-Thomson Filter
%
num1 = [15]; % Numerator coeff of the 3rd order
den1 = [1 6 15 15]; % Denominator coeff of 3rd order
w = logspace(-1,3);
h1 = freqs(num1,den1,w);
f = w/(2*pi);
```

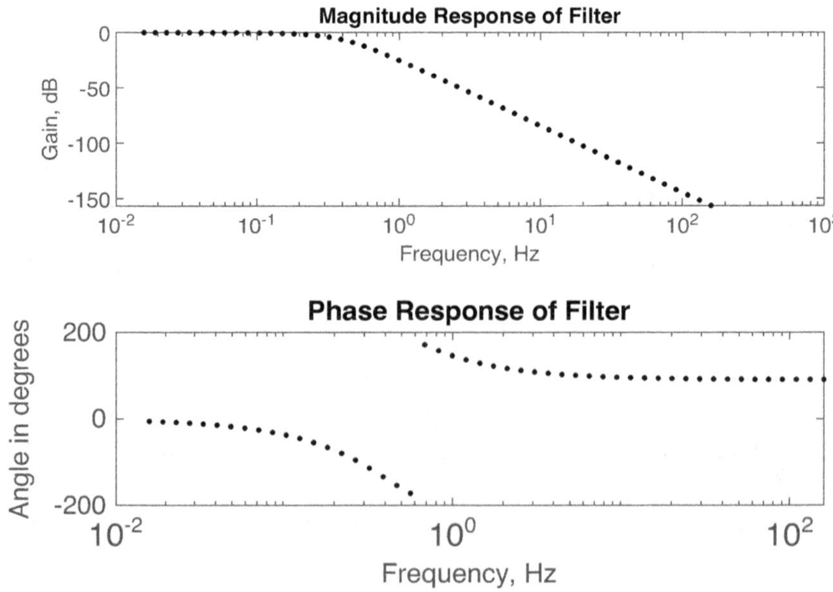

FIGURE 3.5 Amplitude and phases responses of 3rd-order Bessel–Thomson filters.

```
mag1 = 20*log10(abs(h1));
phase1 = angle(h1)*180/pi;
% Plot the response
subplot(211),
semilogx(f, mag1,'k.')
title('Magnitude Response of Filter')
ylabel('Gain, dB')
xlabel('Frequency, Hz')
subplot(212), semilogx(f, phase1,'k.')
title('Phase Response of Filter'),...
xlabel('Frequency, Hz'),
ylabel('Angle in degrees')
```

3.3 FILTER DESIGN FROM TABLES FOR LOW-PASS FILTERS

The circuit structures of a low-pass ladder filter of order n with a resistive load of 1 Ω and a cutoff frequency of 1 rad/s are shown in Figures 3.6 and 3.7, respectively. The element values for the *n*th-order Butterworth low-pass filter are shown in Table 3.5.

The element values L1, C2, L3, C4, …, and L9 are used if Figure 3.6 is used to realize the low-pass filter. However, the element values C1, L2, C3, L4, … and C9 are employed if Figure 3.7 is used to realize the low-pass filter.

Tables 3.6 and 3.7 are element values for low-pass Chebyshev filters with 0.1dB and 1.0dB ripples in the passband, respectively. Figures 3.7 and 3.8 can also be used to realize the filters.

Table 3.8 shows element values for low-pass Bessel–Thomson filters. Figures 3.6 and 3.7 can also be used to realize the filters.

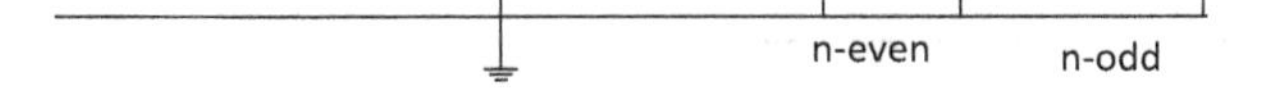

FIGURE 3.6 Structure A of *n*th-order low-pass Butterworth filters.

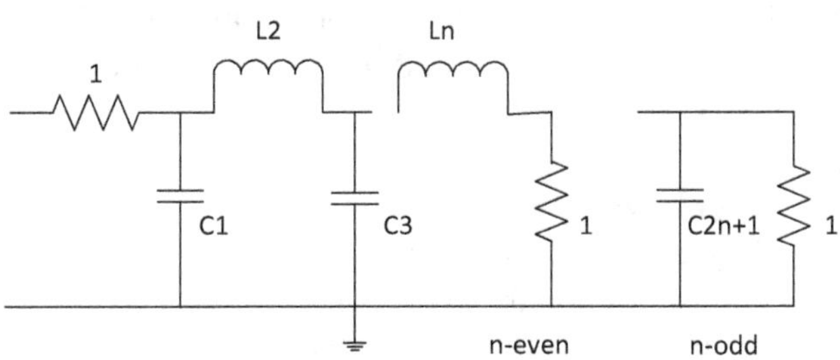

FIGURE 3.7 Structure B of *n*th-order low-pass Butterworth filters.

TABLE 3.5
Element Values for a Low-pass Butterworth Filter with a Load of 1 Ω and Cutoff Frequency of 1 rad/s

n/Figure 3.6 values	L1	C2	L3	C4	L5	C6	L7	C8	L9
2	1.414	1.414							
3	1.000	2.000	2.000						
4	0.765	1.848	1.848	0.765					
5	0.618	1.618	2.000	1.618	0.618				
6	0.518	1.414	1.932	1.932	1.414	0.518			
7	0.445	1.247	1.802	2.000	1.802	1.247	0.445		
8	0.390	1.111	1.663	1.962	1.962	1.663	1.111	0.390	
9	0.347	1.000	1.532	1.879	2.000	1.879	1.532	1.000	0.347
***n*/Figure 3.7 values**	**C1**	**L2**	**C3**	**L4**	**C5**	**L6**	**C7**	**L8**	**C9**

TABLE 3.6
Element Values for a Low-pass Chebyshev Filter with 0.1 dB Equi-Ripple in the Passband, Load of 1 Ω, and Cutoff Frequency of 1 rad/s

n/Figure 3.6 values	L1	C2	L3	C4	L5	C6	L7	C8	L9
3	1.433	1.594	1.433						
5	1.301	1.556	2.241	1.556	1.301				
7	1.262	1.520	2.239	1.680	2.239	1.520	1.262		
9	1.245	1.502	2.222	1.683	2.296	1.683	2.222	1.502	1.245
***n*/Figure 3.7 values**	**C1**	**L2**	**C3**	**L4**	**C5**	**L6**	**C7**	**L8**	**C9**

TABLE 3.7
Element Values for a Low-pass Chebyshev Filter with 1.0-dB Equi-Ripple in the Passband, Load of 1 Ω, and Cutoff Frequency of 1 rad/s

n/Figure 3.6 values	L1	C2	L3	C4	L5	C6	L7	C8	L9
3	2.216	1.088	2.216						
5	2.207	1.128	3.102	1.128	2.207				
7	2.204	1.131	3.147	1.194	3.147	1.131	2.204		
9	2.202	1.131	3.154	1.202	3.208	1.202	3.154	1.131	2.202
***n*/Figure 3.7 values**	**C1**	**L2**	**C3**	**L4**	**C5**	**L6**	**C7**	**L8**	**C9**

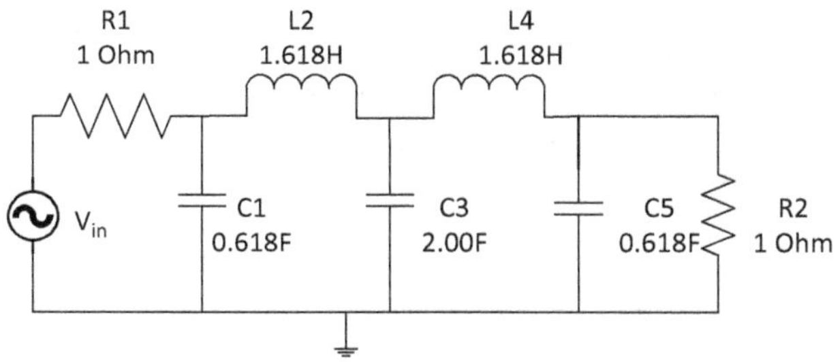

FIGURE 3.8 Element values of a 5th-order low-pass Butterworth filter.

TABLE 3.8
Element Values for a Low-pass Bessel–Thomson Filter with a Load of 1 Ω and Cutoff Frequency of 1 rad/s

n/Figure 3.6 values	L1	C2	L3	C4	L5	C6	L7	C8	L9
2	0.596	2.148							
3	0.337	0.971	2.203						
4	0.233	0.672	1.082	2.240					
5	0.174	0.507	0.804	1.111	2.258				
6	0.137	0.400	0.639	0.854	1.113	2.265			
7	0.110	0.326	o.525	0.702	0.869	1.105	2.260		
n/Figure 3.7 Values	C1	L2	C3	L4	C5	L6	C7	L8	C9

Example 3.4: Design of Normalized Butterworth Low-pass Filters

Determine the element values of 5th-order Butterworth filters with a cutoff frequency of 1 rad/s and load of 1 Ω.

Solution

As from Table 3.5, the element values of the Butterworth low-pass filter that will be realized by Figure 3.7 are as follows:

Using the element values shown in Table 3.9 and Figure 3.7, we get Figure 3.8.

3.4 LOW-PASS TO HIGH-PASS TRANSFORMATION

The low-pass filters discussed in Section 3.3, with a cutoff frequency of 1 rad/s, may be converted into high-pass filters by scaling using 1/s in the transfer function. This means that capacitors become inductors with a value 1/C and inductors become capacitors with a value of 1/L, as illustrated in Figure 3.9. The low-pass filter prototypes can also be transformed to band-pass or band-reject filters, but that is beyond the scope of this book. Example 3.5 shows the design of a high-pass filter from the low-pass element values.

TABLE 3.9
Element Values of a 5th-Order Low-pass Butterworth Filter

n	C1	L2	C3	L4	C5
5	0.618	1.618	2.000	1.618	0.618

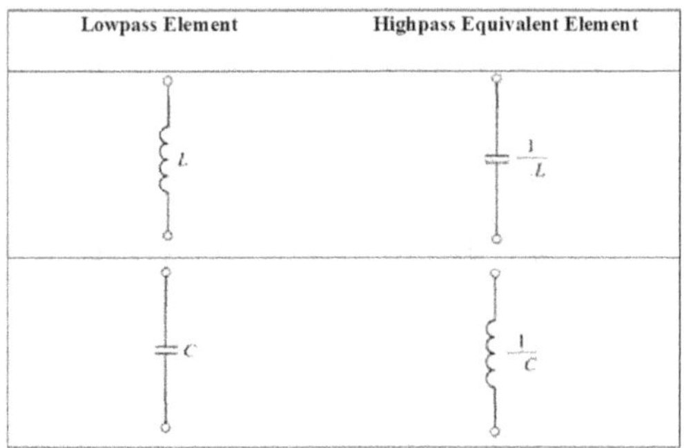

FIGURE 3.9 Low-pass to high-pass filter transformation elements.

TABLE 3.10
Element Values of a 5th-Order High-pass Butterworth Filter

n	L1	C2	L3	C4	L5
5	1.618	0.618	0.500	0.618	1.618

Example 3.5: Design of Normalized High-pass Butterworth Filters

Determine the element values of 5th-order Butterworth high-pass filters with a cutoff frequency of 1 rad/s and load of 1 Ω.

Solution

Taking the reciprocal of the element values of the capacitors and inductors in Table 3.9, we get the element values of the high-pass filter, as given in Table 3.10. The circuit diagram is shown in Figure 3.10.

3.5 FREQUENCY AND MAGNITUDE SCALING

Most filter design data are provided assuming that we have a normalized circuit. A normalized circuit assumes that the cutoff frequency or center frequency is 1 rad/s, and the load and source impedances are 1 Ω. The element values of normalized

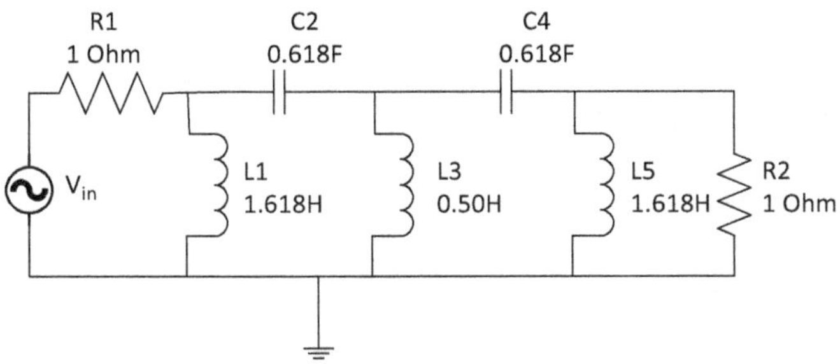

FIGURE 3.10 Element values of a 5th-order high-pass Butterworth filter.

circuits are normally quite unrealistic. Typically, in normalized circuits, the capacitance would be very high (in the Farads range). To obtain cutoff or center frequencies greater than 1 rad/s, frequency scaling can be done. Impedance scaling is normally used to obtain realistic elements and also to increase the impedance level of a filter circuit. In general, frequency and impedance scalings are done to obtain realistic element values.

In frequency scaling, the circuit parameters are changed such that at the new frequency, the impedance of each element remains the same as it was at the original frequency. Resistors are unaffected by frequency scaling, but capacitances and inductances are reduced by the frequency scaling factor. In impedance scaling, the impedance level of a circuit is changed while maintaining constant cutoff frequencies or resonant frequency of the circuit before and after the impedance scaling.

If K_f is the frequency scaling factor and K_m is the impedance scaling factor, then the scaled and unscaled elements are related by the equations

$$R_2 = K_m R_1 \tag{3.12}$$

$$L_2 = \frac{K_m}{K_f} L_1 \tag{3.13}$$

$$C_2 = \frac{C_1}{K_m K_f} \tag{3.14}$$

where

(L_1, C_1, R_1) and (L_2, C_2, R_2) are the element values of the unscaled and scaled circuits, respectively. Example 3.6 shows the applications of magnitude and frequency scaling in the design of filters.

Example 3.6: Design of 4th-order Butterworth Low-pass Filters

Design a 4th-order Butterworth low-pass filter with a cutoff frequency of 5000 Hz and load impedance of 1000 Ω.

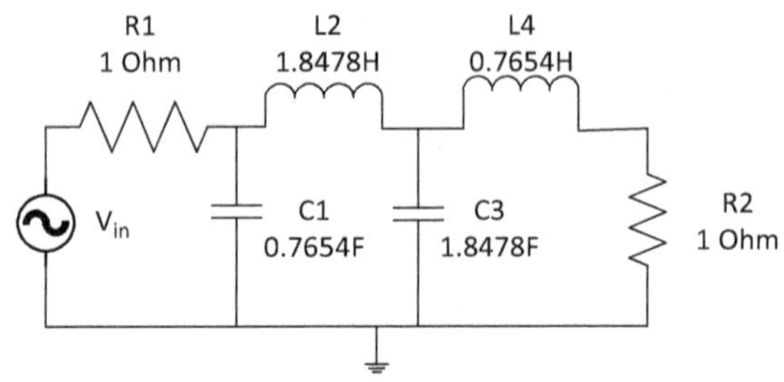

FIGURE 3.11 Normalized 4th-order low-pass Butterworth filter.

Solution

Using the information given in Table 3.5 and Figure 3.7, the circuit of the normalized 4th-order Butterworth lowpass filter is shown in Figure 3.11.

It is desired to redesign the filter such that the cutoff frequency is 5000 Hz and the load impedance is 1000 Ω. The impedance scaling factor is

$$K_m = \frac{1000}{1} = 10^3$$

The frequency scaling factor is given as

$$K_f = \frac{(2\pi)(5000)}{1} = 10{,}000\pi$$

Using Equations (3.12)–(3.14), the element values of the scaled circuit are

$$R' = (1000)1 = 1000\ \Omega$$

$$C1' = \frac{C1}{K_f K_m} = \frac{0.765}{(10000\pi)(1000)} = 24.36\ \text{nF}$$

$$L2' = \frac{K_m L2}{K_f} = \frac{(1.848)(1000)}{10000\pi} = 58.82\ \text{mH}$$

$$C3' = \frac{C3}{K_f K_m} = \frac{1.848}{(10000\pi)(1000)} = 58.82\ \text{nF}$$

$$L4' = \frac{K_m L4}{K_f} = \frac{(0.765)(1000)}{10000\pi} = 24.36\ \text{mH}$$

where

$R', C1', L2', C3', L4'$ are the element values of the scaled circuit. The scaled circuit is shown in Figure 3.12.

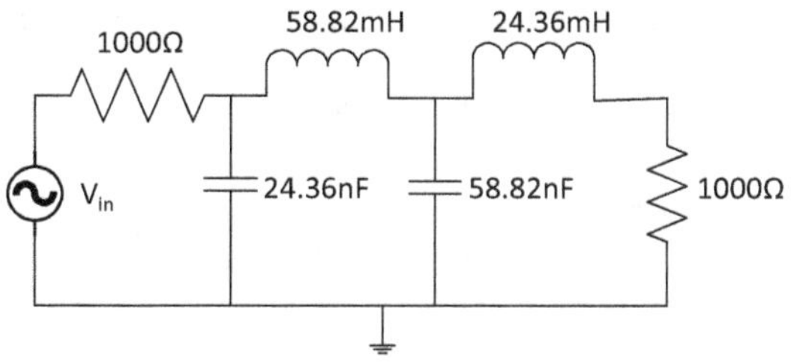

FIGURE 3.12 Fourth-order Butterworth filter with a cutoff frequency of 5000 Hz and load impedance of 1000 Ω.

Example 3.7: Design of 7th-order Chebyshev High-pass Filters

Design a 7th-order Chebyshev high-pass filter with a passband ripple of 1.0 dB with a cutoff frequency of 15.0 kHz and load impedance of 5000 Ω.

Solution

As from Table 3.7, the element values of the 7th-order Chebyshev filter with a passband ripple of 1.0 dB with a cutoff frequency of 1 rad/s are shown in Table 3.11. Taking the reciprocal of the element values of the capacitors and inductors given in Table 3.11, we get the element values of the high-pass filter, given in Table 3.12.

$$K_m = \frac{5000}{1} = 5000$$

The frequency scaling factor is given as

$$K_f = \frac{(2\pi)(15000)}{1} = 30,000\pi$$

TABLE 3.11
Element Values of a 7th-Order Chebyshev Low-pass Filter with 1.0-dB ripple in the Passband

n	L1	C2	L3	C4	L5	C6	L7
7	2.204	1.131	3.147	1.194	3.147	1.131	2.204

TABLE 3.12
Element Values of a 7th-Order Chebyshev High-pass Filter with 1.0-dB Ripple in the Passband

n	C1	L2	C3	L4	C5	L6	C7
7	0.454	0.884	0.318	0.838	0.318	0.884	0.454

Using Equations (3.12)–(3.14), the element values of the scaled circuit are

$$R' = (5000)1 = 5000\ \Omega$$

$$C1' = \frac{C1}{K_m K_f} = \frac{(0.454)}{(30000\pi)(5000)} = 9.63\text{ nF}$$

$$L2' = \frac{K_m L2}{K_f} = \frac{(0.884)(5000)}{(30000\pi)} = 46.9\text{ mH}$$

$$C3' = \frac{C3}{K_m K_f} = \frac{(0.318)}{(30000\pi)(5000)} = 674.7\text{ pF}$$

$$L4' = \frac{K_m L4}{K_f} = \frac{(0.838)(5000)}{(30000\pi)} = 44.45\text{ mH}$$

$$C5' = \frac{C5}{K_m K_f} = \frac{(0.318)}{(30000\pi)(5000)} = 674.7\text{ pF}$$

$$L6' = \frac{K_m L6}{K_f} = \frac{(0.838)(5000)}{(30000\pi)} = 44.45\text{ mH}$$

$$C7' = \frac{C7}{K_m K_f} = \frac{(0.454)}{(30000\pi)(5000)} = 9.63\text{ nF}$$

where:

$C1, L2, C3, L4, C5, L6, C7$ are the element values of the unscaled circuit and $R', C1', L2', C3', L4', C5', L6', C7'$ are the element values of the scaled circuit.

The scaled filter circuit is shown in Figure 3.13.

3.6 ACTIVE AND PASSIVE FILTERS

In general, filters can be classified as passive or active. Passive filters consist of circuits constructed using passive elements, such as capacitors, resistors, and inductors. On the other hand, active filters are networks consisting of resistors, capacitors, and active devices such as op amps and transistors.

There are disadvantages of using inductors in filter networks. Some filter designs might require inductors with large values. Large-valued inductors normally have large size, and they are relatively expensive. In addition, inductors are lossy, and they cause unnecessary power dissipation in filter circuits.

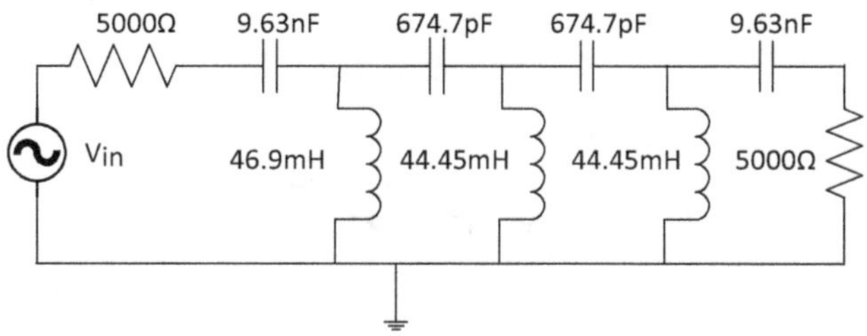

FIGURE 3.13 Element values of a 7th-order Chebyshev high-pass filter.

There are a number of advantages for using active filters as compared to passive filters. Active filters can be designed to use op amps to realize inductors. Thus, active filters can be small in size and less expensive since inductors are not used. In addition, active filters can be designed such that they can easily be cascaded. This is done by providing good isolation between the filter sections. Cascading with isolation amplifiers (voltage follower circuits) allows the filter sections to be designed separately and allows us to easily obtain the total transfer function as the product of the transfer functions of the individual sections. Furthermore, active filters can easily be designed to produce gain to satisfy system or filter requirements.

Active filters have some disadvantages. All active filters require power supply, whereas passive filters do not require power supplies. Active filters, because of their use of operational amplifiers and transistors, have definite signal limits beyond which nonlinear operation of the devices occurs. Furthermore, active filters have low gains at high frequencies. The op amps used in building active filters are inherently low-pass devices, so the active filter (elements and devices) components need to be carefully chosen for medium- or high-frequency operations of the filters.

3.7 ACTIVE FILTER CIRCUITS

The following types of active filters will be discussed in this section: low pass, band pass, high pass, and band-reject. The order of the filter will determine the transition from the passband to stopband.

3.7.1 Low-pass Filters

Low-pass filters pass low frequencies and attenuate high frequencies. The transfer function of a first-order low-pass filter has the general form:

$$H(s) = \frac{k}{s + \omega_0} \tag{3.15}$$

A circuit that can be used to implement a first-order lowpass filter is shown in Figure 3.14.

The voltage transfer function for Figure 3.14 is

$$H(s) = \frac{V_0(s)}{V_{IN}(s)} = \frac{k}{1 + sR_1C_1} \tag{3.16}$$

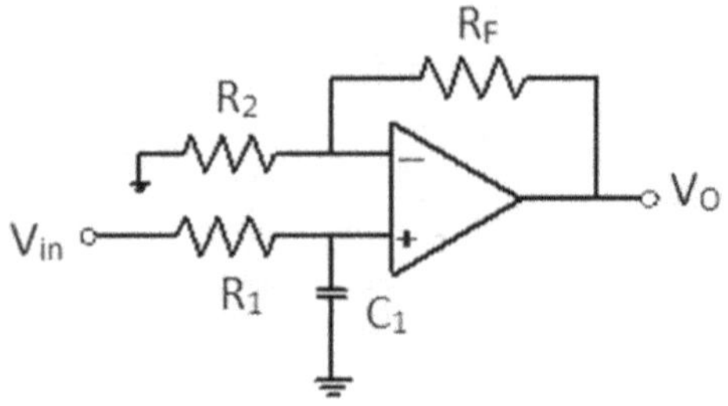

FIGURE 3.14 First-order low-pass filter.

with DC gain, k, given as

$$k = 1 + \frac{R_F}{R_2} \tag{3.17}$$

and the cutoff frequency, f_C

$$f_C = \frac{1}{2\pi R_1 C_1} \tag{3.18}$$

The first-order filter exhibits −20dB/decade roll-off in the stopband. Another low-pass filter circuit is shown in Figure 3.15.

The transfer function of the filter is given as

$$H(f) = \frac{V_O}{V_{in}} = -\frac{Z_f}{Z_{in}} \tag{3.19}$$

$$Z_f = R_2 \left\| \frac{1}{j2\pi fC} = \frac{R_2}{1 + j2\pi fR_2C} \right. \tag{3.20}$$

$$H(f) = -\left(\frac{R_2}{R_1}\right)\left(\frac{1}{1 + j2\pi fR_2C}\right) \tag{3.21}$$

The cutoff frequency, f_C, is given as

$$f_C = \frac{1}{2\pi R_2 C} \tag{3.22}$$

Example 3.8: Low-pass Filter Parameters

For the circuit shown in Figure 3.15, $R_1 = 2$ kΩ, $R_2 = 40$ kΩ, and $C = 0.025$ μF, (i) find gain at DC and (ii) determine the cutoff frequency.

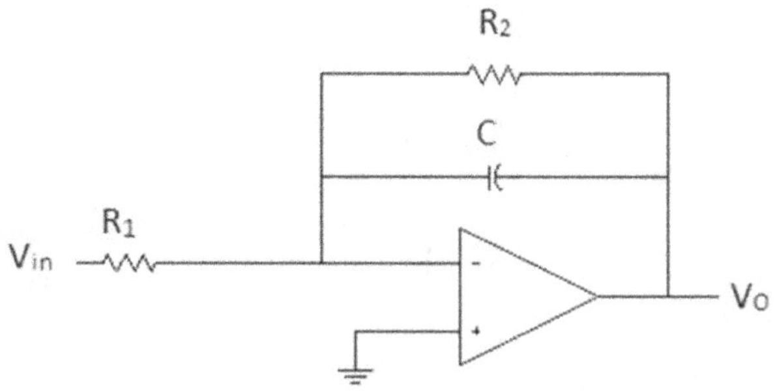

FIGURE 3.15 Low-pass filter.

Solution

i. From Equation (3.21), the DC gain is given as $-\frac{R_2}{R_1} = -\frac{40\,K}{2\,K} = -20$

ii. From Equation (3.22), the cutoff frequency is given as

$$f_C = \frac{1}{2\pi R_2 C} = \frac{1}{2\pi(40\times 10^3)(0.025\times 10^{-6})} = 159.13\text{ Hz}$$

3.7.2 High-pass Filters

High-pass filters pass high frequencies and attenuate low frequencies. The transfer function of the first-order high-pass filter has the general form

$$H(s) = \frac{ks}{s+\omega_0} \tag{3.23}$$

The circuit, shown in Figure 3.16, can be used to implement the first-order high-pass filter.

The voltage transfer function of Figure 3.16 is as follows:

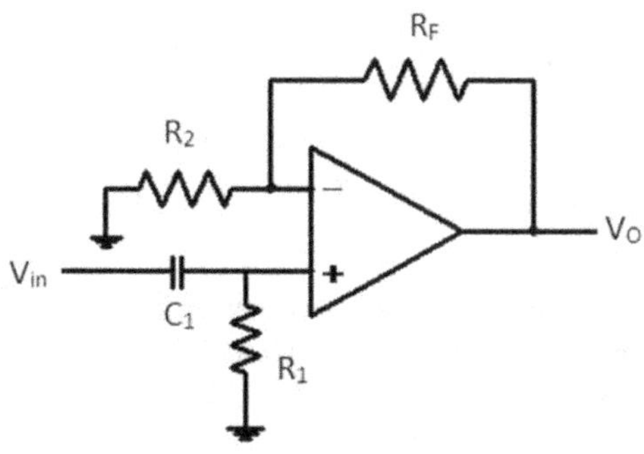

FIGURE 3.16 First-order high-pass filter.

$$H(s) = \frac{V_O}{V_{IN}}(s) = \frac{s}{s + \frac{1}{R_1 C_1}}\left(1 + \frac{R_F}{R_2}\right) \tag{3.24}$$

where

$$k = \left(1 + \frac{R_F}{R_2}\right) \tag{3.25}$$

k = gain at very high frequency
and the cutoff frequency at 3-dB gain is

$$f_C = \frac{1}{2\pi R_1 C_1} \tag{3.26}$$

It should be noted that although the filter, shown in Figure 3.16, passes all signal frequencies higher than f_C, the high-frequency characteristic is limited by the bandwidth of the op amp.

3.7.3 Band-pass Filters

Band-pass filters pass a band of frequencies and attenuate other bands. The filter has two cutoff frequencies f_L and f_H. We assume that $f_H > f_L$. All signal frequencies lower than f_L or greater than f_H are attenuated. The general form of the transfer function of a band-pass filter is

$$H(s) = \frac{k\left(\omega_c / Q\right)s}{s^2 + \left(\omega_c / Q\right)s + \omega_c^2} \tag{3.27}$$

where

k is the passband gain.
w_c is the center frequency in rad/s.

The quality factor Q is related to the 3-dB bandwidth and the center frequency by the expression.

$$Q = \frac{\omega_c}{\mathrm{BW}} = \frac{f_c}{f_H - f_L} \tag{3.28}$$

Band-pass filters with $Q < 10$ are classified as wide bandpass. On the other hand, band-pass filters with $Q > 10$ are considered narrow bandpass. Wide-band pass filters may be implemented by cascading low-pass and high-pass filters. The order of the band-pass filters is the sum of the high-pass and low-pass sections. The advantages of this arrangement are that the fall-off, bandwidth, and midband gain can be set independently.

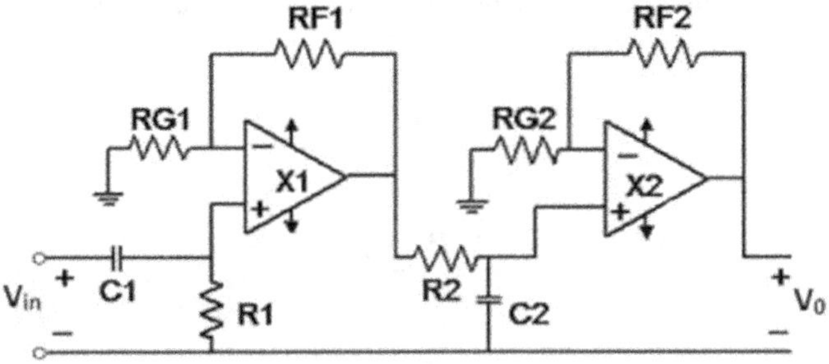

FIGURE 3.17 Second-order wideband-pass filter.

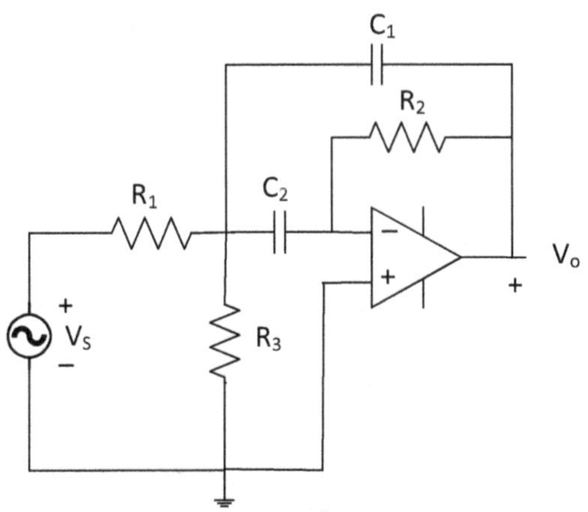

FIGURE 3.18 Multiple feedback band-pass filter.

Figure 3.17 shows the wide-band pass filter, built using first-order high-pass and first-order low-pass filters. The magnitude of the voltage gain is the product of the voltage gains of both the high-pass and low-pass filters.

A circuit that can be used to implement a narrow-band filter is a multiple feedback filter shown in Figure 3.18. The low-pass circuit consists of R_1 and C_1, and the high-pass circuit consists of R_2 and C_2. The feedback paths pass through C_1 and R_2. The center frequency f_0 is given as follows:

$$f_0 = \frac{1}{2\pi\sqrt{(R_1 \| R_3)R_2C_1C_2}} \tag{3.29}$$

The resonant frequency can be tuned by changing the value of resistor R_3. By making $C_1 = C_2 = C$, the resonant frequency becomes

$$f_0 = \frac{1}{2\pi C}\sqrt{\frac{R_1 + R_3}{R_1R_2R_3}} \tag{3.30}$$

The resistor values can be found by using the following equations:

$$R_1 = \frac{Q}{2\pi f_0 C A_0} \tag{3.31}$$

$$R_2 = \frac{Q}{\pi f_0 C} \tag{3.32}$$

$$R_3 = \frac{Q}{2\pi f_0 C(2Q^2 - A_0)} \tag{3.33}$$

where Q is the quality factor. The maximum gain, A_0, that occurs at the center frequency, is given as follows:

$$A_0 = \frac{R_2}{2R_1} \tag{3.34}$$

3.7.4 Band-Reject Filters

A band-reject filter is used to eliminate a specific band of frequencies. It is normally used in communication and biomedical instruments to eliminate unwanted frequencies. The general form of the transfer function of the band-reject filter is

$$H_{BR} = \frac{k_{PB}\left(s^2 + \omega_C^2\right)}{s^2 + \left(\omega_C/Q\right)s + \omega_C^2} \tag{3.35}$$

where

k_{PB} is the passband gain
ω_C is the center frequency of the band-reject filter

Band-reject filters are classified as a wide-band reject filter ($Q < 10$) and narrow-band reject filter ($Q > 10$). Narrow-band reject filters are commonly called notch filters. The wide-band reject filter can be implemented by summing the responses of high-pass section and low-pass section through a summing amplifier. The block diagram of the arrangement is shown in Figure 3.19.

The order of the band-reject filter is dependent on the order of low-pass and high-pass sections. There are two important requirements for implementing the wide-band reject filter using the scheme shown in Figure 3.19.

i. The cutoff frequency, f_L, of the high-pass filter section must be greater than the cutoff frequency, f_H, of the low-pass filter section.
ii. The passband gains of the low-pass and high-pass sections must be equal.

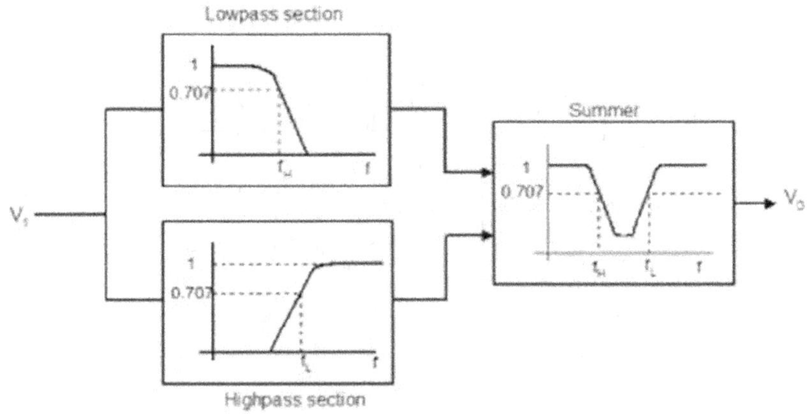

FIGURE 3.19 Block diagram of wideband-reject filter.

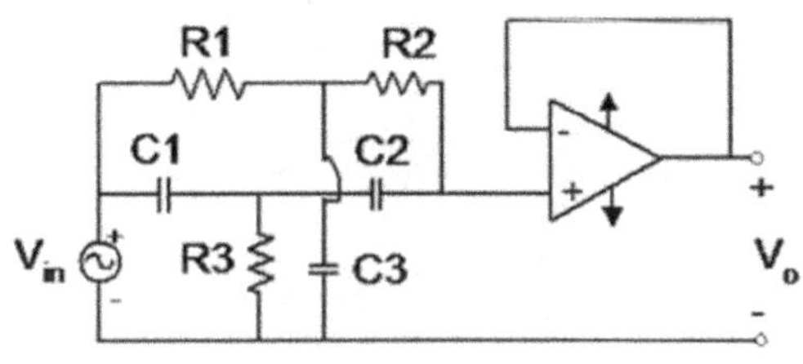

FIGURE 3.20 Narrowband-reject Twin-T network.

The narrow-band reject filter or notch filter can be implemented by using a Twin-T network. This circuit consists of two parallel T-shaped networks, shown in Figure 3.20.

Normally, $R1 = R2 = R$, $R3 = R/2$, $C1 = C2 = C$, and $C3 = 2C$. The $R1$-$C3$-$R2$ section is a low-pass filter with corner frequency $f_C = (4\pi RC)^{-1}$. The $C1$-$R3$-$C2$ is a high-pass filter with corner frequency $f_C = (\pi RC)^{-1}$.

The center frequency or the notch frequency is

$$f_C = \frac{1}{2\pi RC} \tag{3.36}$$

At the notch frequency given by Equation (3.33), the phases of the two filters cancel out.

3.7.5 Sallen–Key Filters

The Sallen–Key filters have high input and low output impedances. The filters implement 2-pole low-pass, high-pass, band-pass, and band-stop responses. The filters do not use inductors, and they can be cascaded with a stage that affects the tuning of the other stages. The Sallen–Key low-pass filter is shown in Figure 3.21.

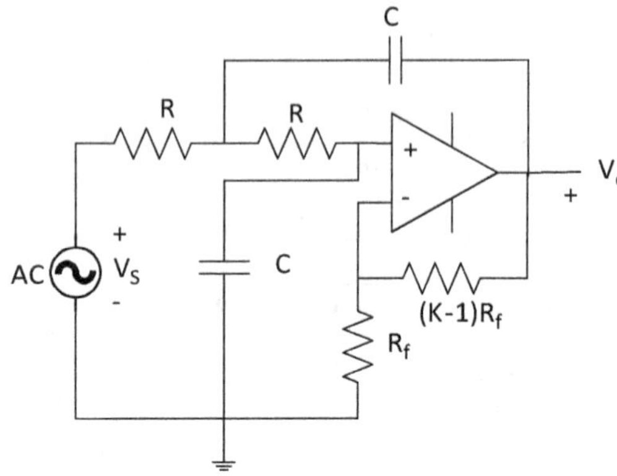

FIGURE 3.21 Sallen–Key low-pass filter.

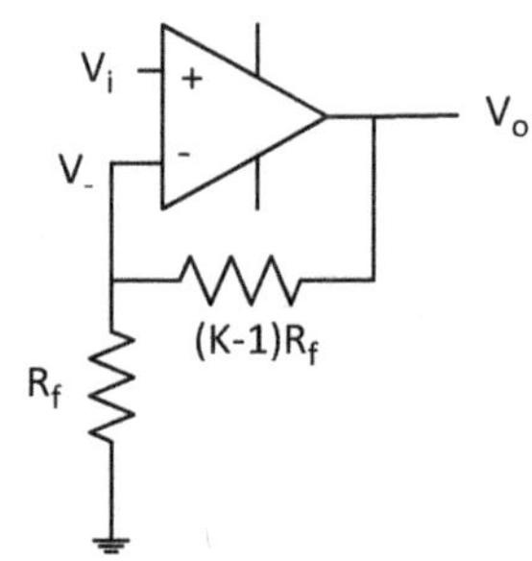

FIGURE 3.22 Non-inverting amplifier with a gain of *K*.

The low cutoff frequency, f_C, is given as follows:

$$f_C = \frac{1}{2\pi RC} \tag{3.37}$$

Figure 3.22 shows that the gain of the Sallen–Key circuit is given as *K*

$$v_- = v_i = \frac{R_f}{R_f + (K-1)R_f} v_O \tag{3.38}$$

$$\frac{v_O}{v_i} = A_V = 1 + \frac{(K-1)R_f}{R_f} = K \tag{3.39}$$

The Sallen–Key low-pass filter, shown in Figure 3.21, can be cascaded to obtain a higher-order low-pass filter. The *K*-values for low-pass Butterworth filters of various orders are shown in Table 3.13.

TABLE 3.13
K-Values of a Low-pass Butterworth Filter of Various Orders

Order	*K*-Values
2	1.586
4	1.152
	2.235
6	1.068
	1.586
	2.483
8	1.038
	1.337
	1.889
	2.610

Example 3.9: Design of Sallen–Key Low-pass Filters

Design a fourth-order low-pass Butterworth filter having a cutoff frequency of 100 Hz

Design:

$$R1 = R2 = R5 = R6 = R$$

$$C1 = C2 = C3 = C4 = C$$

Choose $C = 0.1$ μF. From Equation (3.37), we have

$$R = \frac{1}{2\pi C f_C} = \frac{1}{2\pi(0.1\times 10^{-6})(100)} = 15.92\ \text{k}\Omega$$

From Table 3.13, a fourth-order filter requires *K*-values of 1.152 and 2.235. The DC gain is (1.152)(2.235) = 2.575.

Assume $R4 = R8 = 10\ \text{k}\Omega$

$K\text{-value} =$ 1.152 for the first stage of the 4th-order Butterworth filter (see Table 3.13).

$K\text{-value} =$ 2.235 for the second stage of the 4th-order Butterworth filter (see Table 3.13)

Thus,

$$R3 = 1.52\ \text{k}\Omega, \text{and } R7 = 12.35\ \text{k}\Omega$$

The designed circuit is shown in Figure 3.23.

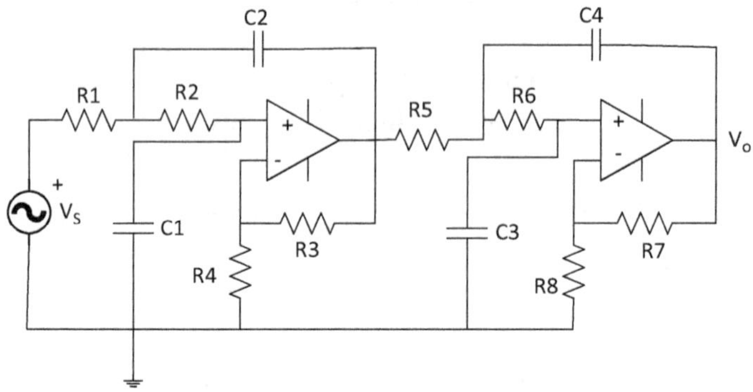

FIGURE 3.23 Design of 4th-order low-pass Butterworth filters.

PROBLEMS

Problem 3.1 Using Table 3.1, the transfer function of the 2nd-order Butterworth filter is

$H_{B2}(s) = \dfrac{1}{s^2 + 1.414s + 1}$ and that of the 4th-order Butterworth filter is

$$H_{B4}(s) = \frac{1}{s^4 + 2.613s^3 + 3.414s^2 + 2.613s + 1}$$

Use MATLAB to plot the magnitude response for the 2nd- and 4th-order Butterworth filters.

Problem 3.2 Using Table 3.3, the transfer function of the fourth-order Chebyshev filter with the 1.0-dB passband ripple is given as follows:

$$H_{C4}(s) = \frac{0.276}{s^4 + 0.953s^3 + 1.454s^2 + 0.743s + 0.276}$$

Use MATLAB to plot the magnitude response for the 4th-order Chebyshev filters.

Problem 3.3 As from Table 3.4, the transfer function of the 5th-order Bessel–Thomson Filter is given as $H_{BT5}(s) = \dfrac{945}{s^5 + 15s^4 + 105s^3 + 420s^2 + 945s + 945}$.

Use MATLAB to plot the magnitude and phase responses of the filter.

Problem 3.4 (a) Design a 5th-order low-pass Butterworth filter with a cutoff frequency of 3 kHz and load impedance of 2 kΩ; (b) use SPICE or other circuit simulator to determine the cutoff frequency of the circuit designed in Part (i).

Problem 3.5 (a) Design a 5th-order low-pass Chebyshev filter with a 1.0-dB passband ripple, cutoff frequency of 5 kHz and load impedance of 10 kΩ; (b) use SPICE or other circuit simulator to determine the cutoff frequency of the circuit designed in Part (a).

Problem 3.6 (a) Design a 7th-order high-pass Butterworth filter with a cutoff frequency of 4 kHz and load impedance of 1 kΩ; (b) use SPICE or other circuit simulator to determine the cutoff frequency of the circuit designed in Part (i).

Problem 3.7 (a) Design a 9th-order high-pass Chebyshev filter with a 0.1-dB passband ripple, cutoff frequency of 15 kHz, and load impedance of 5 kΩ; (b) use SPICE or other circuit simulator to determine the cutoff frequency of the circuit designed in Part (a).

Problem 3.8 For the Figure P3.1, R1 = 1.5 KΩ, R2 = 30 kΩ, and C = 0.005 μF, (i) obtain the transfer function, (ii) plot the magnitude response of the filter, and (iii) determine the cutoff frequency.

Problem 3.9 (a) Design a passive band-pass filter with a center frequency of 40 kHz and quality factor of 2. For the circuit designed in Problem 3.9 (a), use SPICE or other circuit simulator to determine the bandwidth of the filter (Figure P3.2).

Problem 3.10 Design a Sallen–Key 6th-order low-pass filter with the cutoff frequency of 3 kHz.

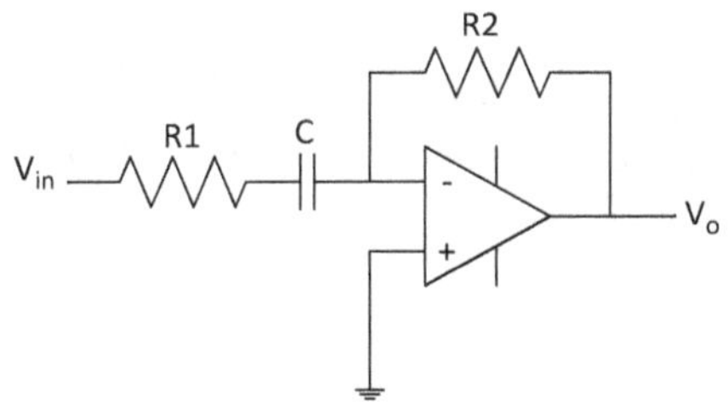

FIGURE P3.1 High-pass filter.

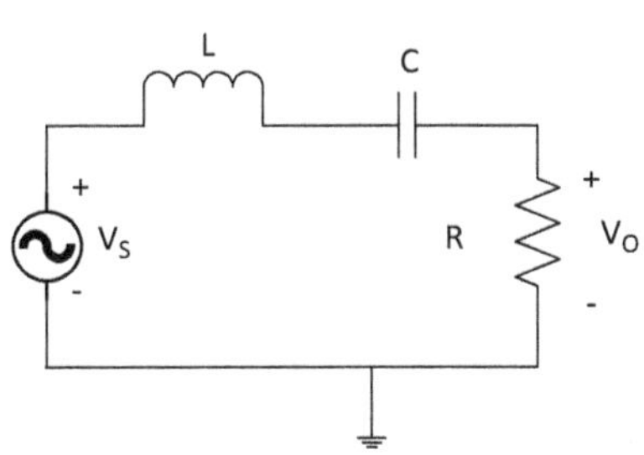

FIGURE P3.2 Passive band-pass filter.

BIBLIOGRAPHY

Alexander, Charles and Matthew N.O. Sadiku, *Fundamentals of Electric Circuit, ISF*, 7th Edition, McGraw-Hill Education, 2020.

Attia, John O., *Electronics and Circuit Analysis Using MATLAB*, CRC Press, 2nd Edition, 2004.

Chapman, Stephen J., *MATLAB Programming for Engineers*, 7th Edition, Cengage Learning, 2024.

Denisenko, Denisenko, Nikolay Prokopenko, and Anna Bugakova, "Low Pass Filter of the Sallen-Key Family with Main Parameters' Independent Adjustment," *2022 IEEE 16th International Conference on Solid-State & Integrated Circuit Technology (ICSICT)*, Nangjing, China, pp. 1–3, 2022.

Dorr, Barry L., "How to Design Analog Filters," in *Ten Essential Skills for Electrical Engineers,* IEEE Press, pp.123–148, 2014.

Etter, Delores, *Engineering Problem Solving with MATLAB*, 3rd Edition, Pearson, 2021.

Hayt, William, Jack Kemmely, Jamie Phillips, and Steven M. Durbin, *Engineering Circuit Analysis*, 10th Edition, McGraw Hill, 2023.

Horowitz, Paul, and Winfield Hill, *The Art of Electronics*, 3rd Edition, 11th Printing with Corrections, Cambridge University Press, 2017.

Irwin, J. David and R. Mark Helms, *Engineering Circuit Analysis*, 12th Edition, Wiley, 2021.

Jaeger, Richard C. and Travis Blalock, *Microelectronic Circuit Design (ISE)*, 6th Edition, McGraw Hill, 2022.

Razavi, Behzad, *Fundamentals of Microelectronics*, 3rd Edition, Wiley, 2021.

Sánchez-Sinencio, Edgar, "Analog filter design: Current design techniques and trends," *2017 IEEE Custom Integrated Circuits Conference (CICC)*, Austin, TX, pp. 1–76, 2017.

Sedra, Adel S. and Kenneth Smith, *Microelectronic Circuits*, 8th International Edition, Oxford University Press, 2020.

Ulaby, Fawwaz, Michael M. Maharbiz, and Cynthia M. Furse, *Circuit Analysis and Design*, Michigan Publishing Services, 2024.

Valentine, Daniel T. and Brian D. Hahn, *Essential MATLAB for Engineers and Scientists*, 8th Edition, Academic Press, 2022.

Vector, Maxwell, *MATLAB Fundamentals for Electrical Engineers*, MATLAB Fundamentals Book Series, 2025.

4 Analog Signal Processing Circuits

In this chapter, some electronic circuits that can be used to perform various operations in electronic instrumentation systems will be discussed. Electronic circuits that will be discussed include integrators, differentiators, limiters, rectifiers, comparators, and peak detectors.

4.1 INTEGRATOR CIRCUITS

The integrator circuit, shown in Figure 4.1, can be used to realize the mathematical operation of an integration. In the frequency domain, we have

$$H(w) = \frac{V_O}{V_{in}} = -\frac{1/jwC}{R_1} = -\frac{1}{jwCR_1} \tag{4.1}$$

In the time domain, we have

$$\frac{V_{in}}{R_1} = I_R \text{ and } I_C = -C\frac{dV_0}{dt} \tag{4.2}$$

Since $I_R = I_C$

$$V_0(t) = -\frac{1}{R_1C}\int_0^t V_{in}(t)\,dt + V_0(0) \tag{4.3}$$

where

R_1C is the integrating time constant.

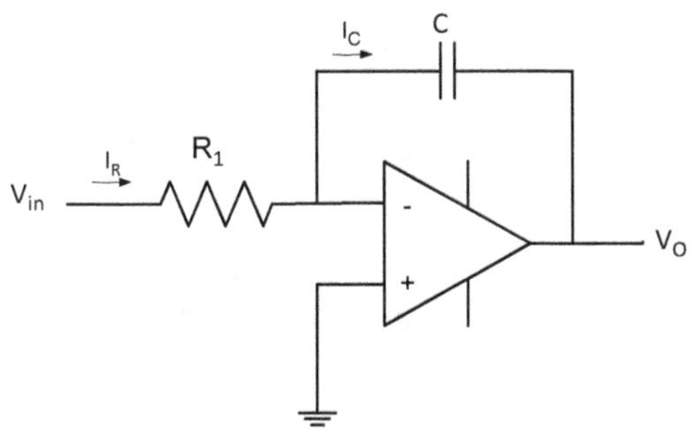

FIGURE 4.1 Op amp inverting integrator.

DOI: 10.1201/9781003765554-4

The above circuit is called the Miller integrator. The output voltage is an inversion of the input.

The Miller integrator behaves as a low-pass filter, passing low frequencies and attenuating high frequencies. However, at DC, the capacitor becomes open-circuited, and there is no longer negative feedback from the output to the input. The output voltage then saturates. To provide finite closed-loop gain at DC, a resistance R_2 is connected in parallel with the capacitor. The circuit is shown in Figure 4.2. The resistance R_2 is chosen such that R_2 is far greater than R_1.

The transfer function of Figure 4.2 is given as

$$H(w) = -\frac{R_2/R_1}{1 + jwR_2C} \tag{4.4}$$

The break frequency, f_b, is

$$f_b = \frac{1}{2\pi R_2 C} \tag{4.5}$$

For

$$f \ll f_b, \quad |H(w)| \simeq \frac{R_2}{R_1} \tag{4.6}$$

and for

$$f \gg f_b, \quad |H(w)| \simeq \frac{1}{wR_2C} \tag{4.7}$$

Equation (4.7) shows that the magnitude response of an AC integrator closely approximates that of a true integrator in the frequency range far above the break frequency. In addition, as from Equation (4.6), the AC integrator behaves as an amplifier with a gain of $\frac{R_2}{R_1}$ at frequencies far below the break frequency. One useful application of

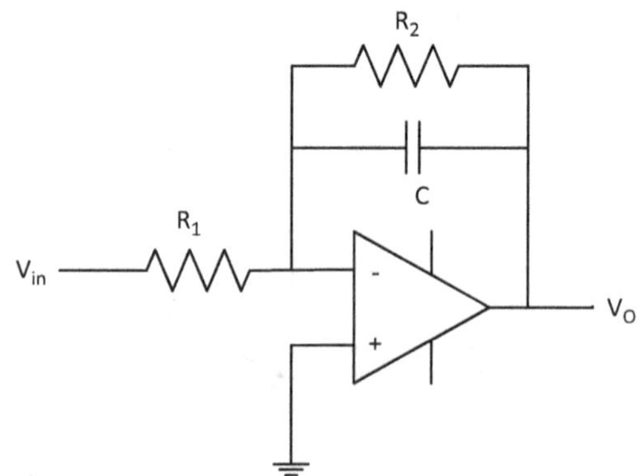

FIGURE 4.2 Miller integrator with finite closed-loop gain at DC.

AC integrators is in wave shaping of periodic waveforms. For example, a triangular wave can be produced by integrating a periodic square wave.

Example 4.1: Output Voltage of an Integrator

For an integrator circuit as shown in Figure 4.1, the voltage across the capacitor at $t = 0$ sec is $V_C = -2.5$ V. A step input voltage $V_{in} = -3$ V is applied at time $t = 0$. Determine the time constant necessary such that the output voltage reaches 9.5 V at time $t = 6$ ms.

Solution

From Equation (4.3), the output voltage is given as follows:

$$V_0(t) = -\frac{1}{R_1C}\int_0^{6\text{ms}} V_{in}(t)\,dt + V_C(0)$$

Given that $V_c(0) = -2.5\text{ V},\quad V_{in}(t) = -3.0\text{ V},\quad V_0(0.006) = 9.5\text{ V}$

Substituting the given values into Equation (4.3), we have

$$9.5 = -2.5 - \frac{(-3)}{R_1C}\int_0^{6\text{ ms}} dt = -2.5 + \frac{3}{R_1C}[6\text{ms}] \tag{4.8}$$

Solving the above Equation (4.8), the time constant, $R_1C = 1.5\text{ ms}$.

Example 4.2: Frequency Response of Miller Integrator

For Figure 4.2, (a) derive the expression for the transfer function $\frac{V_0}{V_{in}}(s)$. (b) If $C = 1$ nF and $R_1 = 1$ kΩ, plot the magnitude response for R_2 equal to (i) 200 kΩ, (ii) 400 kΩ, and (iii) 600 kΩ

Solution

The resistance R_2 and capacitance C_2 are in parallel. Thus,

$$Z_2 = R_2 \| \frac{1}{sC_2} = \frac{R_2}{1 + sC_2R_2}$$

$$Z_1 = R_1$$

The transfer function is given as follows:

$$\frac{V_o}{V_{in}}(s) = \frac{-R_2/R_1}{1 + sC_2R_2}$$

$$\frac{V_o}{V_{in}}(s) = \frac{-1/C_2R_1}{s + 1/C_2R_2} \tag{4.9}$$

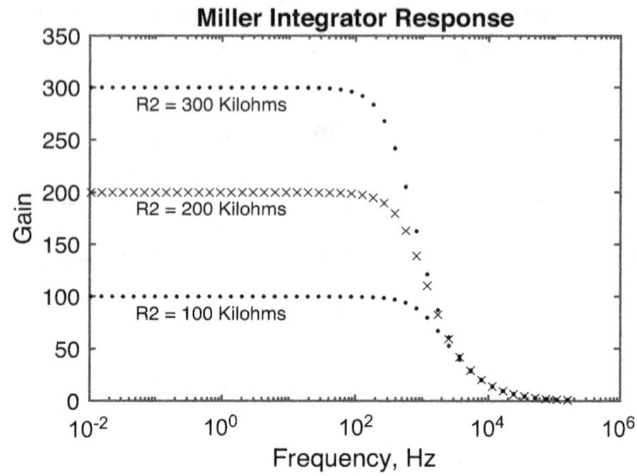

FIGURE 4.3 Frequency response of the Miller integrator with finite closed-loop gain at DC.

MATLAB® CODE

```
% Frequency response of Miller Integrator circuit
c = 1e-9; r1 = 1e3;
r2 = [100e3, 200e3, 300e3];
n1 = -1/(c*r1); d1 = 1/(c*r2(1));
num1 = [n1]; den1 = [1 d1];
w = logspace(-2,6);
h1 = freqs(num1,den1,w);
f = w/(2*pi);
d2 = 1/(c*r2(2)); den2 = [1 d2];
h2 = freqs(num1, den2, w);
d3 = 1/(c*r2(3)); den3 = [1 d3];
h3 = freqs(num1,den3,w);
semilogx(f,abs(h1),'k.',f,abs(h2),'kx',f,abs(h3),'k.')
xlabel('Frequency, Hz')
ylabel('Gain')
axis([1.0e-2,1.0e6,0,350])
text(5.0e-2,85,'R2 = 100 Kilohms')
text(5.0e-2,185,'R2 = 200 Kilohms')
text(5.0e-2,285,'R2 = 300 Kilohms')
title('Miller Integrator Response')
```

Figure 4.3 shows the frequency response of Figure 4.2 for the elements of the problem.

4.2 DIFFERENTIATOR CIRCUITS

Differentiator circuits perform the mathematical operation of differentiation. A circuit diagram of a differentiator circuit is shown in Figure 4.4.

In the frequency domain, we have

$$H(w) = \frac{V_0}{V_{\text{in}}} = -\frac{R_f}{1/jwC} = -jwCR_f \tag{4.10}$$

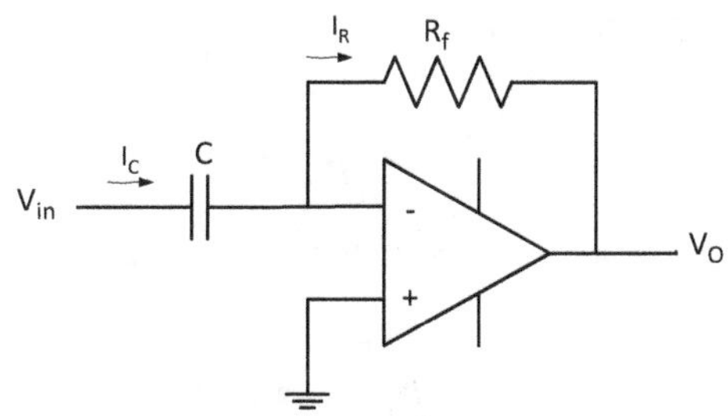

FIGURE 4.4 Op amp differentiator circuit.

In the time domain,

$$I_C = C\frac{dV_{\text{in}}}{dt} \text{ and } V_0(t) = -I_R R_f \tag{4.11}$$

Since

$$I_C(t) = I_R(t)$$

we have

$$V_0(t) = -CR_f \frac{dV_{\text{in}}(t)}{dt} \tag{4.12}$$

Figure 4.4 shows a true differentiator circuit that accentuates sudden changes in the input signal. Noise, which is always present in electronic systems, tends to have sudden abrupt changes. Since the output of the differentiator is proportional to the derivative of the input signal, the sudden changes in the noise signal at the input of the differentiator circuit result in pronounced changes in the output signal. To reduce the differentiation of high-frequency components of noise signal, a low-frequency differentiator, shown in Figure 4.5, is normally used.

The transfer function of Figure 4.5 is given as

$$H(w) = \frac{V_O}{V_{in}}(w) = -\frac{R_f}{R_1 + 1/jwc} = \frac{-jwR_fC}{1 + jwR_1C} \tag{4.13}$$

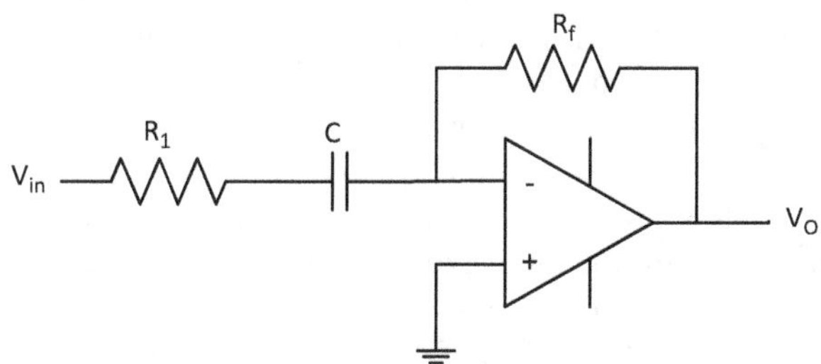

FIGURE 4.5 Low-frequency differentiator.

The break frequency, f_b, is given as

$$f_b = \frac{1}{2\pi R_1 C} \quad (4.14)$$

For frequency $f \gg f_b$, the magnitude response reduces to

$$|H(w)| \simeq \frac{wR_f C}{\sqrt{(wR_1 C)^2}} = \frac{R_f}{R_1} \quad (4.15)$$

Thus, at high frequencies, the circuit behaves as an amplifier and does not accentuate high-frequency noise signals. However, at low frequencies, the circuit approximates that of the true differentiator. The low-frequency differentiator can be used for wave shaping of periodic signals. For example, a triangular waveform may be differentiated to obtain a periodic square wave.

Example 4.3: Output Voltage of a Differentiator

Determine the output voltage of the differentiator circuit shown in Figure 4.4. Assume that the time constant $CR_f = 2.5$ ms and the input voltage is $v_{\text{in}}(t) = 4.5\cos(100\pi t)$ V.

Solution

Using Equation (4.12), we have

$$V_O(t) = -CR_f \frac{dV_{\text{in}}(t)}{dt} = (2.5 \times 10^{-3}) \frac{d[4.5\cos(100\pi t)]}{dt}$$

$$V_O(t) = -(2.5 \times 10^{-3})[(-4.5)(100\pi)(\sin(100\pi t))]$$

$$V_O(t) = 3.53\sin(100\pi t) \text{ V}$$

4.3 RECTIFIERS

A rectifier circuit converts a bidirectional signal into a unidirectional waveform. A simple half-wave rectifier circuit is shown in Figure 4.6. It consists of an alternating current (AC) source, a diode, and a resistor.

In Figure 4.6, when the source voltage, V_{in}, is positive (greater than diode voltage drop), the diode conducts and the output is almost the same as the input. However, if V_{in} is negative, the diode is reversed-biased and $V_O = 0$. It should be noted that the diode does not conduct until the input exceeds the cut-in voltage of the diode, which is approximately 0.55 V. The output voltage V_O will be less than the input voltage by a diode voltage drop. In addition, because the diode has an internal resistance, the voltage across the diode will increase as the input voltage increases.

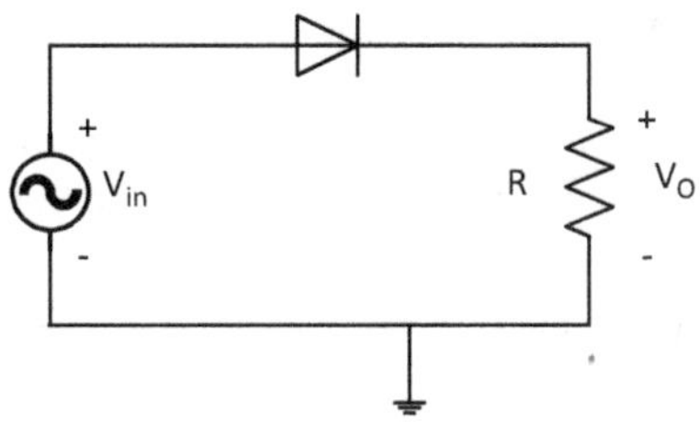

FIGURE 4.6 Half-wave rectifier circuit.

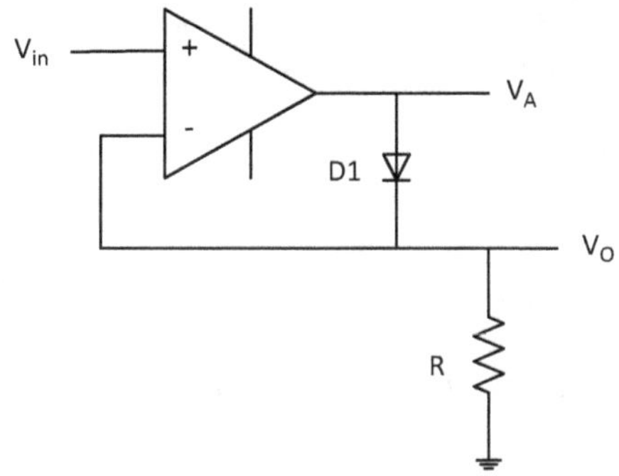

FIGURE 4.7 Precision half-wave rectifier circuit.

To reduce the diode voltage drop at the output and to eliminate the effect of the increase in diode voltage with respect to the current at the output, a precision half-wave rectifier circuit, shown in Figure 4.7, can be used.

As shown in Figure 4.7, when V_{in} is greater than zero, the output of the op amp will go positive and the diode will conduct, thus establishing a negative feedback path between op amp input and output terminals. This negative feedback will make the output voltage equal the input voltage, i.e.

$$V_O = V_{in} \quad \text{for} \quad V_{in} \geq 0 \tag{4.16}$$

When the voltage V_{in} is less than zero, the output of the op amp tries to go negative, and the diode Dl does not conduct. No current flows through the resistor R, and the output is zero.

$$V_O = 0 \quad \text{for} \quad V_{in} < 0 \tag{4.17}$$

The combination of the diode and op amp is called a "superdiode." A superdiode circuit has two disadvantages. When the input voltage is less than 0, the entire negative voltage appears at the two input terminals of the op amp. If this magnitude is greater than a few volts, the op amp may be damaged unless it is equipped with overvoltage protection. In addition, when the input voltage is negative, the op amp saturates. Although, op amp saturation is not harmful to the op amp, the op amp takes more time to get out of saturation.

Example 4.4: Precision Half-Wave Rectifier

As given for Figure 4.7, we assume that the output saturation voltages of the op amp are ±12 V. In addition, we assume that the conduction diode has a voltage drop of 0.7 V.

a. If $V_{in} = -5$ V, find V_O and V_A; (b) If $V_{in} = 2$ V, calculate V_O and V_A.

Solution

a. If $V_{in} = -5$ V, the diode is off, the op amp saturates, and $V_A = -12$ V. Since the diode is off, there is no negative feedback, and $V_O = 0$ V.
b. If $V_{in} = 2$ V, the diode conducts, and negative feedback is established. $V_O = V_{in} = 2$ V. Since the diode conducts, $V_A = V_O + V_D = 2.7$ V.

There are applications in electronic instrumentation systems, where rectification of both halves of AC power is necessary. Figure 4.8 is the circuit diagram of a bridge rectifier.

When $V_{in}(t)$ is positive, diodes $D1$ and $D3$ conduct, but diodes $D2$ and $D4$ do not conduct. The current enters the load resistance R at node A. In addition, when voltage $V_{in}(t)$ is negative, the diodes $D2$ and $D4$ conduct, but diodes $D1$ and $D3$ do not conduct. The current entering the load resistance R enters it through node A. The output voltage is

$$V_O(t) = |V_{in}(t)| - 2V_D \tag{4.18}$$

The circuit is used for full-wave rectification of a signal provided the input voltage is far greater than a diode drop. For input signals that are small, a full-wave precision rectifier using a superdiode can be used. The circuit is shown in Figure 4.9.

When $V_{in} > 0$, diode $D2$ conducts and $D1$ is cut-off and $V_A = -V_{in}$
When $V_{in} < 0$, diode $D1$ conducts and $D2$ is cut-off and $V_A = 0$ V.

The output voltage V_O can be expressed as follows:

For

$$V_{in} > 0,\ V_O = -\frac{R}{R}V_{in} - \frac{R}{R/2}V_A = -V_{in} - 2V_A \tag{4.19}$$

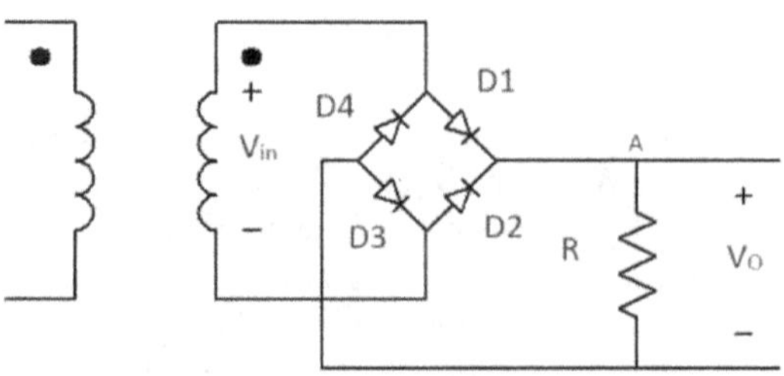

FIGURE 4.8 Bridge rectifier.

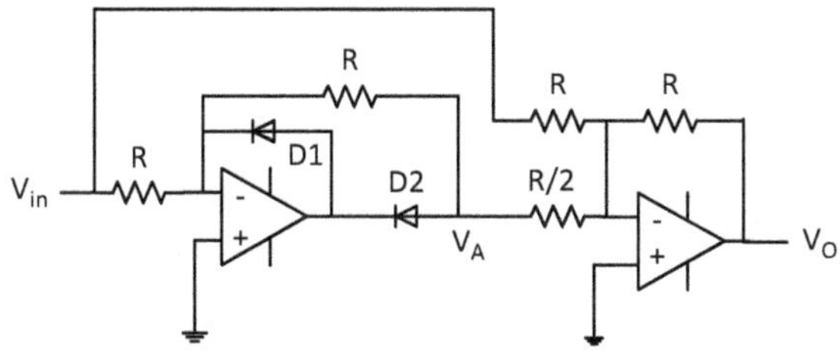

FIGURE 4.9 Full-wave precision rectifier

For

$$V_{in} > 0,\ V_A = -V_{in} \quad \text{and}$$
$$V_O = -V_{in} - 2(-V_{in}) = V_{in} \tag{4.20}$$

For

$$V_{in} < 0 \quad V_A = 0$$
$$V_O = -(\frac{R}{R} V_{in}) = -V_{in} \tag{4.21}$$

For $V_{in} > 0$, $V_O = V_{in}$, and when $V_{in} < 0$, $V_O = -V_{in}$. Thus, for both positive and negative values of the input voltage, V_{in}, the output voltage is given as $V_O = |V_{in}|$. Figure 4.9 is also called an absolute-value circuit. In Figure 4.9, the op amps do not saturate and the op amps do not require an "overvoltage" protection.

4.4 LIMITERS

Limiters are used to confine the level of a signal to a specific range. Most limiters have one or more linear ranges. For an input voltage greater or less than a specified level, the output remains almost constant. Figure 4.10 shows an input–output relationship of a limiter circuit.

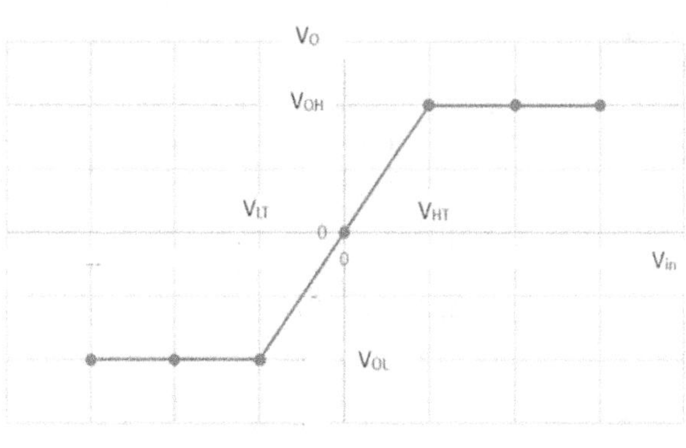

FIGURE 4.10 Input–output relationship of limiters.

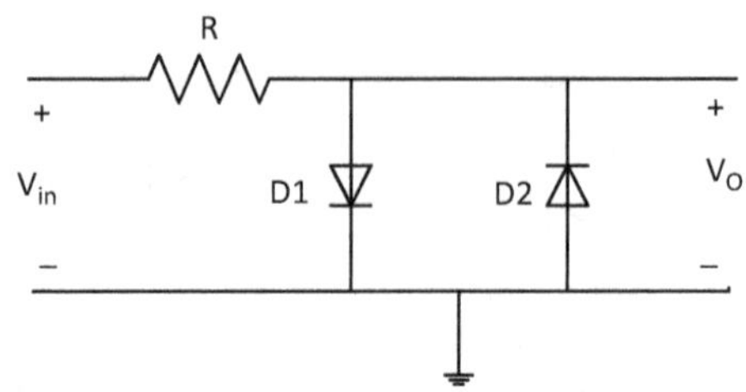

FIGURE 4.11 Diode limiter circuit.

From Figure 4.10, we have the following relationships.

$$
\begin{aligned}
V_O &= kV_{\text{in}} && V_{\text{LT}} \le V_{\text{in}} \le V_{\text{HT}} \\
&= V_{\text{OH}} && V_{\text{in}} \ge V_{\text{HT}} \\
&= V_{\text{OL}} && V_{\text{in}} \le V_{\text{LT}}
\end{aligned}
\tag{4.22}
$$

Where k is a constant.

The limiting action takes place for $V_{\text{in}} > V_{\text{HT}}$ and for $V_{\text{in}} \le V_{\text{LT}}$. Figure 4.11 shows a limiter circuit built using diodes.

When V_{in} is greater than diode drop, Dl conducts and V_O is approximately a diode drop. When the input voltage is less than a diode voltage drop, the diode **D**2 conducts, and the output voltage is a negative voltage of diode drop. For small input voltages, the diodes do not conduct, and the output voltage equals the input voltage. The input–output relationship of the circuit can be written as follows:

$$
\begin{aligned}
V_O &= V_D && V_{\text{in}} > V_D \\
&= V_{\text{in}} && -V_D \le V_{\text{in}} \le V_D \\
&= -V_D && V_{\text{in}} \le -V_D
\end{aligned}
\tag{4.23}
$$

where the voltage V_D is a diode drop. The above relationships assume that the resistance of a conducting diode is zero.

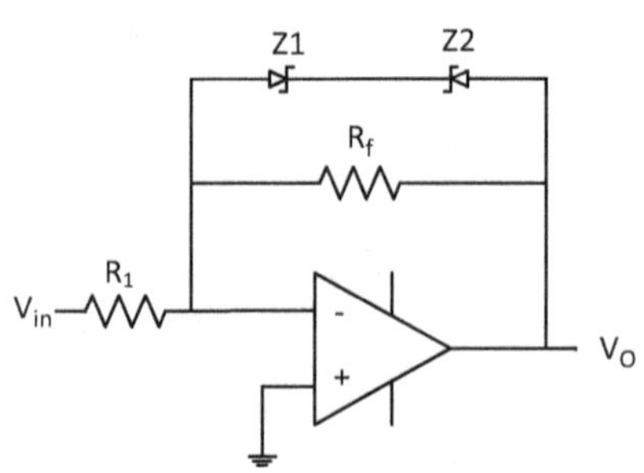

FIGURE 4.12 Limiter circuit with an operational amplifier.

The diode limiter circuit, shown in Figure 4.11, can be used for meter protection. A sensitive meter may have a full scale when the terminal voltage is between 0. 1 V and 0.4 V. The meter is connected at the output of the limiter circuit. If a high voltage is accidentally applied to the meter, the forward-biased diode limits the terminal voltage to about ±0.7 V. Figure 4.12 shows a limiter circuit that uses an operational amplifier.

The input and output voltages are related by the following expressions.

$$\begin{aligned} V_O &= V_{OH} \qquad V_{in} > V_{HT} \\ &= -\frac{R_f}{R_1} V_{in} \qquad V_{LT} \le V_{in} \le V_{HT} \\ &= V_{OL} \qquad V_{in} \le V_{LT} \end{aligned} \tag{4.24}$$

where:

$$\begin{aligned} V_{OL} &= -(V_{D1} + V_{Z2}) \\ V_{OH} &= (V_{D2} + V_{Z1}) \\ V_{LT} &= -\frac{R_1}{R_f}(V_{D2} + V_{Z1}) \\ V_{HT} &= \frac{R_1}{R_f}(V_{D1} + V_{Z2}) \end{aligned} \tag{4.25}$$

V_{D1} and V_{D2} are the forward-biased diode voltages of Zener diodes Z1 and Z2, respectively.

V_{Z1} and V_{Z2} are the Zener breakdown voltages of Zener diodes Z1 and Z2, respectively

The maximum output voltage is obtained when the input is such that the diode Z2 is forward-biased and diode Z1 breaks down. Similarly, the minimum voltage is obtained when the diode Z1 is forward-biased and diode Z2 breaks down. By choosing Zener diodes of different breakdown voltages, the output limiting voltages V_{OH} and V_{OL} can have different voltages.

Example 4.5: Diode Limiter

As given for Figure 4.11, assume that a voltage across a forward-biased conducting diode is 0.7 V and the resistance of a conducting diode is negligible. (a) If V_{in} = −0.4 V, find V_0; (b) If V_{in} = 1.4 V, find V_0.

Solution

a. If $V_{in} = -0.4$ V, both diodes will be off, and $V_0 = V_i = -0.4$ V.
b. If $V_{in} = 1.4$ V, diode D1 will conduct and the output voltage will be $V_0 = 0.7$ V.

4.5 COMPARATORS

A comparator is a circuit that compares the input voltage with a reference voltage and cause the output voltage to change state. Thus, comparators are usually used to compare two signals. There are two forms of comparators: non-inverting and inverting. If an output of a comparator assumes a "high" state when the input voltage is above a reference level, the comparator is said to be non-inverting. However, if the comparator output assumes a "low" state when the input voltage is above a reference level, the comparator is considered inverting.

Op amp can be used as a comparator. Because of the high open-loop gain of an op amp, a very small difference in the input signal in the order of 0.1 μV will result in a large value of the output voltage of the amplifier. As shown in Figure 4.13, if V_{in}, is greater than V_A, the output will switch to the positive saturation voltage of the op amp. If V_{in} is less than the voltage V_A, the output will switch to the negative saturation voltage of the op amp.

Figure 4.14 shows an inverting op amp comparator. If V_{in} is greater than V_A, the output will switch to the negative saturation voltage of the op amp. If V_{in} is less than V_A, the output will switch to the positive saturation voltage of the op amp.

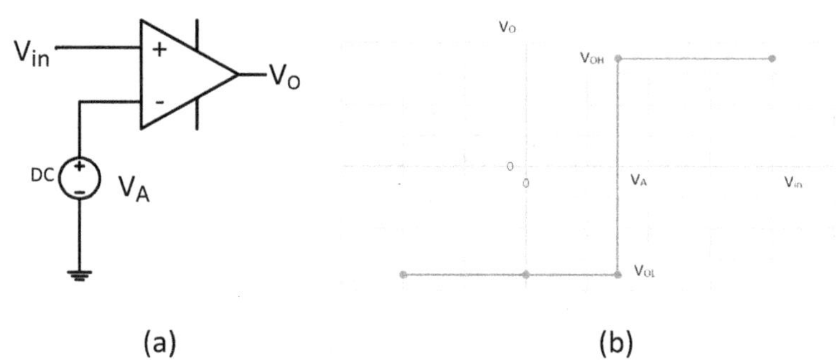

FIGURE 4.13 Op amp non-inverting comparator: (a) circuit; (b) input–output characteristics.

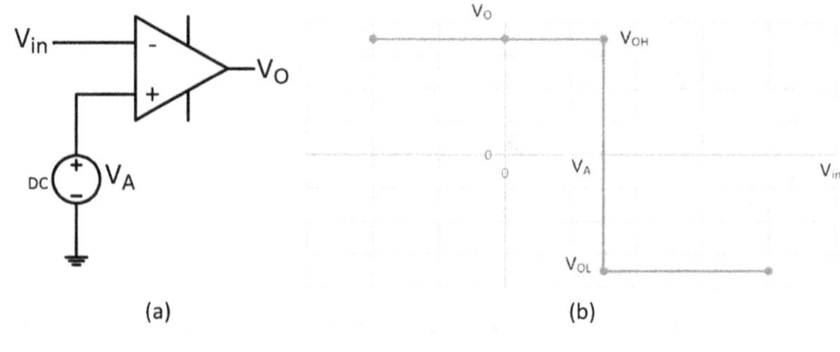

FIGURE 4.14 Op amp inverting comparator: (a) circuit; (b) input–output characteristics.

Simple comparators considered so far are sensitive to false triggers produced by noise or other glitches in a circuit. To minimize false triggering, comparators with hysteresis, often referred to as Schmidt trigger circuit, can be used. Schmitt trigger circuits are operational amplifier circuits with positive feedback.

4.5.1 Non-Inverting Schmitt Circuit

A non-inverting Schmitt trigger, shown in Figure 4.15, has the voltage at the non-inverting input V_i that can be obtained using the Kirchoff's current law as follows:

Using Kirchoff's current law, we have

$$\frac{V_{\text{in}} - V_i}{R_1} + \frac{V_0 - V_i}{R_2} = 0 \tag{4.26}$$

Solving Equation (4.26), we have

$$V_i = \frac{R_2}{R_1 + R_2} V_{\text{in}} + \frac{R_1}{R_1 + R_2} V_0 \tag{4.27}$$

When $V_i > 0$, the output will be high and $V_O = V_{\text{OH}}$. Thus,

$$\frac{R_2}{R_1 + R_2} V_{\text{in}} + \frac{R_1}{R_1 + R_2} V_{\text{OH}} > 0 \tag{4.28}$$

Simplifying Equation (4.28), we have

$$V_{\text{in}} > -\frac{R_1}{R_2} V_{\text{OH}} \tag{4.29}$$

The input low transition voltage, V_{LT}, is given as follows:

$$V_{\text{LT}} = -\frac{R_1}{R_2} V_{\text{OH}} \tag{4.30}$$

When $V_i < 0$, the output will be low and $V_O = V_{\text{OL}}$. From Equation (4.27), we have

$$\frac{R_2}{R_1 + R_2} V_{\text{in}} + \frac{R_1}{R_1 + R_2} V_{\text{OL}} < 0 \tag{4.31}$$

Simplifying Equation (4.31), we have

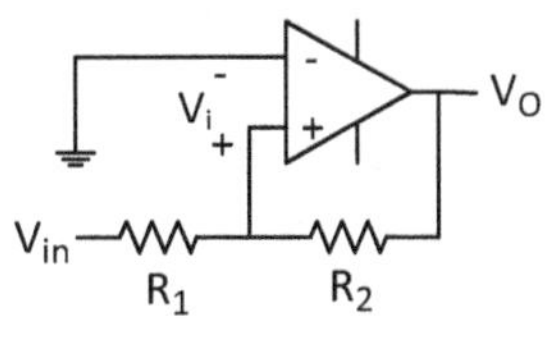

FIGURE 4.15 Non-inverting Schmidt trigger circuit.

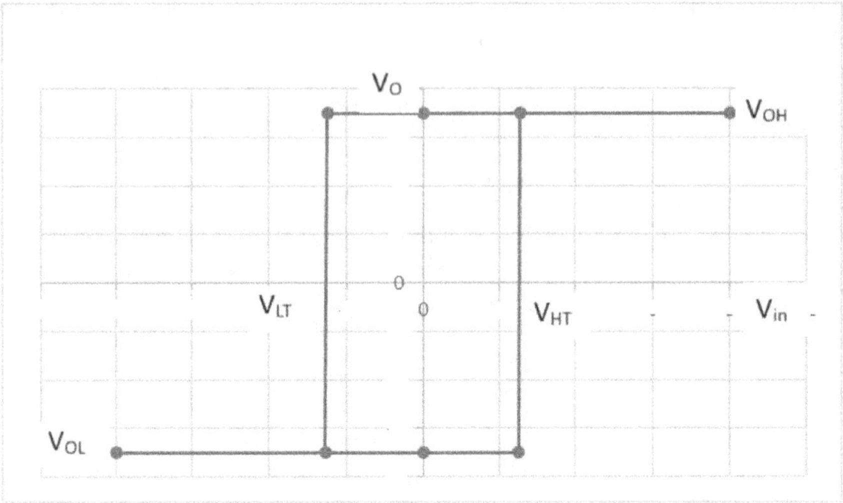

FIGURE 4.16 Non-inverting Schmidt trigger points and hysteresis circuit.

$$V_{\text{in}} < -\frac{R_1}{R_2} V_{\text{OL}} \tag{4.32}$$

The input high transition voltage, V_{HT}, is given as follows:

$$V_{\text{HT}} = -\frac{R_1}{R_2} V_{\text{OL}} \tag{4.33}$$

Hysteresis voltage, V_{HYS}, is the difference between the high and low trip point voltages, and it is given as

$$V_{\text{HYS}} = V_{\text{UT}} - V_{\text{LT}} \tag{4.34}$$

Figure 4.16 shows the Schmitt trigger threshold voltages and low and high output voltages.

4.5.2 Inverting Schmitt Circuit

Figure 4.17 shows an inverting Schmidt trigger circuit.

Since the op amp takes no current at the input, we have

$$V_f = \frac{R_1}{R_1 + R_2} V_0 \tag{4.35}$$

$$V_f = V_{\text{in}} + V_i \tag{4.36}$$

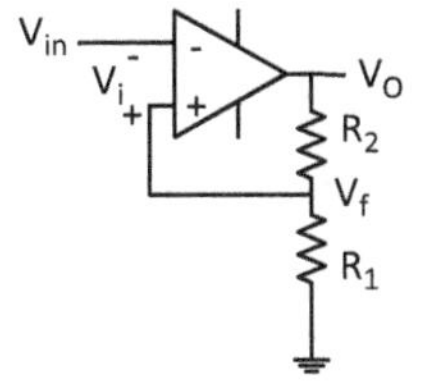

FIGURE 4.17 Inverting Schmidt trigger circuit.

Substituting Equations (4.35)–(4.36), we have

$$V_i = \frac{R_1}{R_1 + R_2} V_O - V_{in} \tag{4.37}$$

When $V_i > 0$, the output will be high and $V_O = V_{OH}$. Thus,

$$V_i = \frac{R_1}{R_1 + R_2} V_{OH} - V_{in} > 0 \tag{4.38}$$

Simplifying Equation (4.38), we have

$$V_{in} < \frac{R_1}{R_1 + R_2} V_{OH} \tag{4.39}$$

The input high transition voltage, V_{UT}, is given as follows:

$$V_{UT} = \frac{R_1}{R_1 + R_2} V_{OH} \tag{4.40}$$

When $V_i < 0$, the output will be low and $V_O = V_{OL}$. From Equation (4.37), we have

$$V_i = \frac{R_1}{R_1 + R_2} V_{OL} - V_{in} < 0 \tag{4.41}$$

Simplifying Equation (4.41), we have

$$V_{in} > \frac{R_1}{R_1 + R_2} V_{OL} \tag{4.42}$$

The input low transition voltage, V_{LT}, is given as follows:

$$V_{LT} = \frac{R_1}{R_1 + R_2} V_{OL} \tag{4.43}$$

Hysteresis voltage, V_{HYS}, is the difference between the high and low trip point voltages, and it is given by Equation (4.34). Figure 4.18 shows the Inverting Schmitt trigger threshold voltages and low and high output voltages.

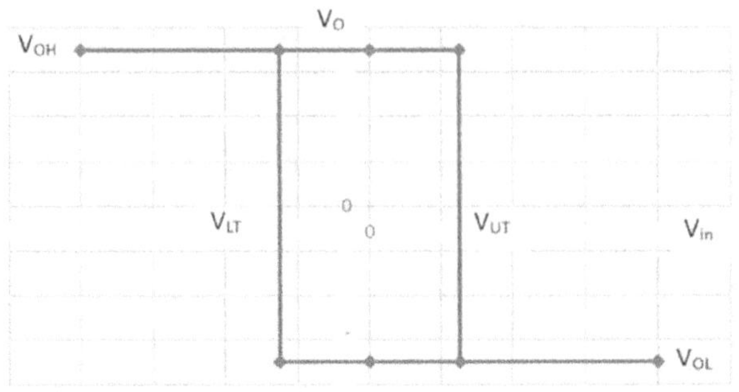

FIGURE 4.18 Inverting Schmidt trigger showing hysteresis.

A Schmitt trigger circuit can be used for wave shaping. A periodic rectangular waveform can be obtained from a triangular wave. The triangular waveform will be connected to the input of the Schmidt trigger, and a rectangular waveform will appear at the output of the Schmidt trigger. By changing the transition points, it is possible to get a rectangular waveform with different duty cycles. In addition, a Schmitt trigger can be used to reject noise interference in a signal. For a signal with noise superimposed on it, there will be multiple zero crossings if we use a zero-crossing comparator circuit. The comparator output will change state a number of times as the signal passes through zero crossings. By using a Schmitt trigger circuit, with transition points V_{UT} and V_{LT}, it is possible to prevent multiple switching at the zero crossings.

Example 4.6: Schmitt Trigger Circuit

For the Schmidt trigger circuit shown in Figure 4.17, $R_1 = 1\ \text{k}\Omega$, $R_2 = 49\ \text{k}\Omega$, $V_{OH} = 13.5$ V, and $V_{OL} = -13.5$ V. What are the Schmitt trigger trip points and hysteresis voltages?

Solution

Using Equations (4.40), (4.43), and (4.34), we have

$$V_{UT} = \frac{R_1}{R_1 + R_2}(V_{OH}) = \frac{1\text{k}\Omega}{50\ \text{k}\Omega}(13.5) = 0.27\ \text{V}$$

$$V_{LT} = \frac{R_1}{R_1 + R_2}(V_{OL}) = \frac{1\text{k}\Omega}{50\text{k}\Omega}(-13.5) = -0.27\,\text{V}$$

$$V_{HYS} = V_{UT} - V_{LT} = 0.27 - (-0.27) = 0.54\ \text{V}$$

Example 4.7: A System with Hysteresis

The output voltage, $v_0(t)$, of a system and the input voltage, $V_{in}(t)$, of a system are related by the expressions

$$
\begin{aligned}
V_O(t) &= -4\ V \quad \text{if} \quad V_{\text{in}}(t) \geq 0.3\ V \\
&= -4\ V \quad \text{if} \quad -0.3\ V < V_{\text{in}}(t) < 0.3\ V \quad \text{and} \quad V_O(t-1) = -4\ V \\
&= +4\ V \quad \text{if} \quad -0.3\ V < V_{\text{in}}(t) < 0.3\ V \quad \text{and} \quad V_O(t-1) = +4\ V \\
&= +4\ V \quad \text{if} \quad V_{in}(t) \leq -0.3\ V
\end{aligned}
$$

If the input voltage, $V_{\text{in}}(t)$, is a noisy signal given as

$$V_{\text{in}}(t) = 0.7\sin(2\pi f_0 t) + 0.5n(t)$$

where

$f_o = 150$ Hz

and

$n(t)$ is a normally distributed white noise,
write a MATLAB program to find the output voltage. Plot both the input and output waveforms of the system with hysteresis.

Solution

MATLAB® CODE

```
% Solution to Example 4.7
% vo is the output voltage
% vin is the input voltage
% Generate the sine voltage
t = 0.0:0.1e-4:5e-3;
fo=150; % frequency of sine wave
len = length(t);
for i =1:len
s(i) = 0.7*sin(2*pi*fo*t(i));
% Generate a normally distributed white noise
n(i) = 0.5*randn(1);
% generate the noisy signal
vin(i) = s(i) + n(i);
end
% calculation of output voltage
%
len1 = len -1;
for i=1:len1
if vin(i + 1) >= 0.3;
vo(i+1) = -4;
elseif vin(i+1) > -0.3 & vin(i+1)< 0.3 & vo(i) == -4
vo(i+1) = -4;
elseif vin(i+1) > -0.3 & vin(i+1 )< 0.3 & vo(i) == +4
vo(i+1) = 4;
else
vo(i+1) = +4;
end
end
```

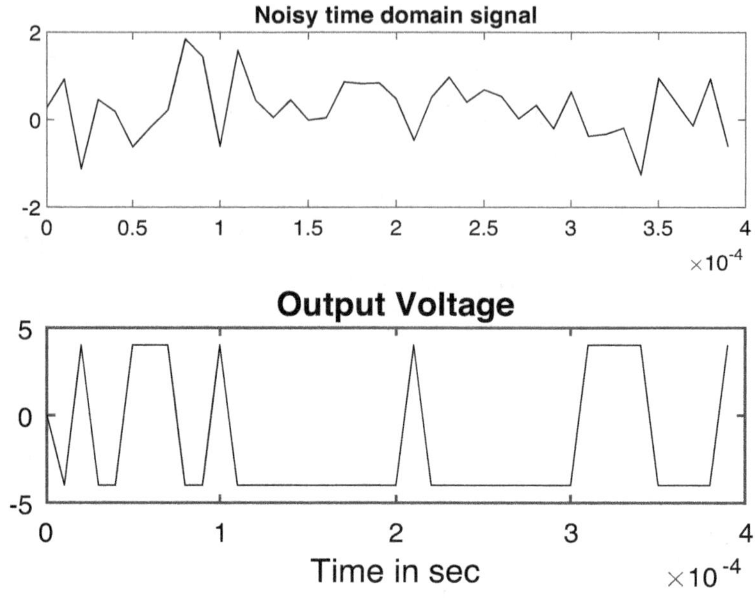

FIGURE 4.19 (a) Input and (b) output voltages.

```
%
% Use subplots to plot vs and vo
subplot (211), plot (t(1:40), vin(1:40),'k')
title ('Noisy time domain signal')
subplot (212), plot (t(1:40), vo(1:40),'k'), axis([0,0.0004, -5,
5])
title ('Output Voltage')
xlabel ('Time in sec')
```

The input and output voltages are shown in Figure 4.19.

4.6 PEAK DETECTOR

A peak detector circuit can be used to detect and hold the peak amplitude of a waveform within a given period of time. A peak detector tracks the input signal and holds the output at the highest peak value found. The input signal is continuously monitored and compared with the stored peak value to determine whether the output should be updated. A positive peak detector circuit is shown in Figure 4.20.

The op amps are connected as voltage followers. When an input voltage V_i appears at non-inverting input, the capacitor $C1$ is charged through the diode D1. The voltage across the capacitor will appear at the output. The capacitor $C1$ can be reset manually or electronically with an FET.

If the input voltage is greater than the capacitor voltage, the capacitor will be charged to the higher input voltage level. Higher positive peaks will increase the output voltage level, while lower input voltage levels will be ignored. By reversing the direction of diode D1, we can detect peak values of negative polarity signals.

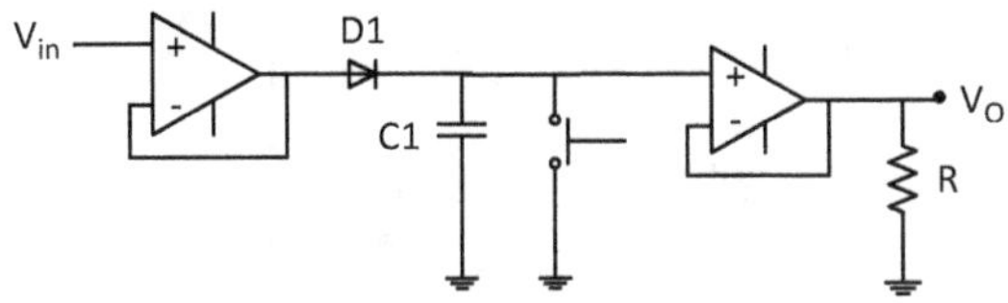

FIGURE 4.20 Peak detector circuit.

To reduce the discharge of the capacitor to the minimum, the capacitor $C1$ should have low leakage. The diode should also have low leakage current. The op amp should have enough current drive to charge the capacitor. In addition, the amplifier should have a high slew rate, low bias current, and high input impedance. By replacing the resistance R by a meter, Figure 4.20 can be used as a peak responding meter.

PROBLEMS

Problem 4.1 For the integrator circuit shown in Figure P4.1, the initial voltage across the capacitor is –2V. A step voltage of –2 V is applied at time $t = 0$. Find the output voltage at time $t = 4$ ms, if $R = 1000\ \Omega$ and $C = 1.0\ \mu F$

Problem 4.2 For the circuit shown in Figure 4.5, $R_1 = 1.5\ k\Omega$, $R_f = 30\ k\Omega$, and $C = 0.005\ \mu F$, (i) what is the voltage gain at a frequency of zero Hz, (ii) what is the voltage gain when the frequency approaches infinity, and (iii) determine the cut-off frequency.

Problem 4.3 For the op amp circuit shown in Figure P4.2, the output saturation voltages of the op amp are ±10 V. In addition, assume that the conduction diode has a voltage drop of 0.7 V. (a) If V_I = - 6 V, find V_0 and V_A. (b) If V_I = 3 V, calculate V_0 and V_A.

Problem 4.4 For the opamp circuit shown in Figure P4.3, the output saturation voltages of the op amp are ±12 V. In addition, assume that the conduction diode has a voltage drop of 0.7 V. (a) If $V_{in} = -2$ V, find V_0 and V_A, (b) If V_{in} = 1 V, calculate V_0 and V_A.

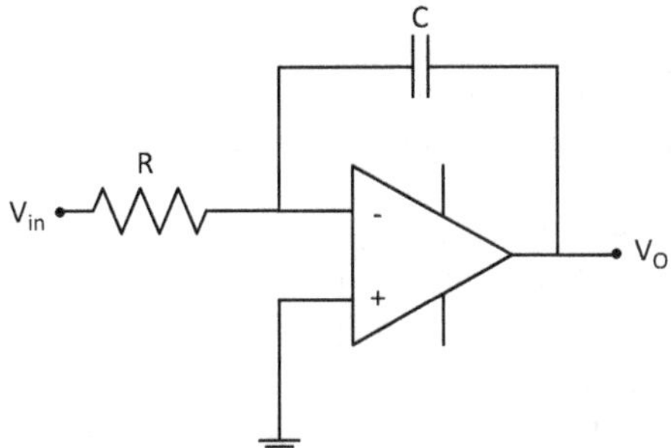

FIGURE P4.1 An integrator circuit.

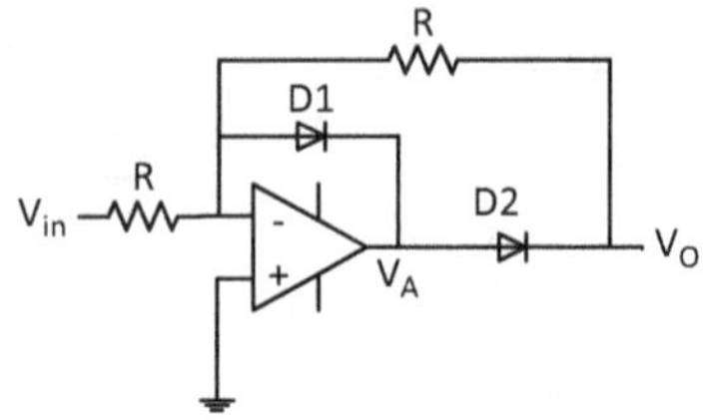

FIGURE P4.2 A rectifier circuit.

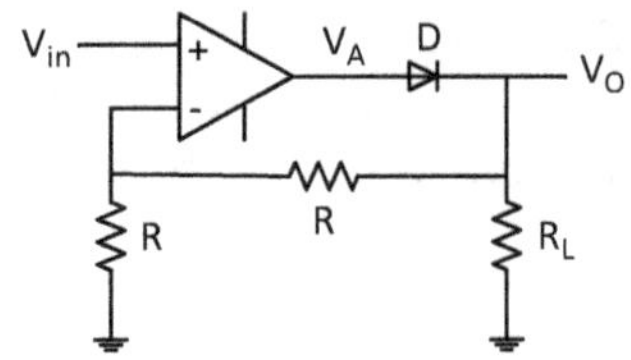

FIGURE P4.3 A superdiode circuit.

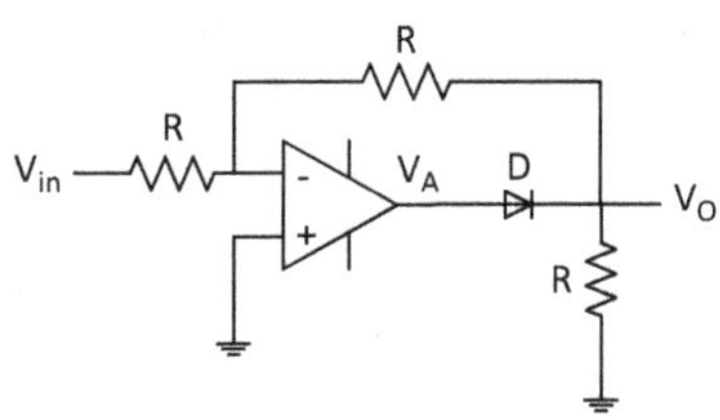

FIGURE P4.4 A circuit with op amp and diode.

Problem 4.5 For the opamp circuit shown in Figure P4.4, the output saturation voltages of the opamp are ±10 V. In addition, assume that the conduction diode has a voltage drop of 0.7 V. (a) If $V_{in} = -3$ V, find V_0 and V_A; (b) If $V_{in} =$ 2 V, calculate V_0 and V_A.

Problem 4.6 For the circuit shown in Figure P4.5, assume that a voltage across a forward-biased conducting diode is 0.7 V. (a) If $V_{in} = 7.0$ V, find V_0; (b) If V_{in} $= -8.0$, find V_0

Problem 4.7 For the circuit shown in Figure P4.6, assume that a voltage across a forward-biased conducting diode is 0.7 V. (a) If $V_{in} = 0.4$ V, find V_0; (b) If V_{in} $= 7.0$, find V_0; and (c) If $V_{in} = -10$ V, find V_0.

Problem 4.8 For the circuit shown in Figure P4.7, assume that a conducting diode has a voltage drop of 0.7 V. (a) If $V_{in} = 0.3$ V, find V_0. (b) If $V_{in} = 8.7$ V, find V_0; (c) If $V_{in} = -10.7$ V, find V_0.

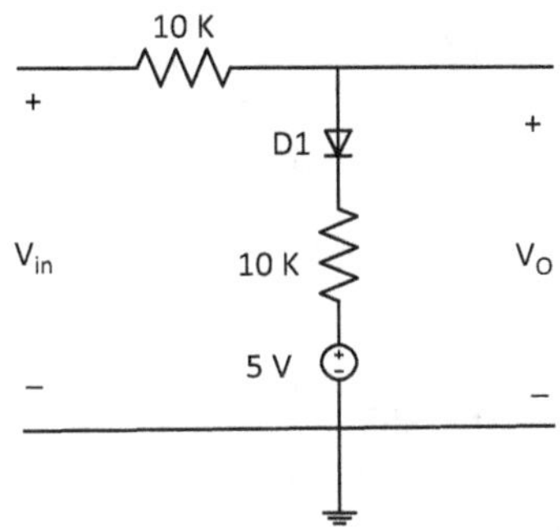

FIGURE P4.5 A limiter circuit with two voltage sources and a diode.

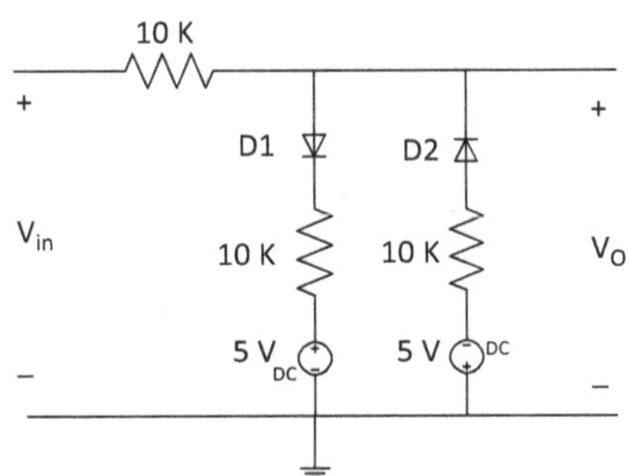

FIGURE P4.6 A limiter circuit with three voltage sources.

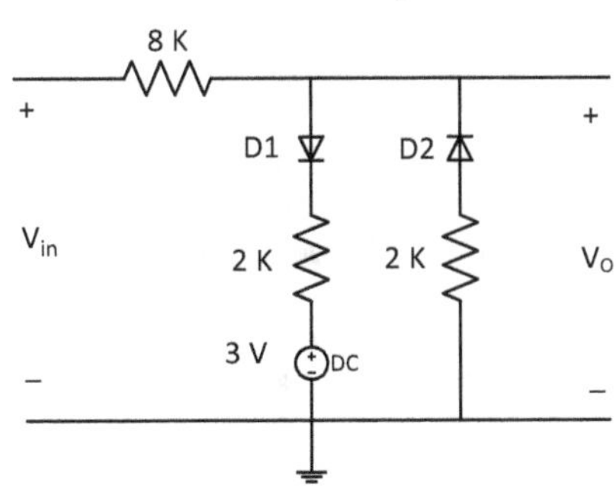

FIGURE P4.7 A limiter circuit with two voltage sources.

Problem 4.9 Design an inverting Schmitt trigger circuit that will have a V_{HYS} of 0.5 V. Assume that the $V_{OH} = |V_{OL}| = 10$ V. Draw the circuit and provide appropriate values for the resistors.

Problem 4.10 For the Schmidt trigger circuit shown in Figure P4.8, $R_1 = 2$ kΩ, $R_2 = 10$ kΩ, $V_{OH} = 10$ V, and $V_{OL} = -10$ V. What are the Schmitt trigger trip points and hysteresis voltages?

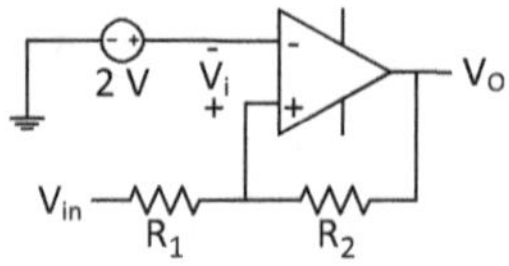

FIGURE P4.8 A Schmitt trigger circuit.

Problem 4.11 In Example 4.7, determine the input and output signals assuming if

a. $V_{in}(t) = 0.7\sin(2\pi f_0 t) + 0.3n(t)$
b. $V_{in}(t) = 0.7\sin(2\pi f_0 t) + 0.6n(t)$

Assume that (i) the hysteresis characteristics are unchanged and (ii) the frequency of the sinusoidal signal is 150 Hz,

Comment on the results obtained in parts (a) and (b).

BIBLIOGRAPHY

Alexander, Charles and Matthew N.O. Sadiku, *Fundamentals of Electric Circuit, ISF*, 7th Edition, McGraw-Hill Education, 2020.

Al-Qadasi, Mohammed, Abdullah Alshehri, Talal Al-Attar, and Hossein Fariborzi, "A Self-Biased Schmitt Trigger for Low Power Applications," *2018 25th IEEE International Conference on Electronics, Circuits and Systems (ICECS)*, Bordeaux, France, pp. 373–376, 2018.

Attia, John O., *Electronics and Circuit Analysis Using MATLAB*, CRC Press, 2nd Edition, 2004.

Attia, John O., *Circuits and Electronics: Hands-on Learning with Analog Discovery*, CRC Press, 2018.

Chapman, Stephen J., *MATLAB Programming for Engineers*, 7th Edition, Cengage Learning, 2024.

Etter, Delores, *Engineering Problem Solving with MATLAB*, 3rd Edition, Pearson, Dec. 20, 2021.

Hayt, William, Jack Kemmely, Jamie Phillips, and Steven M. Durbin, *Engineering Circuit Analysis*, 10th Edition, McGraw Hill, 2023.

Jaeger, Richard C. and Travis Blalock, *Microelectronic Circuit Design (ISE)*, 6th Edition, McGraw Hill, 2022.

Md. Moyeed, Abrar, "Design and Implementation of Schmitt Trigger using Operational Amplifier," *International Journal of Engineering Research and Application*, Vol. 7, Issue 1, pp. 5–9, 2017.

Mele, Filippo, Irisa Dedolli and Giuseppe Bertuccio, "A CMOS Peak Stretcher/Peak Detector for Wide Dynamic Range Spectroscopy Applications With 32-μW Power Consumption," *IEEE Solid-State Circuits Letters*, Vol. 5, pp. 74–77, 2022.

Minaei, Shahram and Erkan Yuce, "A Simple Schmitt Trigger Circuit with Grounded Passive Elements and its Application to Square/triangular Wave Generator," *Circuits, Systems, and Signal Processing*, Vol. 31, Issue 3, pp. 877–888, 2012.

Rana, Sakshi and Poonam Yadav, "Sinusoidal Peak to Peak Voltage Detector," *Journal of Emerging Technologies and Innovative Research (JETIR)*, Vol. 9, Issue 5, pp. a78–a82, 2022.

Razavi, Behzad, *Fundamentals of Microelectronics*, 3rd Edition, Wiley, 2021.

Sedra, Adel S. and Kenneth Smith, *Microelectronic Circuits*, 8th International Edition, Oxford University Press, 2020.

Ulaby, Fawwaz, Michael M. Maharbiz, and Cynthia M. Furse, *Circuit Analysis and Design*, Michigan Publishing Services, 2024.

Valentine, Daniel T. and Brian D. Hahn, *Essential MATLAB for Engineers and Scientists*, 8th Edition, Academic Press, 2022.

Vector, Maxwell, *MATLAB Fundamentals for Electrical Engineers*, MATLAB Fundamentals Book Series, 2025.

5 Analog-to-Digital Conversion

5.1 INTRODUCTION

The majority of variables measured in the world around us are analog in form. Some analog variables are sound, temperature, pressure, and displacement. Processing of these variables is normally done in digital form because of the powerful signal processing abilities of computers, DSP chips, microcontrollers, and microprocessors. Analog signals are converted to digital form using analog–digital converters. After processing the information using digital means, the digital information is converted to analog form using a digital-to-analog converter. Figure 5.1 shows a diagrammatic representation of analog-to-digital (A/D) and digital-to-analog (D/A), as applied to analog signals.

The analog-to-digital conversion process involves three component parts: (i) sampling, (ii) quantizing, and (iii) encoding. Figure 5.2 shows the conversion process. Sampling is done to convert the analog signal into a discrete time signal that will have the same information as the analog signal. Quantization is used to convert the sampled signal into a quantized signal represented through a number of bits. The encoder is used to convert a quantized signal into a digital signal. The processes will be described in the following sections.

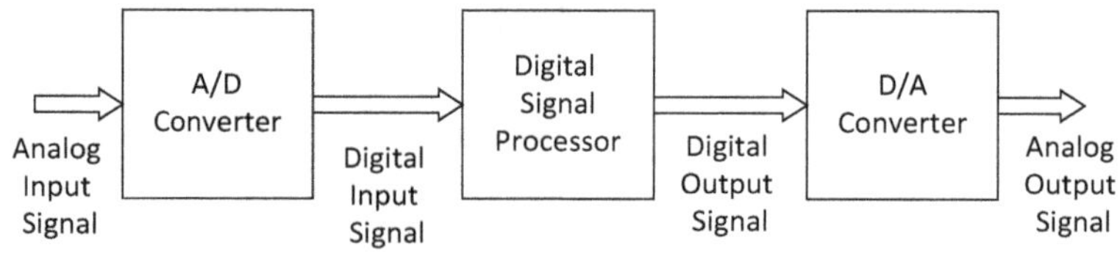

FIGURE 5.1 General representation of the A/D and D/A scheme.

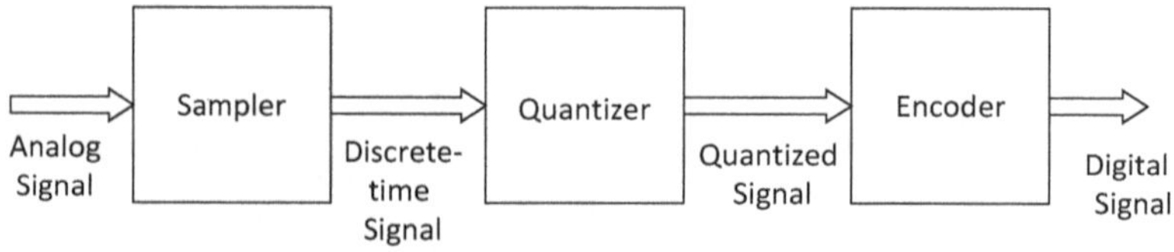

FIGURE 5.2 Analog-to-digital conversion process.

DOI: 10.1201/9781003765554-5

5.2 SHANNON'S SAMPLING THEOREM

It is of fundamental importance to know the minimum rate at which an analog signal should be sampled that will result in no loss of information as a result of the sampling operation. The key to sampling without loss of information is Shannon's Sampling Theorem. The theorem establishes the theoretical basis for sampling analog signals such that there will not be loss of information in the sampled signal.

Shannon's Sampling Theorem states that a real valued band-limited signal, having no spectral components above a frequency of B Hz, is determined uniquely if it is sampled at the minimum rate of 2B Hz. Thus, the sampling frequency, f_S, is given as

$$f_S = 2B \tag{5.1}$$

where B is bandwidth of the band-limited signal.

The theorem assumes a band-limited signal. In addition, recovery of the analog signal from the sampled signal, when the minimum sampling rate is used, necessitates the use of an ideal low-pass filter. In practice, most signals are not band-limited, and second, practical filters have a finite transition band. Thus, the following practical points need to be considered while using the sampling theorem.

First, practical filters, in contradistinction to ideal filters, have a finite transition band. If the minimum sampling rate, dictated by the Sampling theorem is used, unwanted spectral components may be transmitted, and distortion may occur. For this reason, signals are always sampled at a rate greater than $2B$ Hz.

Second, most signals are not band-limited. When a signal which is not band-limited is sampled, there will be some overlap of spectral components. When a low-pass filter, of bandwidth B, is used in an attempt to recover the analog signal, spectral components located above half the sampling frequency will appear as frequency components below half the sampling frequency. This is known as aliasing and results in distortion of the signal. The errors due to the aliasing can be reduced by oversampling the signal.

To use the sampling theorem, the frequency B at which the signal has no spectral components needs to be known. A practice which is commonly used is to pass the analog signal through a low-pass filter before sampling. This filter is called the pre-aliasing filter or anti-aliasing filter. The filter characteristics are chosen to reject all frequency components that are not significant or do not have significant power. Figure 5.3 shows the preprocessing that needs to be performed before sampling.

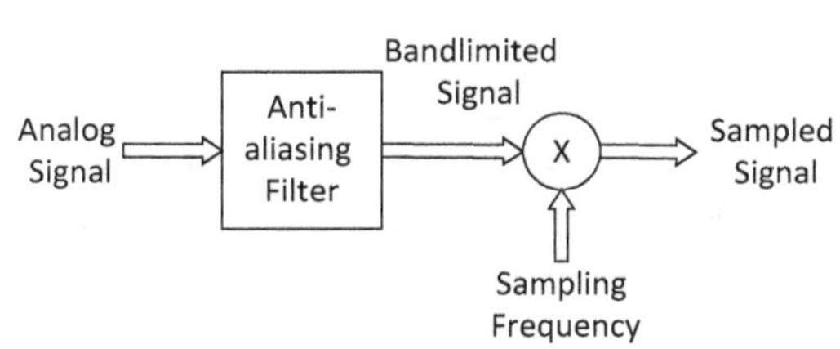

FIGURE 5.3 Preprocessing steps before sampling.

Most of the significant information in human speech is located below 8 kHz, so the minimum sampling rate can be 16 kHz. In telephone recording, the bandwidth is reduced through filtering to 300–3300 Hz, and usually an 8-kHz sampling rate is used. An audio compact disk represents frequencies up to 22 kHz, and the sampling rate of 44 kHz is normally used.

5.3 SAMPLE-AND-HOLD

Sample-and-hold circuits find applications in analog–digital conversion. Sample-and-hold or Track-and-hold is a switched circuit whose output is equal to the input when the circuit is in a sample mode. However, when the circuit is in a hold-mode, the output is almost constant. A sample-and-hold circuit samples a signal voltage and stores it on a capacitor. The voltage stored on the capacitor is maintained almost constant during the time required for the analog-to-digital conversion process. The circuit eliminates the errors that might occur from rapidly changing signals.

A sample-and-hold circuit is shown in Figure 5.4. It consists of two unity gain amplifiers, a capacitor and a switch. When the switch is in the sample mode, there is small impedance between the source and drain of the transistor, and the capacitor is charged to the voltage of the input signal.

The low output resistance of the unity gain amplifier Al and the low on-resistance of the switch provide a small time constant for the charging of the capacitor C to the input voltage. When the circuit is in the hold mode, the switch is turned off. The high input resistance of the buffer amplifier $A2$ causes the discharge of the capacitor voltage to be small, thus making the voltage across the capacitor almost constant during the hold mode of the operation of the circuit.

The value of the capacitor needs to be chosen carefully. When the switch is on, the capacitor should be small if high-speed signals are to be followed accurately. The op amp Al must be able to supply the charging current. In addition, the slew rate of the op amp must be large enough to allow the output of the op amp Al to follow the input without causing distortion of the input signal. During the hold mode, the leakage current in the transistor S1 causes the voltage across the capacitor to decrease or "droop" according to the expression

$$\frac{dV_{\text{in}}}{dt} = \frac{I_{\text{leakage}}}{C} \tag{5.2}$$

The capacitance should be large enough to reduce the droop.

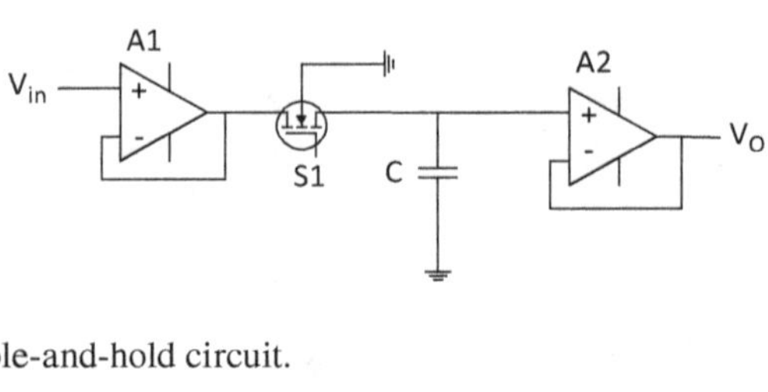

FIGURE 5.4 Sample-and-hold circuit.

5.4 DIGITAL-TO-ANALOG CONVERSION (DAC): ANALOG OUTPUT AND RESOLUTION

5.4.1 Analog Output

Digital-to-Analog Conversion (DAC)is a process of converting a digital word to an analog voltage or current. The emphasis in this section will be in the description of the basic principles of the operation of DAC. An analog output voltage or current, X_a, is related to its related digital input, X_d, through the following expression:

$$X_a = kX_d \tag{5.3}$$

where k is a constant.

Assuming that a digital word X_d consists of n bits, with the following representation:

$$X_d = d_n d_{n-1} ... d_3 d_2 d_1 \tag{5.4}$$

where

d_n is the most significant bit (MSB)
d_{n-1} is the next MSB
d_1 is the least significant bit (LSB)

and

$X_d = d_n, d_{n-1}, \ldots, d_3, d_2, d_1$ are binary numbers with two possible states (0 or 1)
The analog equivalent of Equation (5.4) is given as

$$X_a = d_n(2)^{n-1} + d_{n-1}(2)^{n-2} + \cdots + d_3(2)^2 + d_2(2)^1 + d_1(2)^0 \tag{5.5}$$

Example 5.1: 6-Bit Digital-to-Analog Converter (DAC)

A 6-bit DAC produces an analog voltage of 40 mV for the digital input of 000001. Find the analog voltage for a digital input of 111111.

Solution

The digital input of $000001_2 = 1_{10}$ (decimal equivalent)
The digital input of $111111_2 = 63_{10}$ (decimal equivalent)
From Equation (5.3), $40 \text{ mV} = k(1)$; thus, $k = 0.04 \text{ V}$

Thus, the analog voltage for the digital word of 111111_2 is given as

$$X_a = (0.04)(63) = 2.52 \text{ V}$$

Example 5.2: Analog Equivalent of 8-bit DAC

An 8-bit DAC produces an analog voltage of 0.5 V for a digital input of 00110110; what is the largest output voltage of the 8-bit DAC?

Solution

The digital $00110110_2 = 54_{10}$
The largest digital word for the 8-bit analog digital word $11111111_2 = 255_{10}$
The largest analog output voltage is

$$X_a = (0.5)\frac{255}{54} = 2.361 \text{ V}$$

5.4.2 Resolution

An ideal analog-to-digital converter (ADC) transforms an analog input within a certain range into a limited number of digital output codes. In the process of conversion, an analog signal is converted into a digital signal, and an error called quantization error is introduced. The step between the discrete output levels is called quantum or step size. As the step size decreases, the transfer function of an ADC approaches a straight line, and quantization error decreases to zero.

Resolution of a DAC is defined the smallest change that can occur in the analog output as the result of a change in the digital input. Resolution is always equal to the value of the LSB, which is referred to as the step size. Thus,

$$\text{Re solution} = \text{step size} = \text{LSB} \tag{5.6}$$

It should be noted that the resolution (step size) is the same as the proportional constant k of Equation (5.3).

For an N-bit ADC, the number of possible states is given as

$$\text{Number of Possible States} = 2^N \tag{5.7}$$

The number of steps of an N-bit ADC converter is given as

$$\text{Number of Steps} = 2^N - 1 \tag{5.8}$$

It is normally useful to express the resolution as the percentage of full-scale output. The percentage resolution is given as

$$\text{Percentage Resolution} = \frac{\text{step size}}{\text{full scale (F.S)}}(100\%) \tag{5.9}$$

The LSB, the full-scale range (FSR), and N-bit ADC are related by the following expression:

$$\text{LSB} = \frac{\text{FSR}}{(2^N - 1)} \tag{5.10}$$

The FSR is related to the reference voltage, V_{ref}, by the expression

$$FSR = V_{ref} - LSB \tag{5.11}$$

LSB is given as

$$LSB = \frac{V_{ref}}{2^N} \tag{5.12}$$

Example 5.3: Resolution of a 10-bit DAC

An 8-bit DAC has a step size of 20 mV. Calculate (a) the full-scale output voltage and (b) the percentage resolution.
Solution

With 8 bits, there will be $2^8 - 1 = 256 - 1 = 255$ steps of 20 mV each.

a. The full-scale output will be $(20\,\text{mV})(255) = 5.1\ \text{V}$.
b. The percentage resolution is given as

$$\%\ \text{resolution} = \frac{20\,\text{mV}}{5.1\,\text{V}} \times (100\%) = 0.39\%$$

It should also be noted from Example 5.3 that for an 8-bit ADC, the total number of steps is $2^8 - 1 = 255$. The percentage resolution can also be calculated as follows:

$$\%\ \text{resolution} = \frac{1}{2^8 - 1} \times (100\%) = \frac{1}{255} \times (100\%) = 0.39\%$$

This implies the percentage resolution is determined by the number of bits. Thus, for an N-bit ADC, the percentage resolution is given as

$$\%\ \text{resolution} = \frac{1}{(2^N - 1)} \times (100\%) \tag{5.13}$$

As given by Equation (5.13), the percentage resolution decreases as the number of bits of the ADC increases.

Example 5.4: ADC Specifications Calculations with MATLAB®

Given an 8-bit ADC with a full-scale value of 5 V, write a MATLAB program to determine (i) step size, (ii) LSB, (iii) number of steps, and (iv) percentage resolution.

Solution

MATLAB® CODE

```
% Characteristics of ADC
% NB is the number of bits of ADC
% FS is full scale range value of ADC
```

```
% LSB is the least significant bits
% St_size is the step size
% N_steps is the number of steps of ADC
% p_resol is the percentage resolution
NB = 8;
FS = 5;
N_steps = (2^NB - 1);
LSB = FS/N_steps;
St_size = LSB;
p_resol = LSB*100;
fprintf('Number of bits of ADC is %5d\n', NB)
fprintf('Full scale range value of ADC is  %5d volts\n', FS)
fprintf('Number of steps of ADC is %7.2f\n', N_steps)
fprintf('Least significant bit of ADC is %7.4f\n', LSB)
fprintf('Step size of  ADC is %7.4f volts\n', St_size)
fprintf('Percentage resolution of ADC is %7.3f\n',p_resol)
```

MATLAB® RESULTS

Number of bits of the ADC is 8.
The FSR value of the ADC is 5 V.
Number of steps of the ADC is 255.00.
LSB of the ADC is 0.0196.
Step size of the ADC is 0.0196 V.
Percentage resolution of the ADC is 1.961.

5.5 DAC CIRCUITS

A general way of obtaining an analog signal from a digital word is shown in Figure 5.5. The block diagram has n latches that hold the digital word or binary number that is to be converted to an analog voltage. The output of each latch controls a switch (normally transistor switch). Each switch is associated with turning on or off a particular resistor in the resistor network. Voltage or current reference connected with the resistor network controls the range of the output voltage. The output op amp provides the means of summing up the currents passing through the closed switches of the resistor network. Two circuits for implementing the block diagram shown in Figure 5.5 are (i) weighted resistance D/A converter and (ii) R-2R Ladder D/A converter. The circuits are described in the following sub-sections.

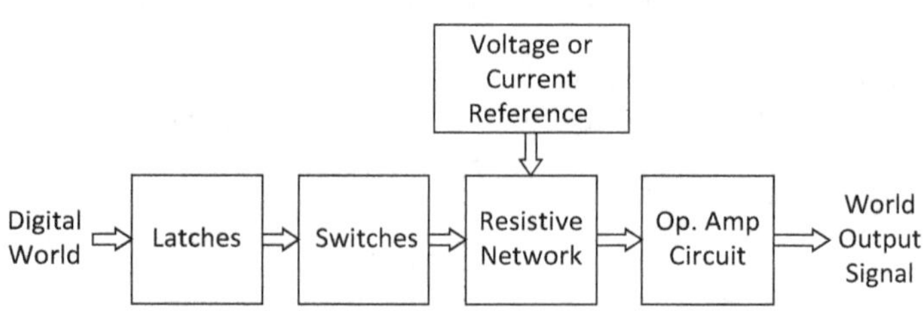

FIGURE 5.5 Block diagram of a typical digital-to-analog converter.

5.5.1 Weighted-Resistance D/A Converter

The circuit diagram of a 5-bit weighted resistance DAC is shown in Figure 5.6. In general, electronic switches are controlled by the logic levels of various bits. The digital word $b_1\ b_2\ b_3 \ldots b_n$ is used to turn the switches on and off. If a bit of the digital word is 1, the switch is closed. However, if the bit is 0, the switch is opened.

The reference voltage V_{ref}, the resistance (R, $2R$, $4R$, … ,$2^{n-1}R$) the switches, R_f and the op amp form a summer amplifier circuit. The output voltage V_O is given as

$$V_O = -(-V_{\text{ref}})\left(b_1 \frac{R_f}{R} + b_2 \frac{R_f}{2R} + b_3 \frac{R_f}{2^2 R} + \cdots + b_n \frac{R_f}{2^{n-1} R} \right) \tag{5.14}$$

where b_i multiplier assumes a value of either 0 or 1.

Equation (5.14) can be simplified to

$$V_O = \frac{2R_f V_{\text{ref}}}{R}\left(b_1 2^{-1} + b_2 2^{-2} + b_3 2^{-3} + \cdots + b_n 2^{-n} \right) \tag{5.15}$$

The factor $\frac{2R_f V_{\text{ref}}}{R}$ is called scaling factor, and it allows normalization of the digital word in such a way that the analog voltage obtained at the output of the op amp will be within the operating limits of the op amp.

It should be noted that the resistor in weighted resistance DAC ranges from R to $2^{n-1}R$. Thus, the highest value of the input resistance is 2^{n-1} times greater than the smallest value of the input resistance. For an 8-bit converter, the ratio is 128. However, for a 16-bit converter, the resistance ratio is 32768. Designing an integrated circuit with large resistance ratios is not easy. In addition, variations in resistance as a result of temperature changes are very difficult to control with large resistance spreads. For large number of bits, this converter circuit is not practicable because of large resistance ratios. In practice, this DAC is limited to applications where a small number of bits (up to eight) are used.

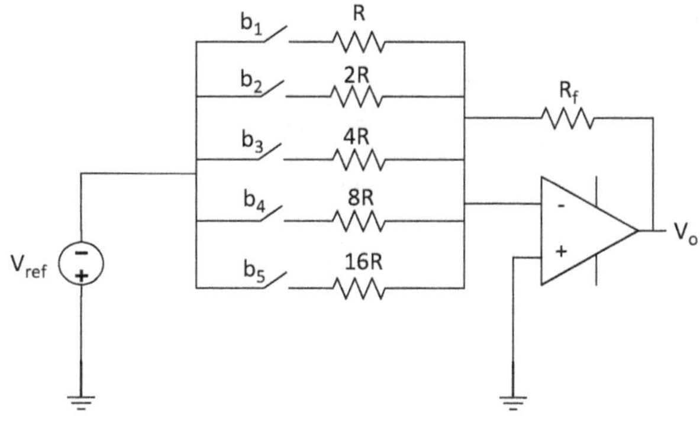

FIGURE 5.6 5-bit weighted-resistance D/A converter.

Example 5.5: 4-bit Weighted-Resistance DAC

In a 4-bit DAC shown in Figure 5.7, $V_{ref} = 5\,V$, $R_f = 1\,k\Omega$, and $R = 2\,k\Omega$. Determine (i) LSB of the DAC, (ii) full-scale output, and (iii) output voltage when $b_1 = b_2 = b_4 = 1$; $b_3 = 0$.

Solution

From Equation (5.14) and assuming that $b_1 = b_2 = b_3 = b_4 = 1$, the output voltage is given as follows:

$$V_O = \frac{R_f V_{ref}}{R}\left(b_1 + b_2 2^{-1} + b_3 2^{-2} + b_3 2^{-3}\right) = \frac{R_f V_r}{R}\left(1 + 2^{-1} + 2^{-2} + 2^{-3}\right) \quad (5.16)$$

i. From Equation (5.12), the LSB of the DAC $=$ LSB $= \frac{V_{ref}}{2^N} = \frac{5}{2^4} = 0.3125$ V

ii. The full-scale output occurs when $b_1 = b_2 = b_3 = b_4 = 1$. Thus, from Equation (5.16), the output voltage

$$V_O = \frac{R_f V_r}{R}\left(1 + 2^{-1} + 2^{-2} + 2^{-3}\right) = \left(\frac{1\,\text{K}}{2\,\text{K}}\right)(5)(1 + 0.5 + 0.25 + 0.125) = 4.6875\,\text{V}$$

iii. When $b_1 = b_2 = b_4 = 1$; $b_3 = 0$; from Equation (5.16), we have

$$V_O = \frac{R_f V_r}{R}\left(1 + 2^{-1} + 2^{-3}\right) = \left(\frac{1\,\text{K}}{2\,\text{K}}\right)(5)(1 + 0.5 + 0.125) = 4.0625\,\text{V}$$

Example 5.6: 8-bit Weighted-Resistance DAC

A weighted resistor 8-bit DAC has a reference voltage of 5 V, the most significant bit resistance is $10\,k\Omega$, and feedback resistance of the operational amplifier is $R_f = 5\,k\Omega$. Determine (i) the output voltage corresponding to the LSB, (ii) the output voltage that corresponds to the MSB, (iii) the maximum value of the output

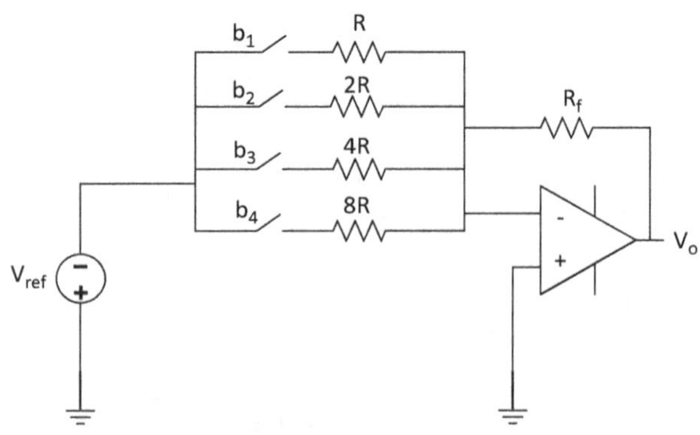

FIGURE 5.7 4-bit weighted-resistance DAC.

voltage, (iv) the resolution, and (v) the output voltage corresponding to the binary input 10101000.

Solution

For $n = 8$ and using Equation (5.14), we have

$$V_O = \frac{2R_f V_r}{R}\left(b_1 2^{-1} + b_2 2^{-2} + b_3 2^{-3} + b_4 2^{-4} + b_5 2^{-5} + b_6 2^{-6} + b_7 2^{-7} + b_8 2^{-8}\right)$$

i. For the LSB, the binary input is 00000001. Thus, the output voltage is

$$V_O = \frac{2(5\text{ K})(5)}{(10\text{ K})}\left(b_8 2^{-8}\right) = (5)\left(2^{-8}\right) = 0.0195\text{ V}$$

ii. For the MSB, the binary input is 10000000. Thus, the output voltage is

$$V_O = \frac{2(5\text{ K})(5)}{(10\text{ K})}\left(b_1 2^{-1}\right) = (5)(2^{-1}) = 2.5\text{ V}$$

iii. The maximum value of the digital input is 11111111. Thus, the output voltage is

$$V_O = \frac{2(5\text{ K})(5)}{(10\text{ K})}\left(2^{-1} + 2^{-2} + 2^{-3} + 2^{-4} + 2^{-5} + 2^{-6} + 2^{-7} + 2^{-8}\right) = 4.98\text{ V}$$

iv. The resolution is the LSB voltage. It is given as

$$V_O = \frac{2(5\text{ K})(5)}{(10\text{ K})}\left(b_8 2^{-8}\right) = (5)\left(2^{-8}\right) = 0.0195\text{ V}$$

v. The output voltage corresponding the binary input 10101000 is given as

$$V_O = \frac{2R_f V_r}{R}\left(b_1 2^{-1} + b_3 2^{-3} + b_5 2^{-5}\right)$$

$$V_O = \frac{2(5\text{ K})(5)}{(10\text{ K})}\left(2^{-1} + 2^{-3} + 2^{-5}\right) = (5)(0.656) = 3.28\text{ V}$$

5.5.2 R - 2R Ladder D/A Converter

To avoid the large resistance ratios found in the weighted resistance DAC, the R-2R ladder DAC can be used. The values of resistances found in the circuit are R and 2R. The circuit diagram for a 5-bit R-2R DAC is shown in Figure 5.8.

By using successive application of Thevenin's Theorem, the input R-2R ladder network can be simplified to the circuit shown in Figure 5.9

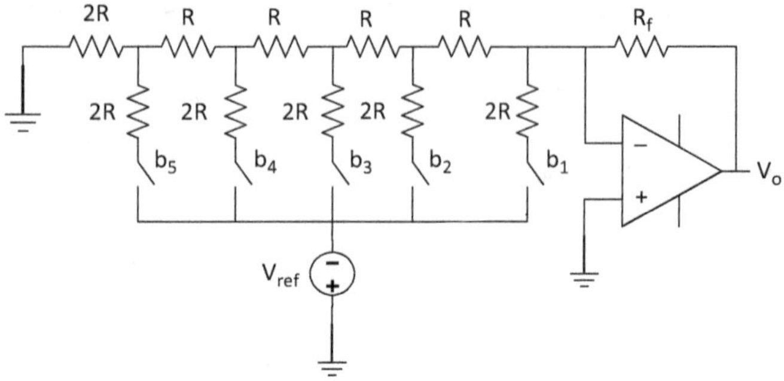

FIGURE 5.8 5-bit R-2R ladder D/A converter.

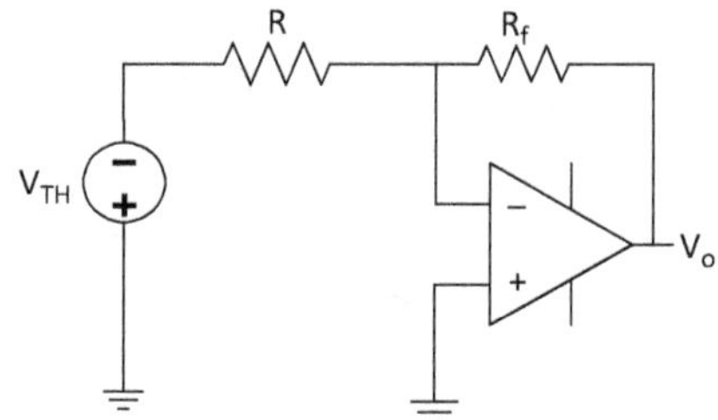

FIGURE 5.9 Simplified circuit of an R-2R ladder converter.

As in Figure 5.9, the Thevenin voltage V_{TH} is given as

$$V_{TH} = -V_{ref}\left(b_1 2^{-1} + b_2 2^{-2} + \cdots + b_{n-1} 2^{n-1} + b_n 2^{-n}\right) \tag{5.17}$$

The output voltage, V_O, is given as

$$V_O = \frac{R_f V_{ref}}{R}\left(b_1 2^{-1} + b_2 2^{-2} + b_3 2^{-3} + \cdots + b_n 2^{-n}\right) \tag{5.18}$$

The factor $\frac{R_f V_{ref}}{R}$ is the scaling factor, which can be changed to prevent t, the output of the op amp, from saturation. As given in Equation (5.18), the analog signal at the output, V_O, is proportional to the reference voltage V_{ref}. R-2R ladder DAC circuits are the most widely used conversion circuits.

While many DACs employ a fixed stabilized reference voltage on the chip, some converters allow the user to apply an external reference voltage for V_{ref}. Such a unit is designated as multiplying DAC (MDAC).

Example 5.7: 4-bit R - 2R Ladder DAC

For the 4-bit DAC circuit shown in Figure 5.10, $V_{ref} = 5\,\text{V}$, $R_f = 1\,\text{k}\Omega$, and $R = 2\,\text{k}\Omega$. Determine (i) full-scale output voltage and (ii) output voltage when $b_1 = b_2 = b_3 = 1$; $b_4 = 0$.

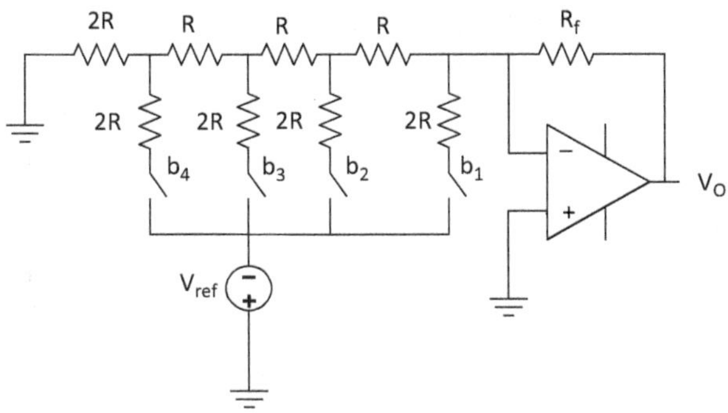

FIGURE 5.10 4-bit ADC circuit with R-2R resistors.

Solution

i The full-scale output occurs when $b_1 = b_2 = b_3 = b_4 = 1$. Thus, from Equation (5.18), the output voltage

$$V_O = \frac{R_f V_r}{R}\left(1 + 2^{-1} + 2^{-2} + 2^{-3}\right) = 5\left(\frac{1}{2}\right)(1 + 0.5 + 0.25 + 0.125) = 4.6875 \text{ V}$$

ii When $b_1 = b_2 = b_3 = 1; b_4 = 0$, we have

$$V_O = \frac{R_f V_r}{R}\left(1 + 2^{-1} + 2^{-2}\right) = 5\left(\frac{1}{2}\right)(1 + 0.5 + 0.25) = 3.5 \text{ V}$$

5.6 ANALOG-TO-DIGITAL CONVERSION TECHNIQUES

ADC is a process wherein the sample of an analog signal is converted to a digital word. In some of the DACs, the digital output has to be compared with the analog signal before a bit in the digital output is changed. Since the digital signal can only take a finite number of levels and the analog signal has an infinite number of levels, we need a quantizer to convert the continuous analog signal into a discrete amplitude level. The quantizer will divide the continuous value analog signal into digital levels. The most common quantizer is the uniform quantizer that provides equal-sized intervals of the range of the analog signal. Figure 5.11 shows a mid-tread staircase quantizer.

In ADC, the analog input is compared with the reference voltage to produce a digital output. The digital value V_d is given as

$$V_d = \left(\frac{V_a}{(V_{\text{ref}} - V_{\text{gnd}})}\right)\left(2^n - 1\right) \tag{5.19}$$

where

X_a is the analog input voltage
V_{ref} is the reference voltage used by the ADC for conversion

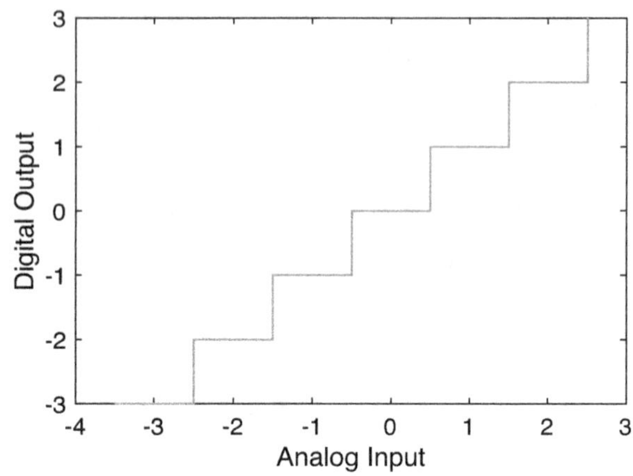

FIGURE 5.11 Mid-tread staircase quantizer.

V_{gnd} is the analog ground voltage
n is the number of digital bits output by the ADC

An ADC with a n-bit digital output has 2^n digital values, which include both 0 and $(2^n - 1)$ levels.

Example 5.8: Digital Output for Analog Input

Write a MATLAB program that will convert an analog signal into a digital value, given the reference voltage and the number of bits of the ADC. Test your program for the following: (a) find the digital value for a 4-bit ADC, reference voltage of 5 V, analog ground voltage of 0 V, and input analog voltage of 3.2 V; (b) find the digital value for an 8-bit ADC, reference voltage of 3.3 V, analog ground voltage of 0 V, and input analog voltage of 1.2 V.

Solution

MATLAB® CODE

```
% Calculation of digital value  from analog voltage of ADC
% NB is the number of bits of ADC
% VREF is reference voltage of ADC
% VGND is the voltage of the ground
% VLOG is the analog voltage
% VDIG is the digital value equivalent analog voltage VLOG
% NSTP is the number of steps of ADC
% Get the number of bits of the ADC
VGND = 0.0;
NB = input('Enter the number of bits of ADC  ')
VREF = input('Enter the reference voltage of ADC  ')
VLOG = input('Enter the analog voltage  ')
FS = VREF - VGND;
```

```
NSTP = (2^NB - 1);
LSB = FS/NSTP;
VD = VLOG*NSTP/FS;
VDIG =floor(VD);
fprintf('Number of bits of ADC is %5d\n', NB)
fprintf('Number of steps is %7d \n', NSTP)
fprintf('Least significant bit is %7.2f \n', LSB)
fprintf('Reference Voltage of DC is  %7.2f volts\n', VREF)
fprintf('Analog input voltage of ADC is %7.2f volts\n', VLOG)
fprintf('Digital equivalent of Analog input voltage is %5d \n',
VDIG)
```

Answers

a.

Number of bits of the ADC is 4.
Number of steps is 15.
LSB is 0.33.
Reference voltage of the DC is 5.00 V.
Analog input voltage of the ADC is 3.20 V.
Digital equivalent of an analog input voltage is 9.

b.

Number of bits of the ADC is 8.
Number of steps is 255.
LSB is 0.01
Reference voltage of the DC is 3.30 V.
Analog input voltage of the ADC is 1.20 V.
Digital equivalent of the analog input voltage is 92

ADCs are available for various applications that may require one to consider the number of bits, resolution, and conversion time. To meet the various needs, different conversion techniques are used. An application might require high-speed conversion with low resolution, whereas another application might need high resolution but low conversion speed.

The objective in this section is to describe the various conversion techniques. There are four commonly used ADC techniques: (i) Successive approximation converter, (ii) flash converter, (iii) double scope integration converter, and (iv) delta–sigma converter (oversampling). The conversion techniques are described in the following sub-sections. It should be noted that ADCs employ sample-and-hold circuits at the input to hold a sampled value of an analog signal during the conversion process.

5.6.1 Successive Approximation Converter

Successive approximation A/D converters are the most popular converters in general use. In successive approximation ADC, conversion is completed in n steps, where n is the resolution of the converter in bits. The block diagram of the successive approximation A/D converter is shown in Figure 5.12.

When the converter receives a signal to start conversion, the successive approximation register turns on the MSB to logic 1 (with all the other bits set to zero). The DAC converts the digital word into analog and compares with the analog input, V_{in}.

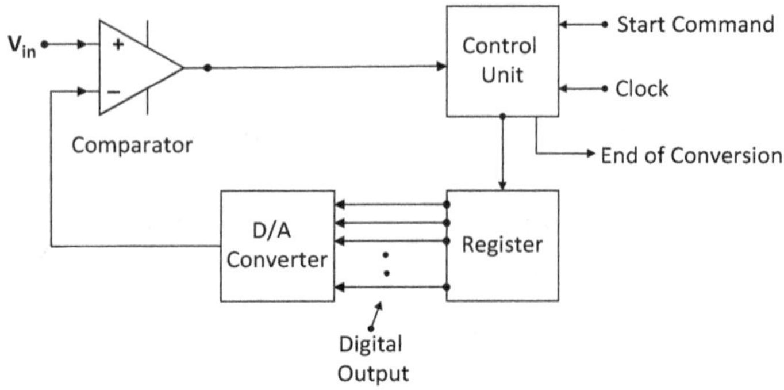

FIGURE 5.12 Successive approximation ADC.

If the latter is less than the DAC output, the MSB of 1 is rejected and replaced by 0. However, if the input signal V_{in} is greater than or equal to the DAC output, the assumed bit of 1 at the MSB is retained.

Attention is next focused on the second MSB. This value is changed to bit 1. The digital value is converted to analog form and compared with analog signal V_{in} . Once again, if the analog signal V_{in} is less than the DAC output, the second MSB is rejected and replaced by 0. However, if the input signal V_{in} is greater or equal to the DAC output, the assumed bit of 1 is retained for the second MSB. This process continues until all bits are tested and the nearest approximation to the input signal is obtained. The result is then passed to the *n*-bit latch. For *n*-bit ADC, *n* conversion steps (i.e., DAC conversion and comparisons) are required before the digital output is obtained.

In Figure 5.12, a reference voltage is required for the DAC circuit. Control logic starts the conversion process if a "start conversion" signal is received. A flag can be set by the control logic to indicate the end of the conversion process. An oscillator provides a clock signal for the analog-to-digital conversion process. The digital output is available in parallel format (i.e., all the bits are available at once on *n* separate output lines). Sometimes, the digital output is available in serial format (i.e., n successive output bits, with MSB being the first).

It should be noted that while the successive approximation conversion process is in operation, the input signal must be held constant using a sample-and-hold circuit in front of the comparator.

Successive approximations converters with a resolution from 8 to 16 bits are available with conversion rates from 100 µS to below 1 µS. The following example illustrates the iterative process used in the successive approximation ADC method.

Example 5.9: Successive Approximation of a 4-bit ADC

For a 4-bit ADC, with a minimum voltage of 0 V and a reference voltage of 16 V, describe the algorithm for successive approximation to obtain a voltage of 11 V.

Solution

Step 1: We make the MSB $(b_3) = 1$ and $b_2 = b_1 = b_0 = 0$, the digital word $1000_2 = 8$ V, which is compared to the analog input of 11. Since the analog input is greater than the digital word, the MSB (b_3) maintained at 1.

TABLE 5.1
Successive Approximation Process for an Analog Voltage of 11 V

Clock	SAR Content at Beginning of Clock (in Digital Word $b_3b_2b_1b_0$)	Output Voltage of DAC, V_{DAC}	Analog Voltage, V_a	Is, $V_a \geq V_{DAC}$?	SAR Content at the End of Clock (in Digital Word $b_3b_2b_1b_0$)	Comment
1	1000	8	11	yes	1000	Keep $b_3=1$, make $b_2=1$
2	1100	12	11	no	1000	Make $b_2=0$ and $b_1=1$
3	1010	10	11	yes	1010	Keep $b_1=1$, make $b_0=1$
4	1011	11	11	yes	1011	Keep $b_0=1$

Step 2: We make the next significant bit $b_2=1$, and $b_1=b_0=0$. Now the digital word $1100_2=12$ V is greater than the analog input of 11; we reset $b_2=0$.

Step 3: We make the next significant bit $b_1=1$ and make $b_0=0$. We have the digital word $1010_2=10$ V. Since the digital word is less than the analog input of 11, we keep $b_1=1$.

Step 4: We make $b_0=1$. We have the digital word $1011_2=11$ V, which is the same as the analog input voltage of 11 V.

The resulting digital word is 1011_2. The steps are summarized in Table 5.1. SAR means successive approximation register.

5.6.2 Flash (Parallel) ADC

Flash or parallel ADC is one of the fastest A/D converters available. The ADC uses one comparator for every quantization level. Thus, for an N-bit converter, flash ADC requires 2^N-1 comparators to compare the input signal level with each of the 2^N-1 possible quantization levels. The flash ADC uses 2^N resistors to form a ladder voltage divider network that divides the reference voltage into 2^N voltage levels. The series resistor chain and the voltage reference provide reference voltages for the comparators. The comparators examine the input analog signal simultaneously and make an immediate conversion. A block diagram of flash ADC is shown in Figure 5.13. Each comparator has one input connected to the input signal, and the other input is connected to a potential divider network. The trip-points of the comparators (the reference voltages for the comparators) are spaced one quantization level apart.

For a given analog input voltage, all the comparators biased below the input voltage turn on, and all those biased above remain off. Since the logic output of the comparators is not in binary form, an encoder is needed to convert the output of the comparators into a binary form. The binary number is outputted through a latch.

It is interesting to note that all the comparators change state simultaneously. Thus, the "quantization" process is a one-step process. Flash A/D converters have the shortest conversion time. However, for an N-bit converter, a flash ADC requires 2^N-1

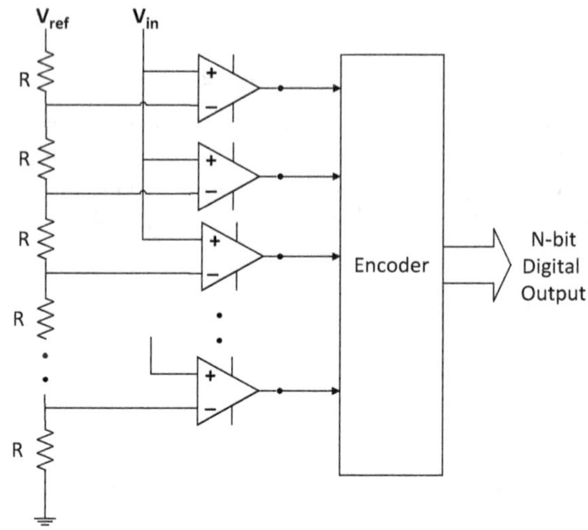

FIGURE 5.13 Flash (parallel) ADC.

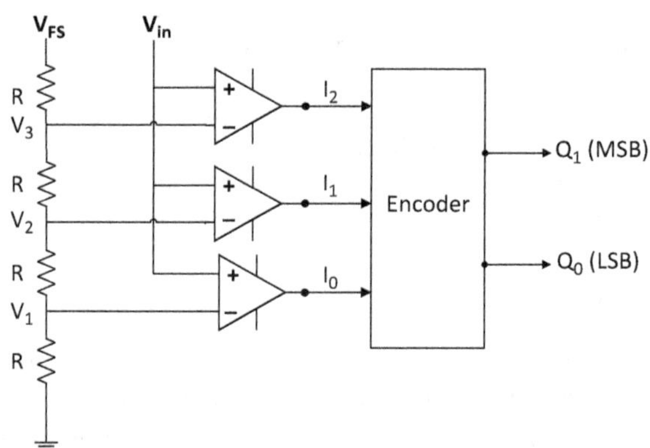

FIGURE 5.14 2-bit flash converter.

comparators. For example, we require 255 comparators for an 8-bit ADC and 65535 comparators for a 16-bit converter. For high-resolution ADC, the flash converter requires the use of a large portion of the chip area for comparators and resistor ladder network.

Flash converters are available from 4- to 10-bit resolution ADC with sampling rates in the gigahertz range. Flash converters are used in applications where high-speed conversion is required, as in video, radar, and digital oscilloscope applications.

Example 5.10: 2-bit Flash Converter

For the 2-bit flash converter circuit shown in Figure 5.14, if the full-range voltage V_{FS} of the converter is 5 V and $R = 1\ \text{k}\Omega$, (i) determine the comparator reference

TABLE 5.2
The Truth Table for the Input and Output of the Encoder

	Input			Output	
Voltage Ranges	I_2	I_1	I_0	Q_1	Q_0
$V_{in} \leq 1.25$	0	0	0	0	0
$1.25\text{ V} < V_{in} \leq 2.5\text{ V}$	0	0	1	0	1
$2.5\text{ V} < V_{in} \leq 3.75\text{ V}$	0	1	1	1	0
$3.75\text{ V} < V_{in} \leq 5.0\text{ V}$	1	1	1	1	1

voltages V_1, V_2, and V_3. (ii) Find the output of the comparators for the following input voltage to the comparators ranges: (1) $V_{in} \leq 1.25$ V; (2) $1.25\text{ V} < V_{in} \leq 2.5$ V; (3) $2.5\text{ V} < V_{in} \leq 3.75$ V; and (4) $3.75\text{ V} < V_{in} \leq 5.0$ V (iii) Complete a truth table for the relation between the input and output of the encoder (Table 5.2).

Solution

Using voltage division, we have

$$V_1 = 5 * \left(\frac{1\text{ K}}{4\text{ K}}\right) = 1.25\text{ V}$$

$$V_2 = 5 * \left(\frac{2\text{ K}}{4\text{ K}}\right) = 2.5\text{ V}$$

$$V_3 = 5 * \left(\frac{3\text{ K}}{4\text{ K}}\right) = 3.75\text{ V}$$

The logical expressions for the input versus output of the encoder are given as follows:

$$Q_0 = I_0(I_2 I_1 + \bar{I}_2 \bar{I}_1)$$

$$Q_1 = I_1 I_0$$

5.6.3 Dual-Slope ADC

The dual-slope ADC consists of an integrator, counter, clock, control logic, comparator, and electronically controlled switches. A block diagram of the converter is shown in Figure 5.15. Prior to the start of the conversion, the switch S2 is closed to discharge the capacitor, thus ensuring that the output of the capacitor is 0. When the conversion starts, a one-shot sends a signal to open a switch S1. In addition, the analog input signal (which is assumed negative) is connected to the input of the integrator, with switch S2 opened. Simultaneous to the start of the conversion, the counter is

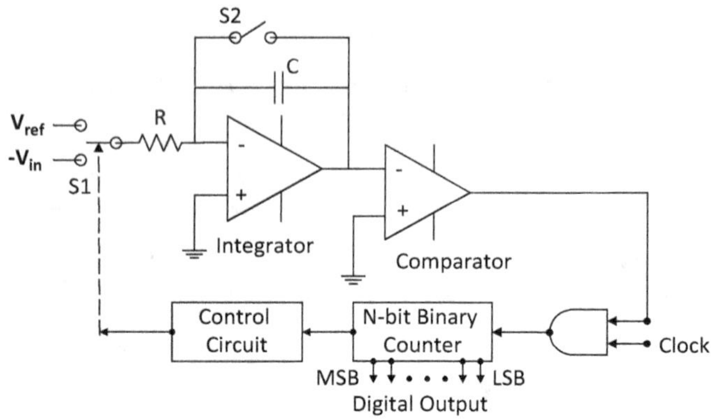

FIGURE 5.15 Dual-slope ADC circuit.

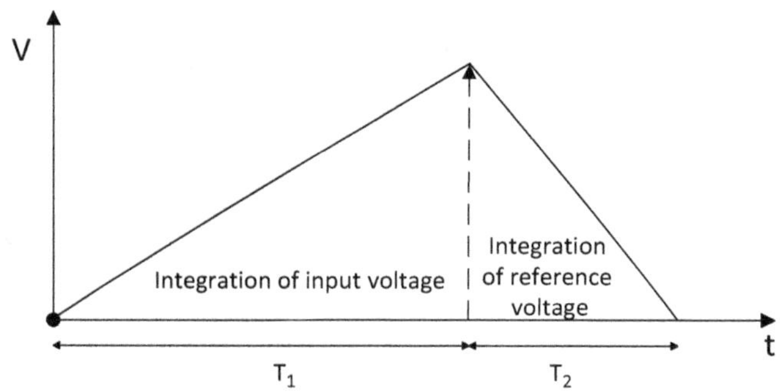

FIGURE 5.16 Integrator output for a dual-slope ADC circuit.

enabled, which counts the pulses from a fixed-frequency clock. The input voltage is integrated for a fixed interval of time T_1 which corresponds to the maximum count of the counter, $n_{c\max}$. Usually for an N-bit converter, $n_{c\max} = 2^N$. At the end of this interval, the counter overflows and causes the control logic to connect the integrator input to a negative reference voltage, V_{ref}, through a switch S1. The integrator output decreases linearly until it reaches 0, wherein the comparator output changes state (from logic state 1 to logic state 0). The latter causes the counter to stop. The output of the one-shot has a duration equal to the maximum conversion time. After the conversion, switch S1 is closed and the capacitor C is discharged, resetting the integrator.

Figure 5.16 shows the output voltage of the integrator. The time T_1 represents the time required for the counter to count through all its possible states. The time T_2 is the time taken for the integrator to go from its maximum value to 0. Since the charge gained by the integrator capacitor during the time T_1 is equal to the charge lost by the capacitor during the time T_2, we have

$$T_1 V_{in} = T_2 V_{ref} \tag{5.20}$$

From Equation (5.20), we have

$$V_{in} = V_{ref} \frac{T_2}{T_1} \tag{5.21}$$

From Equation (5.21), the digital word corresponding to the analog sample V_{in} is proportional to the time T_2. Since T_1 and T_2 are almost equally affected by long-term drift, the ratio T_2/T_1 is relatively insensitive to slow drifts in the clock frequency. It should also be noted that the conversion accuracy is independent of the values of the resistor and capacitor of the integrator. However, the accuracy of conversion is dependent on the accuracy of the reference voltage.

The dual-slope ADC has excellent noise-rejection characteristics. Since the input voltage is integrated over a long period of time, any high-frequency noise riding on the input signal is filtered out. A dual-slope ADC has low conversion speeds. However, it has high resolution. A 10 to 16 bit resolution is possible with this ADC. A dual-slope ADC is used in digital panel meters, digital multimeters, weigh scales, and in similar low-sample-rate applications.

5.6.4 Sigma–Delta Converter

The delta–sigma converter has high resolution at reasonable conversion speeds. The input analog input is oversampled at a rate far above the Nyquist rate. The oversampled signal is passed through the sigma–delta modulator. The latter provides a stream of zeros and ones to a digital filter that processes the signal bit stream into word length of *N*-bits.

A block diagram of a sigma–delta converter is shown in Figure 5.17. It consists of two major blocks: sigma–delta modulator and a digital filter. The sigma–delta modulator consists of a l-bit DAC, difference node, an analog integrator, a comparator, and a D-flip flop. The output of the 1-bit DAC is subtracted from the input sample to produce a delta output (hence "delta") at the difference node. The difference signal is integrated (hence "sigma") and fed to a comparator. The comparator is connected to

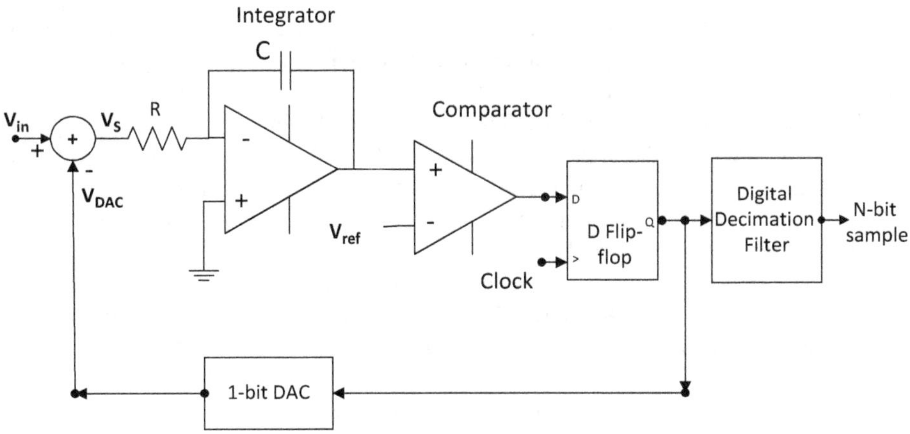

FIGURE 5.17 Sigma–delta converter.

a D-flip–flop. The latter output is a bit stream of zeros and ones that are fed to digital filter and to 1-bit DAC. The output of the 1-bit DAC is summed with the input signal at the difference node. The feedback provides the mechanism for approximating the input signal.

A positive input voltage will appear as a charge on the capacitor and voltage at the output of the integrator. This voltage, sensed by the comparator, produces a train of ones at its output and a train of pulses of opposite polarity to the integrator input voltage through the 1-bit DAC and summing node. When the pulses from the 1-bit DAC have reduced the charge on the capacitor to zero and begin charging the capacitor to the opposite polarity, negative voltage appears at the input of the comparator, and the latter produces zeros. The more positive the input voltage, the greater the ratio of ones to zeros in the bit stream conveyed to the digital filter. Similarly, the more negative the input, the greater the ratio of zeros to ones. The negative feedback loop operates to make the output of the 1-bit DAC equal to the input signal. The digital decimation filter averages the zero and one bit streams to output an n-bit sample.

Oversampling reduces the level of noise in the band of interest by spreading the quantization noise over a much wider bandwidth. In addition, sigma–delta oversampling reduces the noise in the bandwidth of interest because of the low-pass filtering property of the integrator loop.

Sigma–delta converters require only a simple RC filter in the input for anti-aliasing because the ratio of the sampling frequency to maximum frequency of the input signal is high. In addition, sigma–delta converters do not require sample-and-hold circuits. Furthermore, sigma–delta converters are easily integrated into digital integrated circuits technology.

In practical systems, the converters usually have several stages of sigma–delta modulators. Higher-order modulators reduce the noise power in the baseband. The 8 to 24-bit resolution sigma–delta converters are possible. Sigma–delta converters are used in digital audio and in telecommunication speech processing applications.

5.7 CONVERSION TIME OF ADC

The conversion time of ADC is the necessary time required to convert an analog level to digital level. It is dependent on the type of ADC, system sampling rate, and the type of converter. It is generally expressed as conversions per second. The flash ADC has the fastest conversion time. The dual-slope ADC has low conversion speed, as described in Section 5.6.3. Also, the sigma–delta ADC has low conversion speed. The analog signal is oversampled, and the conversion performs an averaging operation to produce a high-resolution digital signal. Table 5.3 shows the relative conversion time for several types of ADCs.

An ADC with n-bit output should be able to resolve the input analog signal to within one LSB during its conversion. It can be shown that the maximum permissible rate of change of the input signal is given as follows:

$$\left.\frac{dV}{dt}\right|_{\max} = \frac{V_{\mathrm{FS}}}{2^n T_{\mathrm{convert}}} \tag{5.22}$$

TABLE 5.3
Relative Speed and Typical Resolution of ADCs

ADC Type	Relative Speed	Typical Resolution (Bits)	Relative Cost
Flash	Fast	6–12	High
Dual-slope	Slow	12–22	Medium
Successive approximation	Medium	8–16	Low
Sigma–delta	Slow	16–24	Low

where

n is the number of bits in the digital output.
V_{FS} is the full-scale output voltage of the ADC.
$T_{convert}$ is the time required for a single conversion.

If the input signal is a full-scale sine wave given as

$$V(t) = V\sin(2\pi ft) \tag{5.23}$$

Then,

$$\left.\frac{dV}{dt}\right|_{max} = 2\pi fV_{FS} \tag{5.24}$$

Equating Equations (5.22) and (5.24), we have the maximum frequency signal that may be applied to the ADC and still maintain full resolution, given as

$$f_{max} = \frac{1}{2\pi(2)^n T_{convert}} \tag{5.25}$$

Sometimes, the maximum rate of voltage variation is greater than that specified in Equation (5.22). For applications with fast signal variations, an ADC with very short conversion times needs to be chosen to maintain the high-resolution capability of that ADC.

Example 5.11: Maximum Rate of Change

Given a 10-bit integration-type A/D converter with $T_{convert} = 0.1$ s and $V_{FS} = 3.3$ V, find the maximum rate of change that an analog input signal may have in order that the converter can resolve the input signal into a 10-bit number in a single conversion.

Solution

From Equation (5.22), the maximum rate of change of the input signal is as follows:

$$\left.\frac{dV}{dt}\right|_{\max} = \frac{V_{FS}}{2^n T_{\text{convert}}} = \frac{3.3}{(2)^{10}\,0.1} = \frac{33}{(2)^{10}} = 0.032 \text{ V/s}$$

Example 5.12: Maximum Frequency of Operation

Given an 8-bit, 5V full-scale, successive approximation A/D converter that has 25-μs conversion time and is used without a sample-and-hold circuit, find the maximum input frequency (assuming a full-scale sine wave input signal) that will still allow the A/D to operate at full resolution.

Solution

From Equation (5.25), the maximum frequency, $f_{\max}$, is given as follows:

$$f_{\max} = \frac{1}{2\pi(2)^n T_{\text{convert}}} = \frac{1}{2\pi(2)^8(25)(10)^{-6}} = 24.86 \text{ Hz}$$

5.8 DAC SPECIFICATIONS

5.8.1 Accuracy

Two ways of specifying the accuracy are (i) full-scale error and (ii) linearity error. The full-scale error is the maximum deviation of the output of the DAC from its expected or ideal value, normally expressed as a percentage of the full scale. For example, if a DAC has an accuracy of $\pm 0.02\%$ FS, with a full scale value of 5V, then the percentage is equal to $(\pm 0.02\%)(5) = \pm 1.0$ mV. This means that the output of the DAC can be off by as much 1.0 1 mV from its expected value.

Linearity error is the maximum deviation in the step size from the expected step size. If the expected step size of an 8-bit DAC is 19.5 mV and the linearity error is $\pm$ 0.02% FS, then this means that the actual step size can vary as much as

$$0.0195 * (2^8 - 1) * \frac{0.02}{100} = 0.0995 \text{ mV}$$

Example 5.13: Full-Scale Error of a 7-bit DAC

A certain 7-bit DAC has a full-scale voltage of 3.3 V and full-scale error of $\pm 0.1\%$ FS. What is the range of possible outputs for an input of 1000000_2.

Solution

$$\text{step size} = \frac{3.3}{2^7 - 1} = \frac{3.3}{127} = 25.98 \text{ mV}$$

The input $1000000_2 = 2^6 = 64_{10}$

The expected output for the binary input 1000000_2 is given as

$$(64)(25.98\text{mV}) = 1.663\text{ V}$$

The error can $(\pm 0.1\%)(3.3\text{ V}) = \pm 3.3\text{ mV}$

Thus, the output of the DAC may deviate by 3.3 mV, which is

$$1663 \pm 3.3\text{ mV} = 1659.7\text{ to }1666.3\text{ mV}$$

5.8.2 Offset Error

Offset error is the difference between the actual analog output and the zero analog output when the binary code has only 0s. For ADC, the offset error is the mid-step value when the digital output is 0. The offset error affects all codes by the same amount and can usually be compensated for by a trimming process.

5.8.3 Settling Time

The settling time is the time required for the output of a DAC to reach a specified error band when a step change in the input is applied. The fixed-error band is generally expressed as a fraction of full scale. In particular, for DAC, the settling time is usually defined as the time taken for the converter to settle after a full-scale code change to within an analog equivalent of $\pm$LSB.

Normally, the settling time of current-output DAC is quite fast (<200 ns), whereas the settling time of voltage-output DAC is strongly dependent on the settling time of the op amp. A proper choice of op amp can significantly reduce the settling time.

5.8.4 Gain Error

The gain error is defined as the difference between the actual analog output and the theoretical (nominal) full-scale output, after the offset error has been corrected to 0. The error represents the difference in the slope of the actual and ideal transfer functions. At each step, the error corresponds to a percentage of the maximum gain error obtainable when all the bits are turned on. This error can also be adjustable to 0 by trimming.

5.8.5 Integral Nonlinearity

Integral nonlinearity error (sometimes called as linearity error) is the maximum deviation of the actual transfer function from the best straight line. The best straight line can be (i) the line drawn so as to minimize the deviations between the straight line and the values on the actual transfer characteristics or (ii) the line between the end points of the transfer function once the gain and offset errors have been corrected to 0.

5.8.6 Differential Nonlinearity

For an ideal converter, any two adjacent codes shared produce an output value exactly equal to 1 LSB. In practice, the measured step deviates from the ideal difference, and this produces an error. The differential nonlinearity error is the difference between the actual step width (for an ADC) or the step height (for a DAC) and the ideal value of 1 LSB. The differential linearity is 0 if the step width is exactly 1 LSB, and the converter can become non-monotonic (i.e., an increase in the input voltage results in a corresponding increase in the output voltage. In addition, if there is differential linearity in ADC, there will be a possibility that one or more of the 2^n possible codes do not show at the output of the ADC (i.e., missing codes occur at the output).

PROBLEMS

Problem 5.1 A 5-bit DAC produces an analog voltage of 20 mV for the digital input of 00001. Find the analog voltage for a digital input of 10110.

Problem 5.2 If a 10-bit DAC produces an analog voltage of 0.3 V for a digital input of 0011001101, what is the largest output voltage of the 10-bit digital-to-analog converter?

Problem 5.3 A 10-bit DAC has a step size of 2 mV. Calculate the full-scale output voltage and the percentage resolution.

Problem 5.4 A 12-bit DAC has a step size of 1 mV. Calculate the full-scale output voltage and the percentage resolution.

Problem 5.5 Given an ADC with N-bits and a given FSR, FS, write a MATLAB program to calculate the following: (i) LSB, (ii) step size, (iii) number of steps, and (iv) percent resolution. Test your program for (i) a 10-bit ADC with an FSR of 5V and (ii) 12-bit ADC with an FSR of 5 V.

Problem 5.6 For the 3-bit DAC shown in Figure P5.1, $V_{ref} = 5$ V, $R_f = 1$ kΩ, and $R = 2$ kΩ. Determine (i) the resolution of the DAC; (ii) output voltage when $b_1 = b_2 = 1$; $b_3 = 0$; and (iii) the maximum current that can flow through the feedback resistor.

Problem 5.7 A weighted resistor 7-bit DAC has a reference voltage of 10 V, the MSB resistance of 15 kΩ, and a feedback resistance of the operational amplifier is $R_f = 7.5$ kΩ. Determine (i) the output voltage corresponding to the LSB; (ii) the output voltage that corresponds to the MSB, (iii) the

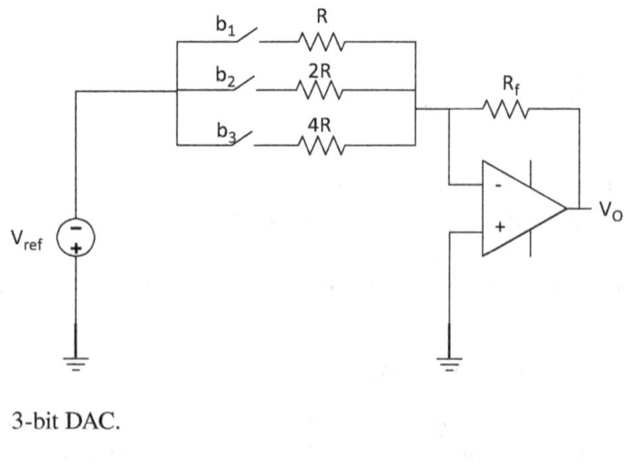

FIGURE P5.1 3-bit DAC.

maximum value of the output voltage, (iv) the resolution, (v) full-scale output voltage, and (v) the output voltage corresponding the binary input 1011100.

Problem 5. 8 An 8-bit R- 2R ladder DAC, similar to Figure 5.8, has $V_{ref} = 5\,V$, $R_f = 2\,k\Omega$ and $R = 4\,k\Omega$. Find the analog output voltage for the input 00101010.

Problem 5.9 A 10-bit R-2R ladder DAC, similar to Figure 5.8, has the full-scale output voltage of 4 V and a reference voltage of 5 V; determine the values of R_f and R.

Problem 5.10 For a 4-bit ADC, with a minimum voltage of 0 V and a reference voltage of 16 V, describe the algorithm for successive approximation to obtain a voltage of 13 V. Construct a table similar to Table 5.1.

Problem 5.11 For a 4-bit ADC, with a minimum voltage of 0 V and a reference voltage of 8 V, describe the algorithm for successive approximation to obtain a voltage of 3.5 V. Construct a table similar to Table 5.1.

Problem 5.12 A 3-bit flash ADC is shown in Figure P5.2. If the full-range voltage of the converter is 4 V, (i) determine the comparator reference voltages

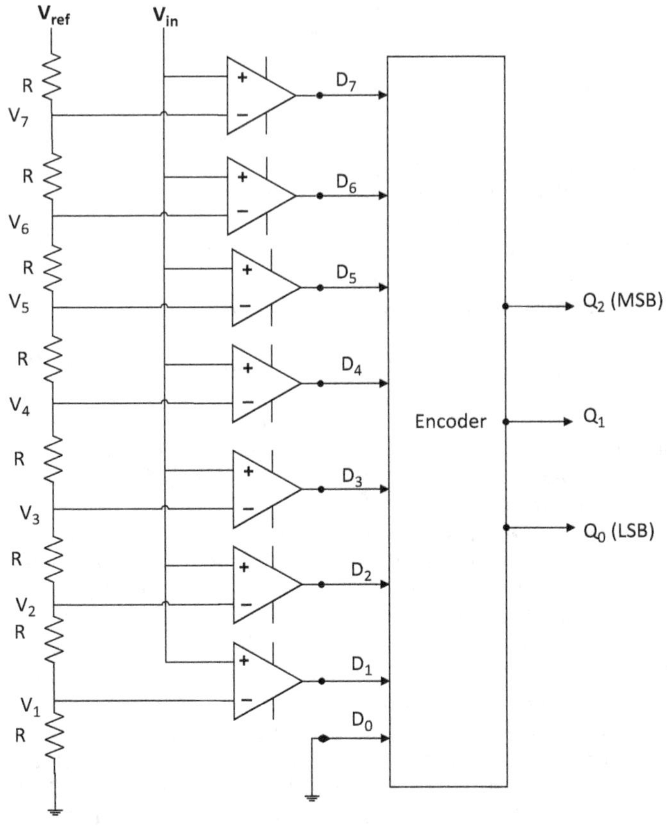

FIGURE P5.2 3-bit flash converter.

V_1, V_2, … and V_7. (ii) Find the output of the comparators for the following range of input voltages to the comparators: $V_X \leq 0.5$ V; $0.5\text{ V} < V_X \leq 1.0$ V, $1.0\text{ V} < V_X \leq 1.5$ V, $1.5\text{ V} < V_X \leq 2.0$ V, $2.0\text{ V} < V_X \leq 2.5$ V, $2.5\text{ V} < V_X \leq 3.0$ V, $3.0\text{ V} < V_X \leq 3.5$ V and $3.5\text{ V} < V_X \leq 4.0$ V (iii) Complete a truth table for the relation between the input and output of the encoder.

Problem 5.13 Given a 10-bit, 5 V full-scale, successive approximation A/D converter that has 0.015 s conversion time and is used without a sample-and-hold circuit, (a) find the maximum rate of change of the input signal $\left.\frac{dV}{dt}\right|_{max}$ and (b) the maximum input frequency (assuming a full-scale sine wave input signal) that will still allow the A/D to operate at full resolution.

Problem 5.14 Given an 8-bit A/D converter with $T_{convert} = 0.1$ s and $V_{FS} = 3.3$ V, find the maximum rate of change that an analog input signal may have in order that the converter can resolve the input signal into an 8-bit number in a single conversion.

Problem 5.15 A certain 9-bit DAC has a full-scale voltage of 5.0 V and full-scale error of ±0.2% FS. What is the range of possible outputs for an input of 100000011_2?

BIBLIOGRAPHY

Alexander, Charles and Matthew N. O. Sadiku, *Fundamentals of Electric Circuit, ISF*, 7th Edition, McGraw-Hill Education, 2020.

Arafa, Kawther, Dina Ellaithy, Abdelhalim Zekry, M. Abouelatta-Ebrahim, and Heba Shawkey, "Successive Approximation Register Analog-to-Digital Converter (SAR ADC) for Biomedical Applications," *Active and Passive Electronic Components*, Vol. 5, pp. 1–29, 2023.

Attia, John O., *Electronics and Circuit Analysis Using MATLAB*, CRC Press, 2nd Edition, 2004.

Attia, John O., *Circuits and Electronics: Hands-on Learning with Analog Discovery*, CRC Press, 2018.

Bashir, Saima, Samiya Ali, Suhaib Ahmed and Vipan Kakkar, "Analog-to-Digital Converters: A Comparative Study and Performance Analysis," *IEEE International Conference on Computing, Communication and Automation (ICCCA2016)*, Galgotias University in Greater Noida, India, pp. 999–1004, 2016.

Beev, Nikolai, "Analog-to-Digital Conversion Beyond 20 Bits: Applications, Architectures, State of the Art, Limitations, and Future Prospects," *IEEE International Instrumentation and Measurement Technology Conference (I2MTC)*, Houston, TX, 14–17 May 2018.

Chapman, Stephen J., *MATLAB Programming for Engineers*, 7th Edition, Cengage Learning, 2024.

Chen, Chia-Hung and Gabor C. Temes, "Fundamentals of Analog-to-Digital Data Converters (ADCs)," In Chia-Hung Chen and Gabor C. Temes (Eds.), *Chapter 1 of Book: Incremental Data Converters for Sensor Interfaces*, John Wiley and Sons, pp. 1–16, 2024.

Dalmia, Hemlata and Sanjeet K. Sinha, "Analog to Digital Converters (ADC): A Literature Review," *E3S Web of Conferences,* Hyderabad, India, Vol. 184, p. 01025, 2020.

Etter, Delores, *Engineering Problem Solving with MATLAB*, 3rd Edition, Pearson, 2021.

Hayt, William, Jack Kemmely, Jamie Phillips, and Steven M. Durbin, *Engineering Circuit Analysis*, 10th Edition, McGraw Hill, 2023.

Jaeger, Richard C. and Travis Blalock, *Microelectronic Circuit Design (ISE)*, 6th Edition, McGraw Hill, 2022.

Patel, Jignesh V. and Hetal Bhatt, "Performance Evaluation of Different Types of Analog to Digital Converter Architecture," *International Journal of Engineering Research & Technology (IJERT)*, Vol. 1, Issue 10, pp. 1–6, 2012.

Razavi, Behzad, *Fundamentals of Microelectronics*, 3rd Edition, Wiley, 2021.

Sedra, Adel S. and Kenneth Smith, *Microelectronic Circuits*, 8th International Edition, Oxford University Press, 2020.

Ulaby, Fawwaz, Michael M. Maharbiz, and Cynthia M. Furse, *Circuit Analysis and Design*, Michigan Publishing Services, 2024.

Valentine, Daniel T. and Brian D. Hahn, *Essential MATLAB for Engineers and Scientists*, 8th Edition, Academic Press, 2022.

Vector, Maxwell, *MATLAB Fundamentals for Electrical Engineers*, MATLAB Fundamentals Book Series, 2025.

6 Temperature Sensors

In this chapter, four basic types of temperature sensors will be discussed: pn junction thermometers, resistance thermal detectors (RTD), thermistors, and thermocouples.

6.1 PN JUNCTION THERMOMETERS

In the forward-biased region of a diode, shown in Figure 6.1, the current flowing through a diode is related to the voltage across the diode by the following diode equation:

$$I_D = I_S\left[\exp\left(\frac{qV_D}{2kT}\right)\right] \tag{6.1}$$

where

I_D = current flowing through a diode.
V_D = voltage across a diode.
I_S = reverse saturation current.
k = Boltzmann's constant.
q = electronic charge.

For high-injection conditions of the operation of a pn junction diode, the saturation current I_S can be expressed as

$$I_S = g(T)T^{3/2}\exp\left(-\frac{E_g}{2kT}\right) \tag{6.2}$$

where

E_g = energy band for silicon at 0°K.

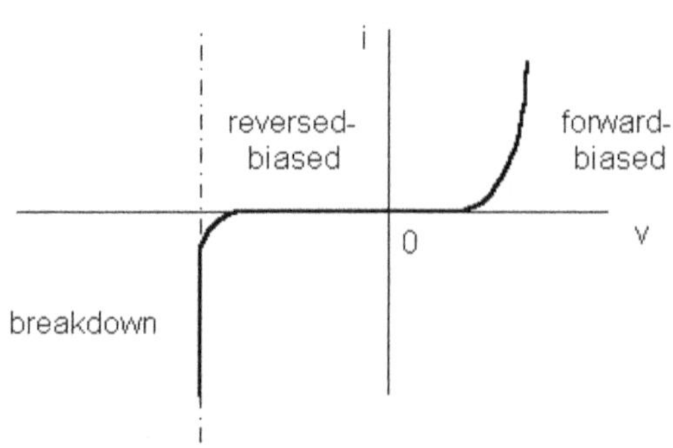

FIGURE 6.1 Diode current versus voltage characteristics.

DOI: 10.1201/9781003765554-6

The function $g(T)$ is approximately given as

$$g(T) = AT^{2/3} \tag{6.3}$$

where A is a constant.

Substituting Equations (6.2) and (6.3) into Equation (6.1), we get

$$V_D = \frac{E_g}{q} + \frac{2kT}{q}\left[\ln(I_D) - \ln(A)\right] \tag{6.4}$$

From Equation (6.4), it can be seen that the diode voltage is linearly proportional to the temperature, T, provided the current, I_D, is constant. Normally, a constant current source is used to obtain a constant current, thus allowing the diode voltage to be proportional to temperature. For silicon diodes operated at 1 mA, the sensitivity is approximately –2.0 mV/°C. From Equation (6.4), the sensitivity of the diode thermometer is given as

$$\frac{dV_D}{dT} = \frac{2k}{q}\left[\ln(I_D) - \ln(A)\right] \tag{6.5}$$

The diode thermometer is limited to temperature measurements below 200°C. It requires a constant current source. It is slow, and the diode heats up when the current flowing through it is large. When used in magnetic fields greater than 1 to 2 Tesla, the diode thermometer is less accurate. Between temperatures −40°C and 150°C, diode thermometers are more sensitive and more linear than thermocouples and resistance thermometers.

If the diode characteristic is unknown, the temperature versus voltage of the diode can be calibrated over the desired temperature range. This can be done by recording the temperature and the corresponding voltage of the diode. A MATLAB function **polyfit** that will be used in this chapter to characterize the voltage or current relationship with temperature will be discussed in the following section.

6.2 MATLAB® FUNCTION POLYFIT

The MATLAB® **polyfit** function is used to compute the best fit of a set of data points to a polynomial with a specified degree. The general form of the function is

$$\mathbf{poly\_xy = polyfit(\ x,\ y,\ n)}$$

where

x and **y** are the data points.
n is the n-th degree polynomial that will fit the vectors x and y.
poly_xy is a polynomial that fits the data in vector y to x in the least-square sense. **poly_xy** returns (n+1) coefficients in descending powers of x.

An application of the **polyfit** function is illustrated by Example 6.1.

Example 6.1: Diode Thermometer

Table 6.1 shows the diode voltage versus temperature of a silicon diode at 1mA. (i) Determine the line of best fit for the current, (ii) estimate the diode voltage at a temperature of 90°C, and (iii) calculate the change in voltage versus temperature for the diode.

Solution

MATLAB® CODE

```
% Example 6.1
% Diode Thermometer
T = [ -40 -20  0  20  40  60  80 ];
V = [ 770  735  705  670  630  600 560];
n = length(T);
%
% Linear equation is y = mx + b
% coefficient
vfit = polyfit (T, V, 1);
b = vfit(2)
m = vfit(1)
%
vfit = m*T + b;
% Plot V versus T and best fit  model
plot (T, vfit, 'k', T, V, 'xk')
xlabel ('Temperature in degree Centigrade')
ylabel('Voltage in mV')
title('Diode Voltage versus Temperature')
v90 = m*90+b
```

ANSWERS

b=701.9643
m=−1.7411
v90=545.2679

TABLE 6.1
Diode Voltage versus Temperature at a Diode Current of 1 mA

Diode Temperature, °C	Voltage (mV) at a Diode Current of 1 mA
−40	770
−20	735
0	705
20	670
40	630
60	600
80	560

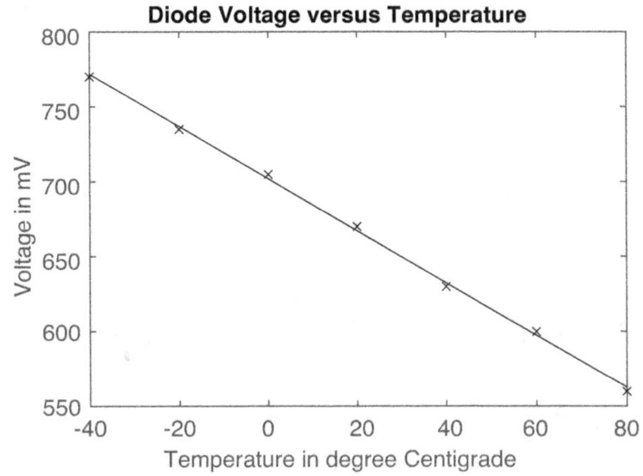

FIGURE 6.2 Voltage versus temperature of a diode thermometer.

i. The diode voltage versus temperature is shown in Figure 6.2. The equation of the line of best fit between voltage and temperature is

$$V_D(T) = -1.741T + 701.9 \text{ mV}.$$

ii. At 90°C, the diode voltage is 545.3 mV.
iii. The change in voltage with respect to temperature is –1.74 mV/°C.

6.3 INTEGRATED-CIRCUIT SENSOR

The base-emitter junction of a transistor can be used to measure temperature. For a transistor, the collector current is related to the base-emitter voltage by the expression

$$I_C = I_S \exp\left(\frac{V_{\text{BE}}}{2kT}\right) \tag{6.6}$$

In the case of the Integrated-Circuit (IC) thermometer, I_S can be canceled out by using two similar transistors. A simple IC thermometer is shown in Figure 6.3. From Figure 6.3, we get

$$\frac{I_{C1}}{I_{C2}} = \exp\left[q\frac{(V_{\text{BE1}} - V_{\text{BE2}})}{2kT}\right] \tag{6.7}$$

and

$$V_{\text{BE1}} - V_{\text{BE2}} = \frac{2kT}{q}\ln\left(\frac{I_{C1}}{I_{C2}}\right) \tag{6.8}$$

For large β transistors,

$$V_{\text{BE1}} - V_{\text{BE2}} = I_{E2}R_E \simeq I_{C2}R_E \tag{6.9}$$

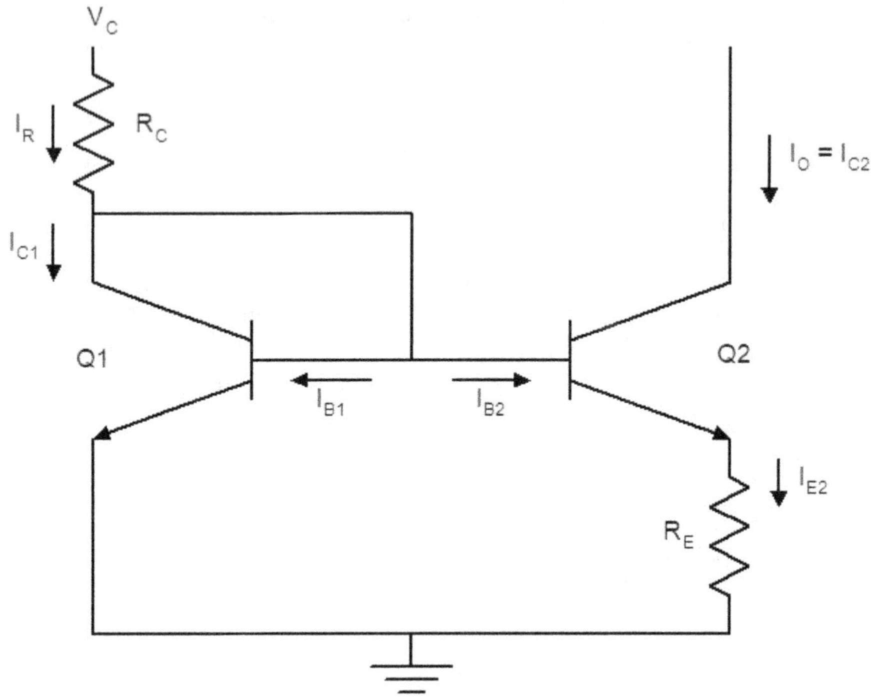

FIGURE 6.3 IC Sensor using a Widlar current mirror.

Equating Equations (6.8) and (6.9), we have

$$I_{C2}R_E = \frac{2kT}{q}\ln\left(\frac{I_{C1}}{I_{C2}}\right) \tag{6.10}$$

Thus, the voltage across R_E will be proportional to temperature, provided the ratio $\frac{I_{C1}}{I_{C2}}$ has a constant value. The latter can be achieved by using a current mirror circuit.

If the β of the transistors is not large, we use the Kirchoff's voltage law to get the following:

$$V_{BE1} - V_{BE2} = (I_{C2} + I_{B2})R_E \tag{6.11}$$

$$V_{BE1} - V_{BE2} = \left(I_{C2} + \frac{I_{C2}}{\beta}\right)R_E = I_{C2}\left(1 + \frac{1}{\beta}\right)R_E \tag{6.12}$$

From Equations (6.6) and (6.12) and assuming the similar reverse saturation current, I_S, for the transistors, we obtain

$$\frac{kT}{q}\ln\left(\frac{I_{C1}}{I_S}\right) - \frac{kT}{q}\ln\left(\frac{I_{C2}}{I_S}\right) = I_{C2}\left(1 + \frac{1}{\beta}\right)R_E \tag{6.13}$$

Simplifying Equation (6.13), we get

$$I_{C2} = \frac{kT}{q}\ln\left(\frac{I_{C1}}{I_{C2}}\right)\left(\frac{1}{\left(1+\frac{1}{\beta}\right)R_E}\right) \tag{6.14}$$

But,

$$I_R = \frac{V_{CC} - V_{BE1}}{R_C} \tag{6.15}$$

Also,

$$I_R = I_{C1} + I_{B1} + I_{B2} = I_{C1}\left(1+\frac{1}{\beta}\right) + \frac{I_{C2}}{\beta} \tag{6.16}$$

Equating Equations (6.15) and (6.16), we have

$$\left(\frac{V_{CC} - V_{BE1}}{R_C}\right) = I_{C1}\left(1+\frac{1}{\beta}\right) + \frac{I_{C2}}{\beta} \tag{6.17}$$

Rewriting Equation (6.17), we have

$$I_{C1} = \left(\left(\frac{V_{CC} - V_{BE1}}{R_C}\right) - \frac{I_{C2}}{\beta}\right)\left(\frac{\beta}{\beta+1}\right) \tag{6.18}$$

If β is very large, Equations (6.14) and (6.18) become

$$I_{C2} = \frac{kT}{q}\ln\left(\frac{I_{C1}}{I_{C2}}\right)\left(\frac{1}{R_E}\right) \tag{6.19}$$

$$I_{C1} = \left(\frac{V_{CC} - V_{BE1}}{R_C}\right) \tag{6.20}$$

For a given R_E, R_C, and V_{CC}, we can solve for I_{C2} by using Equations (6.19) and (6.20) through an iterative method, illustrated in Example 6.2.

Example 6.2: Widlar Current Source

For the Widlar current source shown in Figure 6.3, determine the output current I_{C2} versus temperature if $R_C = 50\ \text{k}\Omega$, $V_{CC} = 5\ \text{V}$, $V_{BE1} = 0.7\,\text{V}$, $R_E = 25\,\text{k}\Omega$, and β =100.

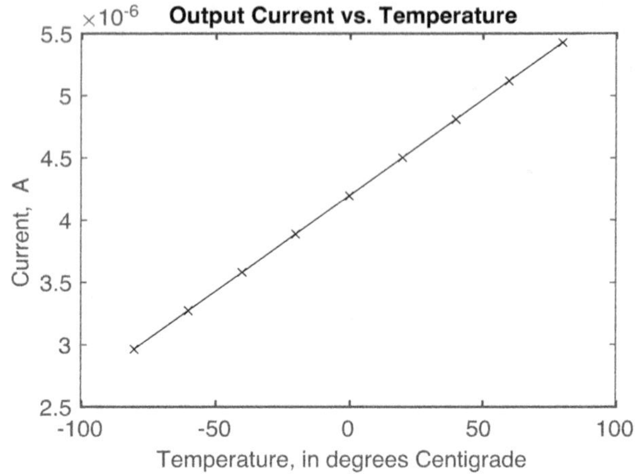

FIGURE 6.4 Output current versus temperature of a Widlar current mirror circuit.

MATLAB® CODE

```
% Solution to Example 6.2
% Determination of Ic2
kB = 1.38e-23; % Boltzmann's constant
q = 1.60e-19; % electronic charge
beta = 100; %  Beta of transistors
re = 25e3; % value of resistance RE
vbe1 = 0.7;
vcc = 5;  % voltage source
rc = 50e3; % value of resistance RC
idiff = 1;
reltol = 1e-30;
ibeta = 1/beta;
kont = (vcc - vbe1)/rc;
T = -100:20:80;  %Temperature in degree centigrade
nT = length(T);  % total number of temperatures
for k=1:nT
    TK(k) = T(k)+273;
    ic2(1)= 1.0e-6; % initialize ic2
    ic2(2) = ((kB*TK(k))/(q*re))*(log(kont/ic2(1)));
    i = 2;
    while idiff > reltol
      ic2(i+1) = ((kB*TK(k))/(q*re))*(log(kont/ic2(i))); %
Equation (6.20)
    idiff = ic2(i+1)-ic2(i);
    i = i+1;
    end
    io(k) = ic2(i); % output current
end
% plot the output current versus temperaturen
plot(T(2:k), io(2:k), 'k', T(2:k), io(2:k),'xk')
title('Output Current versus Temperature')
xlabel('Temperature, in degrees Centigrade')
ylabel('Current, A')
```

Figure 6.4 shows the current versus temperature.

6.4 RESISTANCE TEMPERATURE DETECTORS

The RTD is a temperature sensor made of a metal with a positive temperature coefficient. The variation of resistance with temperature for RTDs is given as follows:

$$R_T = R_0\left(1+\alpha(T-T_0)\right) \tag{6.21}$$

where

R_T = resistance at temperature T.
R_0 = resistance at T_0 (normally 0°C).
α is the temperature coefficient.

Some metals that are commonly used for RTD are platinum, nickel, and copper. Platinum RTD is the more commonly used because of its high temperature range, stability, and low sensitivity to chemical attack. Table 6.2 shows the range of operation of the platinum, copper, and nickel RTDs and their temperature coefficients.

A more accurate equation for the changes in resistance with the temperature of RTD is given as

$$R_T = R_0\left(1+AT+B(T-T_0)^2\right) \tag{6.22}$$

where A and B are constants.

A more accurate relationship between resistance and temperature is given by Callendar–Van Dusen equations shown below:

For temperatures $-200°C < T \le 0°C$,

$$R_T = R_0\left(1+AT+BT^2+C(T-100)T^3\right) \tag{6.23}$$

For temperatures $0°C < T \le 661°C$,

$$R_T = R_0\left(1+AT+BT^2\right) \tag{6.24}$$

TABLE 6.2
RTD Types, Range of Operation, and Temperature Coefficients

RTD Type	Temperature Range of Operation (°C)	Temperature Coefficient
Platinum (Standard DIN 43760, IEC 751, named European Industrial Standard)	−270 to 1000	0.003850
Platinum (Pure Pt. 99.999%, named US Industrial Standard)	−270 to 1000	0.003926
Copper	−200 to 260	0.00427
Nickel	−200 to 430	0.00672

where A, B, and C are known as Callendar–Van Dusen constants, defined by the following equations:

$$A = \alpha + \frac{\alpha\delta}{100} \tag{6.25}$$

$$B = \frac{-\alpha\delta}{100^2} \tag{6.26}$$

$$C = \frac{-\alpha\beta}{100^4} \tag{6.27}$$

α, β, and δ are defined by the following equations.

$$\alpha = \frac{R_{100} - R_0}{100 + R_0} \tag{6.28}$$

β is constant if $T < 0°C$ (it is found experimentally); else it is zero.

$$\delta = \frac{R_0[1+\alpha(260)] - R_{200}}{4.16R_0\alpha} \tag{6.29}$$

Table 6.3 shows the different coefficients for platinum RTD

There are several approaches for obtaining electronic readouts of RTDs. The Wheatstone bridge circuit, shown in Figure 6.5, is good for small temperature changes wherein the change in resistance is linearly related to the change in temperature. If R_b is resistance near balance, then

$$R_T = R_b(1 + k\Delta T) \tag{6.30}$$

The bridge output voltage is given as

$$V_X = \frac{V_{\text{ref}}R_1}{2R_1} - \frac{V_{\text{ref}}R_b}{R_b + R_T} \tag{6.31}$$

TABLE 6.3
Coefficients of Platinum RTD

Coefficient	Platinum (Standard DIN 43760, IEC 751, named European Industrial Standard)	Platinum (Pure Pt. 99.999%, named US Industrial Standard)
α	0.003850	0.003926
δ	1.4999	
β	0.10863	
A	3.9083e-3	3.9848e-3
B	−5.775e-7	−5.870e-7
C	−4.18301e-12	−4.000e-12

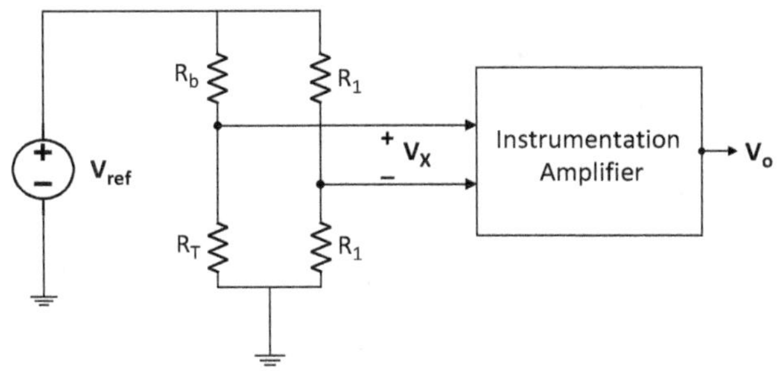

FIGURE 6.5 Circuit for obtaining readout from an RTD using a Wheatstone bridge with an instrumentation amplifier.

$$V_X = \frac{V_{\text{ref}}}{2} - \frac{V_{\text{ref}} R_b}{R_b + R_b(1 + k\Delta T)} = V_{\text{ref}}\left(\frac{1}{2} - \frac{1}{2 + k\Delta T}\right) \tag{6.32}$$

$$V_X = V_{\text{ref}}\left(\frac{1}{2} - \frac{1}{2 + k\Delta T}\right) = V_{\text{ref}}\left(\frac{1}{2} - \frac{1/2}{1 + k\Delta T/2}\right) \tag{6.33}$$

$$V_X = V_{\text{ref}}\left(\frac{1}{2} - \frac{1}{2 + k\Delta T}\right) = V_{\text{ref}}\left(\frac{1}{2} - \frac{1/2}{1 + k\Delta T/2}\right) \tag{6.34}$$

Assuming $\frac{k\Delta T}{2} < 1$, then

$$V_X = V_{\text{ref}}\left(\frac{k\Delta T}{2} - \left(\frac{k}{2}\right)^2 \Delta T^2\right) \simeq \frac{k\Delta T}{2} V_{\text{ref}} \tag{6.35}$$

Equation (6.35) shows that the voltage V_X is proportional to the change in temperature. An instrumentation amplifier can be used to amplify the output of the bridge circuit. The op amp used for the instrumentation amplifiers should have low offset voltage and low drift. For large temperature changes that lead to a large change in resistance, the output is nonlinearly proportional to temperature.

To convert the resistance of the RTD directly to voltage, Figure 6.6 can be used. The output voltage is

$$V_0 = -\frac{R_T}{R_1} V_{\text{ref}} \tag{6.36}$$

As from Figure 6.6, if we set $R_1 = R_0$ and use Equations (6.21) and (6.36), we get

$$V_0 = -V_{\text{ref}}\left(1 + \alpha(T - T_0)\right) \tag{6.37}$$

From Equation (6.37), it can be seen that the output voltage is proportional to temperature. The op amp for the circuit should have low offset voltage and low drift.

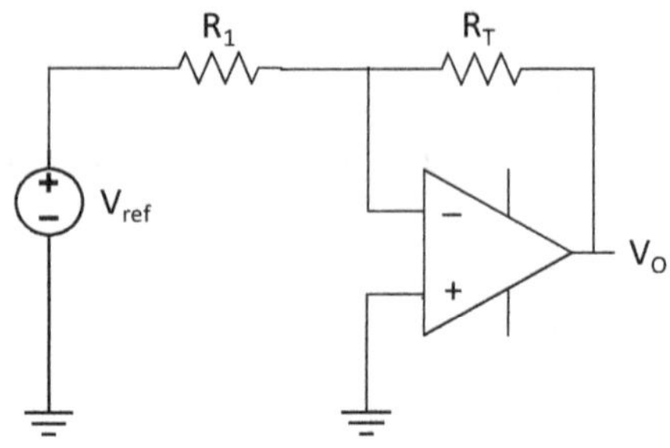

FIGURE 6.6 Circuit for obtaining readout from an RTD using a resistance-to-voltage converter.

The reference voltage, V_{ref}, must be limited to avoid the internal heating of the RTD and the saturation of the op amp.

Example 6.3: Platinum Resistance Thermometer

For a platinum resistance thermometer with the American Alloy, the resistance is related to the temperature by the following expression:

$$R = a + bT + cT^2 \tag{6.38}$$

where R is the resistance in ohms and T is temperature in °C. Using the data shown in Table 6.4, determine the coefficients a, b, and c.

Solution

MATLAB® CODE

```
% Example 6.3
% Platinum Resistance thermometer
T = [0 20  40  60  80  100];
R = [200  207.79  215.54  223.24  230.89  238.5];
n = length(T);
%
% coefficient
```

TABLE 6.4
Resistance versus Temperature of an RTD

T, °C	*R*, Ω
0	200
20	207.79
40	215.54
60	223.24
80	230.89
100	238.5

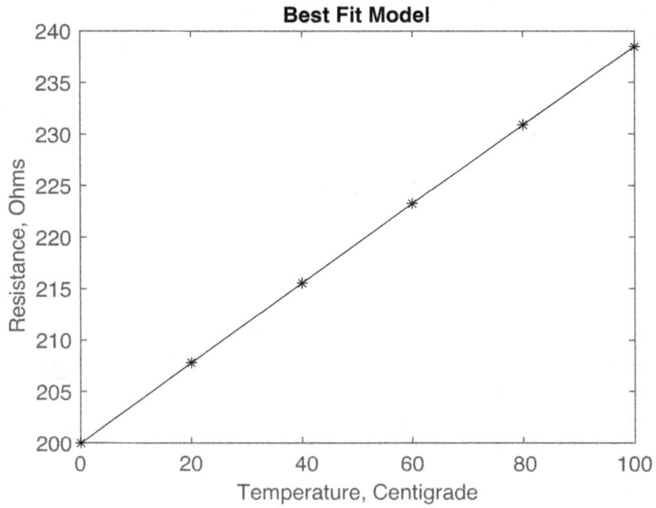

FIGURE 6.7 Resistance versus temperature of an RTD.

```
pfit = polyfit (T, R, 2);
% Equation is of the form R = c*T*T + b*T + a
a = pfit(3)
b = pfit(2)
c = pfit(1)
%
for i = 1:n
    Rfit(i) = c*T(i)*T(i) + b*T(i) + a;
end
% Plot R versus T and best fit  model
plot (T, Rfit, 'k', T, R, '*k')
xlabel ('Temperature, Centigrade')
ylabel('Resistance, Ohms')
title('Best Fit Model')
```

Figure 6.7 shows the plot of resistance versus temperature of the platinum thermometer.

The values of a, b, and c are as follows:

$a = 199.9993$.

$b = 0.3908$.

$c = -5.8036\text{e-}05$.

6.5 THERMISTORS

6.5.1 General Features

Thermistor resistances change with temperature, and they are normally manufactured from semiconducting materials. They can be formed into small beads, rods, and disks and are normally coated for electrical insulation and mechanical protection.

Thermistors can have either positive or negative temperature coefficients depending on the materials used to construct the thermistor. Most thermistors have a negative temperature coefficient (NTC).

The resistance versus temperature of a thermistor can be expressed as

$$R_T = \mathrm{R}_0 \exp\left[\beta\left(\frac{1}{T}-\frac{1}{T_0}\right)\right] \tag{6.39}$$

where

R_T = the resistance of the thermistor at temperature T.
R_0 = resistance of the thermistor at a reference temperature (usually 0°C or room temperature 25°C).
T = thermistor temperature in degrees Kelvin.
T_0 = reference temperature.
β = constant that depends on the thermistor material.
R_0, the resistance of the thermistor at a reference temperature may vary 10 per cent or more from unit to unit. Thus, each thermistor requires calibration before being used.

The thermistor temperature coefficient, α, relates the change of resistance with temperature. It is a measure of thermistor sensitivity. It is defined as

$$\alpha = \frac{1}{R_T}\frac{dR_T}{dT} \tag{6.40}$$

Differentiating Equation (6.39) with respect to temperature and dividing by resistance R_T, we obtain

$$\alpha = -\frac{\beta}{T^2} \tag{6.41}$$

It should be noted that α depends on temperature, and it decreases with increasing temperature. Typical value of α ranges from 3% to 5% per °C.

Equation (6.39) is typically not more than ±5°C over a useful range of the NTC thermistor. The best approximation of thermistor resistance versus temperature is given by the Steinhart–Hart formula. It is given as

$$\frac{1}{T} = A + B.\ln(R_T) + C.(\ln(R_T))^3 \tag{6.42}$$

where

T = thermistor temperature in degrees Kelvin.
R_T = the resistance of the thermistor at temperature T.
A, B, and C are constants obtained from experimental measurement through curve-fitting.

The Steinhart–Hart formula is known to be typically accurate to around ±0.15°C over a range of −50°C to 150°C. It should be noted that if C = 0 in the Steinhart–Hart formula, Equation (6.42) becomes

$$\frac{1}{T} = A + B.\ln(R_T) \tag{6.43}$$

which is similar to Equation (6.39).

At low currents, corresponding to low power dissipation, there is a linear relationship between the voltage and current. Also at low currents, the thermistor temperature is equal to the ambient temperature, and there is hardly any self-heating of the thermistor. At higher currents, self-heating occurs in the thermistor, and the thermistor temperature will be higher than the ambient temperature. Because of the negative temperature coefficient of an NTC thermistor, the incremental resistance of the thermistor decreases. When operating the thermistor at high currents, the current flowing through the thermistor should be limited to prevent thermal destruction.

6.5.2 Voltage Readout of Thermistors

The resistance of a thermistor can be converted to voltage using a resistance-to-voltage converter shown in Figure 6.6. To limit self-heating, the voltage of the precision-regulated voltage source needs to be limited.

When a good accuracy over a small temperature change is required, a bridge circuit can be used. This is shown in Figure 6.5. The output voltage of the bridge V_0 is proportional to the change in temperature, provided we have small temperature change. Once again, the reference voltage V_{ref} should be chosen such that self-heating of the thermistor is prevented.

To convert the resistance of the RTD directly to voltage, a resistance-to-voltage converter, shown in Figure 6.6, can be used. The output voltage is given by Equation (6.36). Using Equations (6.36) and (6.39), we get

$$V_0 = -\mathrm{R}_o \exp\left[\beta\left(\frac{1}{T} - \frac{1}{T_0}\right)\right] * \left(\frac{V_{ref}}{R_1}\right) \tag{6.44}$$

If we set $R_1 = R_0$, then Equation (6.44) becomes

$$V_0 = -V_{ref} \exp\left[\beta\left(\frac{1}{T} - \frac{1}{T_0}\right)\right] \tag{6.45}$$

Thus, the output voltage is dependent on temperature. The op amp for the circuit should have low offset voltage and low drift. In addition, the reference voltage, V_{ref}, must be carefully selected to avoid the internal heating of the RTD and the saturation of the op amp.

Example 6.4: Resistance versus Temperature of a Thermistor

The relationship between resistance and temperature of a thermistor is approximated by Equation (6.39). If $R_0 = 25{,}000\ \Omega$, use the values in Table 6.5 to obtain the constant β and the reference temperature T_0.
Solution

Equation (6.39) can be re-written as follows:

$$\ln\left(\frac{R_T}{R_0}\right) = \beta\left(\frac{1}{T}\right) + \left(-\frac{\beta}{T_0}\right)$$

Thus, a plot of $\ln\left(\frac{R_T}{R_0}\right)$ versus $\left(\frac{1}{T}\right)$ will result in a straight line with a slope of β and y-intercept of $\left(-\frac{\beta}{T_0}\right)$

The plot of resistance versus temperature is shown in Figure 6.8.

The MATLAB code for obtaining β and T_0 is as follows:

MATLAB® CODE

```
% Example 6.4
% Thermistor Temperature vrs. Resistance
T = 300:10:400;
R = [2.25e4  1.46e4 9.77e3 6.69e3 4.68e3 3.35e3 2.44e3 1.80e3
1.36e3 1.04e3 8.03e2];
figure(1)
plot (T, R, 'k*',T, R,'k')
xlabel ('Temperature in degree Kelvin')
ylabel('Resistance in Ohms')
title('Temperature versus Resistance')

n = length(T);
R0 = 25000;
```

TABLE 6.5
Thermistor Resistance versus Temperature

Temperature (°K)	Resistance of Thermistor, Ohms
300	2.25E+04
310	1.46E+04
320	9.77E+03
330	6.69E+03
340	4.68E+03
350	3.35E+03
360	2.44E+03
370	1.80E+03
380	1.36E+03
390	1.04E+03
400	8.03E+02

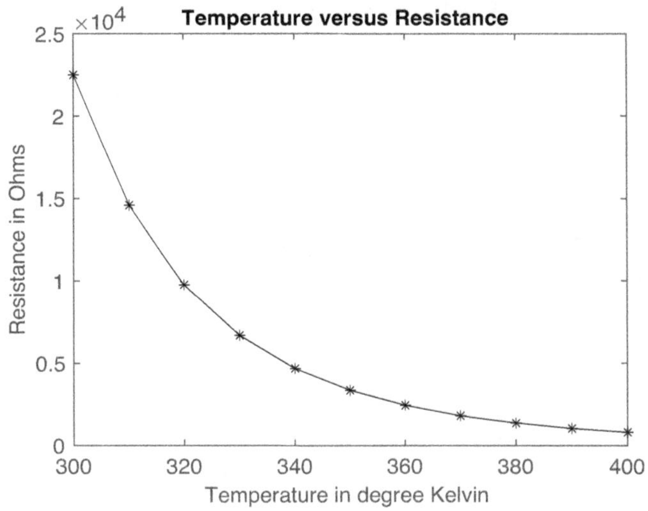

FIGURE 6.8 Resistance versus temperature of a thermistor.

```
% calculate reciprocal of temperature
for i=1:n
    Tinv(i) = 1/T(i);
end
% calculate the natural log of (R/R0)
for i=1:n
    lnRR0(i)= log10(R(i)/R0);
end
% Linear equation is y = mx + b
% coefficient
vfit = polyfit (Tinv, lnRR0, 1);
b = vfit(2);
m = vfit(1);
% calculate beta (slope) and To
beta = m
T0 = -b/beta
%
vfit = m*Tinv + b;
% Plot V versus T and best fit  model
figure(2)
plot (Tinv, vfit, 'k', Tinv, lnRR0, 'k*')
xlabel ('Inverse Temperature in inverse degree Kelvin')
ylabel('Log of Resistance Ratio, ln(R_R0)')
title('Inverse Temperature versus Resistance Ratio')
```

The plot for obtaining β and T_0 is shown in Figure 6.9.

MATLAB® Results

beta = 1.7360e+03
T0 = 0.0034

Thus,

$\beta = 1.7360e+03$.
$T_0 = 0.0034$.

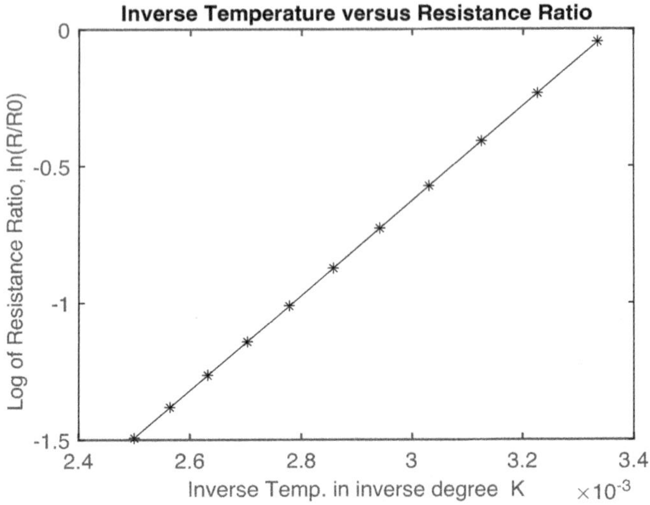

FIGURE 6.9 Log of resistance ratio versus inverse temperature.

6.6 THERMOCOUPLES

A thermocouple is a sensor consisting of two different conductors that produces a voltage that is proportional to a temperature difference between the two ends of the conductors. The voltage produced by a thermocouple is known as the Seebeck voltage, which is nonlinear with respect to temperature. For small changes in temperature, the Seebeck voltage is approximately linear, and it is given as follows:

$$\Delta V = S(\Delta T) \tag{6.46}$$

where

ΔV is the change in voltage.
S is the Seebeck coefficient.
ΔT is the change in temperature.

Figure 6.10 shows two dissimilar conductors: copper and constantan joined at two ends. If the temperatures at the ends are different, a current will flow. The

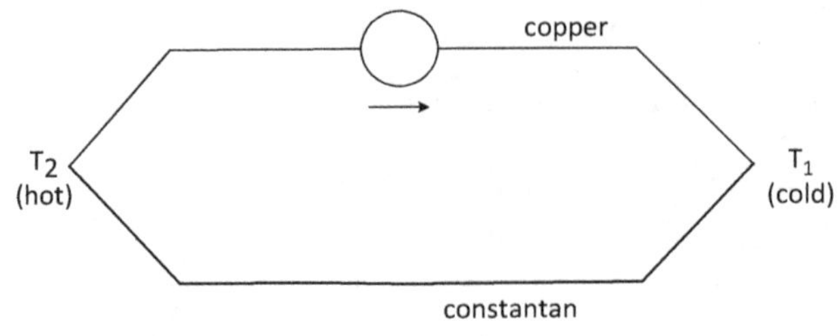

FIGURE 6.10 Basic circuit for Seebeck effect.

electromotive force (emf) producing the current is called Seebeck thermal emf. The thermoelectric laws will be described in the following sub-sections.

6.6.1 Thermoelectric Laws

The measurement of temperature by thermocouples is based on the application of the thermoelectric laws, which are as follows:

1. *Law of Interior Temperatures*: The Seeback thermal emf is dependent on the temperature at the ends of the dissimilar conductors welded together, and the thermal emf is independent on temperatures maintained along the homogeneous conductors. As shown in Figure 6.11, the emf produced by the thermocouple is dependent on the difference in temperature T_1 and T_2 and the emf is independent of temperatures T_3 and T_4.
2. *Law of Intermediate Metals*: A third conductor inserted into one of the dissimilar conductors forming the thermocouple will have no effect on the Seebeck effect emf provided the junctions of the third conductor are maintained at the same temperature. As shown in Figure 6.12, the J_1 and J_2 junctions of metal X are maintained at the same temperature T_3. The resulting emf will be the same as that when X is not present. The Law of Intermediate Metal allows us to attach lead wires to thermocouples, provided that junctions J_1 and J_2 are at the same temperature.

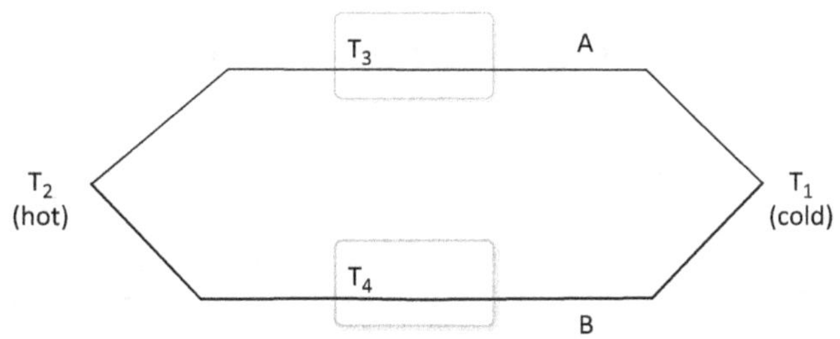

FIGURE 6.11 Figure showing the Law of Interior Temperatures.

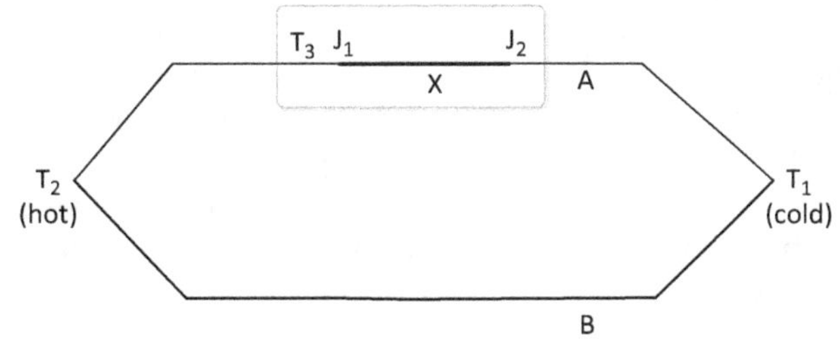

FIGURE 6.12 Figure showing the Law of Intermediate Metal

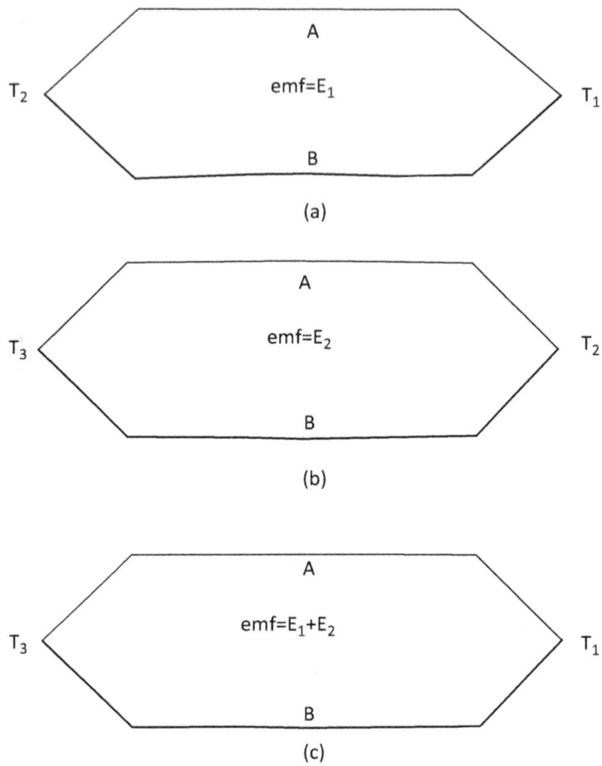

FIGURE 6.13 Figure showing the Law of Intermediate Temperature.

3. *Law of Intermediate Temperature*: For a thermocouple, the generated emf will add provided the intermediate temperature is the same. Specifically, Figure 6.13 shows two dissimilar conductors that produce an emf of E_1 when the junction temperatures are T_1 and T_2 and also produce an emf of E_2 when the junction temperatures are T_2 and T_3. Now if the thermocouple has junction temperatures of T_1 and T_3, the generated emf is $E_1 + E_2$. This law allows us to use thermocouple reference tables, even if the reference junction of our application is not 0°C.
4. *Law of Additive EMF*: If a thermocouple TC_1 constructed from conductors A and C generates an emf of E_{AC} and a thermocouple TC_2 constructed from conductors B and C generates an emf of E_{BC}, then a thermocouple constructed from conductors A and B will generate an emf $E_{AC} + E_{BC}$. This is illustrated in Figure 6.14. This law allows us to create a nonstandard thermocouple from combinations of conductors and still use a thermocouple reference table.

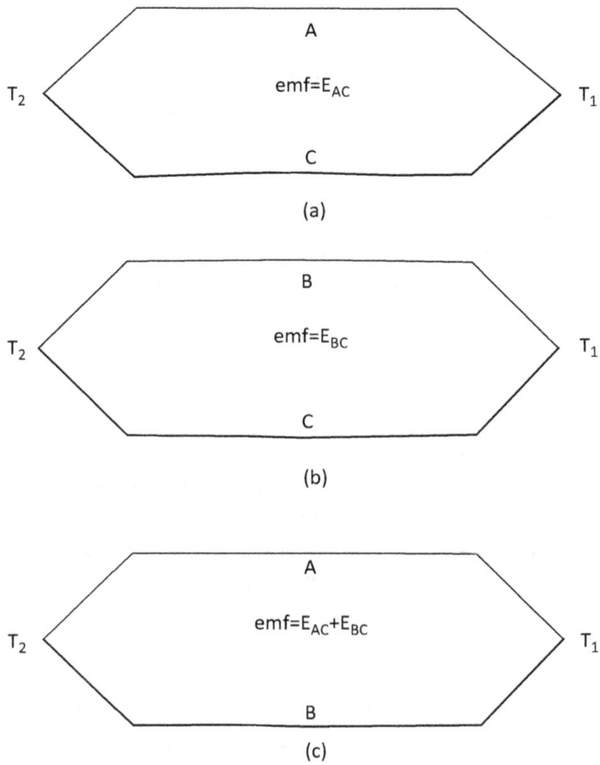

FIGURE 6.14 Figure showing the Law of Additive EMF.

Example 6.5: Seebeck Voltage of an Iron–Copper Thermocouple

For a thermocouple made of copper and iron that is connected at their ends, if the Seebeck coefficient of copper is 1.5 μV/°K and the Seebeck coefficient of iron is 18 μV/°K, the temperature at the hot junction is 350°K, while the temperature at the cold junction is 250°K. (i) Determine the overall Seebeck coefficient of the iron–copper thermocouple and (ii) calculate the generated voltage.

Solution

i. The overall Seebeck coefficient of the iron–copper thermocouple is obtained by subtracting the coefficient of copper from that of iron. Thus,

$$S_{\text{total}} = S_{\text{FE}} - S_{\text{CU}}$$

$$S_{\text{total}} = (18 - 1.5)\ \mu\text{V}/°\text{K} = 16.5\ \mu\text{V}/°\text{K}$$

ii. The temperature difference $\Delta T = (350 - 250)°\text{K} = 100°\text{K}$. Using Equation (6.46), the Seebeck voltage is given as

$$\Delta V = S_{\text{total}}(\Delta T) = (16.5\ \mu\text{V}/°\text{K})(100°\text{K}) = 1650\ \mu\text{V}$$

TABLE 6.6
Temperature Ranges of Common Thermocouples

Type	Temperature Range (°C)	Materials	Seebeck Coefficient (μV/°C)	Color Code (ANSI/ ASTM E230)
K	−270 to 1370	Chromel/Alumel	39.48	Red(−)/Yellow(+)
J	−210 to 1200	Iron/Constantan	50.37	Red(−)/White(+)
E	−270 to 1000	Chromel/Constantan	58.70	Red(−)/Purple(+)
T	−200 to 400	Copper/Constantan	38.74	Red(−)/Blue(+)

6.6.2 Thermocouple Types

Thermocouples operate at different temperature ranges, and some thermocouples may be subjected to corrosion and oxidation. The common thermocouple metals, also called base materials, are types T, E, J, and K. Specialized designed metals, also called noble metals, are types R and S. Thermocouples, constructed from these noble metals, are able to operate at high temperatures and are less susceptible to oxidation and corrosion. However, the sensitivity of noble metal thermocouples is much less than those of base–metal thermocouples. Table 6.6 shows the temperature ranges of common thermocouples.

Type K is the most common thermocouple type because of its large temperature range of operation and its durability. In addition, type K thermocouples are less chemically reactive compared to Type T and Type J thermocouples. Although, Type K thermocouples have output voltages lower than those of Types J and E, the output of Type K thermocouples is higher than that of Type N.

The voltage-to-temperature conversion relationship of a thermocouple is nonlinear. The relation can be expressed using a power-series polynomial as follows:

$$T = a_0 + a_1V + a_2V^2 + a_3V^3 + \cdots + a_nV^n \tag{6.47}$$

where

V = thermocouple output voltage.
T = temperature.
$a_0, a_1, a_2 \ldots a_n$ = polynomial coefficients that depend on a thermocouple type.
n = maximum order of the polynomial.

Table 6.7 lists National Bureau of Standards (NIST) polynomials for voltage-to-temperature conversion for some thermocouples.

The thermocouple voltage can also be obtained as a function of temperature as follows:

$$V = c_0 + c_1T + c_2T^2 + c_3T^3 + \cdots + c_nT^n \tag{6.48}$$

where

V = thermocouple output voltage.
T = temperature.

TABLE 6.7
NIST Polynomial Coefficients for Voltage-to-Temperature Conversion of Thermocouples ($T = a_0 + a_1V + a_2V^2 + a_3V^3 + \ldots + a_nV^n$)

	Type K	Type J	Type E	Type T
Range	0°C to 500°C	0°C to 760°C	0°C to 1000°C	−160°C to 400°C
a_0	0	0	0	0
a_1	2.508355E-2	1.978425E-2	1.7057035E-2	2.592800E-2
a_2	7.860106E-8	−2.001204E-7	−2.33017597E-7	−7.602961E-7
a_3	−2.503131E-10	1.036969E-11	6.5435585E-12	4.637791E-11
a_4	8.315270E-14	−2.549687E-16	−7.3566274E-17	−2.165394E-15
a_5	−1.228034E-17	3.585153E-21	−1.7896001E-21	6.048144E-20
a_6	9.804036E-22	−5.344285E-26	8.4036165E-26	−7.7293422E-25
a_7	−4.4130330E-26	5.099890E-31	−1.3735879E-26	
a_8	1.057734E-30		−1.0629823E-23	
a_9	−1.052755E-35		−3.2447087E-41	

The voltage is in μV and temperature in °C.

TABLE 6.8
NIST Polynomial Coefficients for Temperature-to-Voltage Conversion of Thermocouples ($V = c_0 + c_1T + c_2T^2 + c_3T^3 + \cdots + c_nT^n$)

	Type K	Type J	Type E	Type T
Range	0°C to 1372°C	−210°C to 760°C	0°C to 1000°C	−160°C to 400°C
c_0	−17.600413686	0	0	0
c_1	38.921204975	50.38118782	58.665508710	38.748106364
c_2	1.855877E-2	3.047583693E-2	4.503227558E-2	3.32922279E-2
c_3	−9.9457593E-5	−8.56810657E-5	2.890840721E-5	2.06182434E-4
c_4	3.1840945E-7	1.322819530E-7	−3.30568967E-7	−2.18822568E-6
c_5	−5.607284E-10	−1.7052958E-10	6.50244033E-10	1.09968809E-8
c_6	5.607059E-13	2.09480907E-13	−1.9197496E-13	−3.0815759E-11
c_7	−3.202072E-16	−1.2538395E-16	−1.2536600E-15	4.54791353E-14
c_8	9.7151147E-20	1.56317257E-20	2.14892176E-18	−2.7512902E-17
c_9	−1.210472E-23		−1.4388042E-21	
c_{10}	NOTE A		3.59608995E-25	

NOTE A: The equation for a Type K thermocouple is $V = c_0 + c_1T + c_2T^2 + c_3T^3 + \cdots + c_9T^9 + 118.597e^{(-1.183432E-4)(T-126.9686)^2}$.
The voltage is in μV, and temperature in °C.

$a_0, a_1, a_2 \ldots a_n$ = polynomial coefficients that depend on a thermocouple type.
n = maximum order of the polynomial.

Table 6.8 shows the NIST thermocouple coefficients for Types K, J, E, and T thermocouples for temperature-to-voltage conversion.

6.6.3 Measuring Thermocouple Temperatures

If a voltmeter is used to obtain the thermal emf of a thermocouple, the copper cable of the voltmeter when connected to a thermocouple will create additional junctions, which will create additional emfs. Figure 6.15 shows a voltmeter with copper lead connected to an iron–constantan thermocouple (Type J). Two junctions J_2 and J_3 will be created. J_2 is an iron-to-copper junction, and it creates a thermal emf V_2. Similarly, J_3 is a copper-to-constantan junction, and it also creates a thermal emf V_3. If the junctions J_2 and J_3 are in opposition, the voltage $V_2 - V_3$ will be approximately zero, and the voltmeter will indicate moderately accurate measurements.

If the iron–constantan thermocouple (Type J) is replaced by the copper–constantan thermocouple (Type T), then the junction J_2 in Figure 6.16 creates no emf since we have a copper-to-copper junction. The junction J_3, a copper-to-constantan junction, creates a thermal emf V_3. Since V_2 is zero, the voltage at the voltmeter is given as

$$\Delta V = V_1 - V_3 \tag{6.49}$$

The voltmeter reading ΔV will be proportional to the temperature difference T_1 and T_3. Thus,

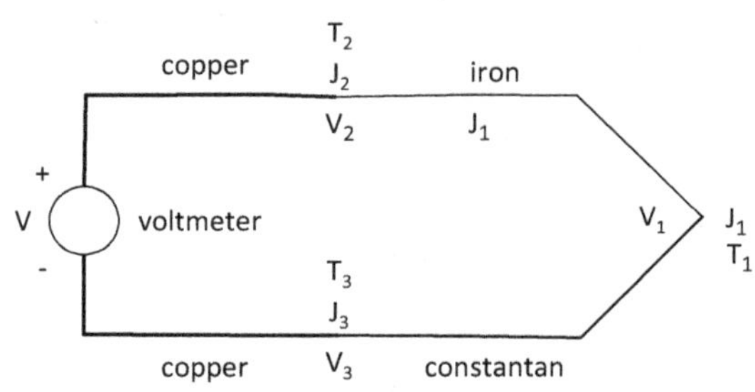

FIGURE 6.15 Voltage measurement on a thermocouple (Type J).

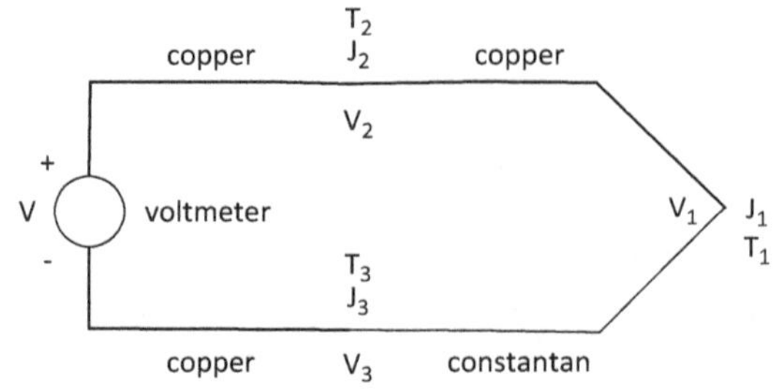

FIGURE 6.16 Voltage measurement on a thermocouple (Type K).

$$\Delta V = k(T_1 - T_3) \tag{6.50}$$

where k is a constant.

If we put junction J_3 in an ice bath, the T_3 is zero, and the voltmeter reading is given as

$$\Delta V = kT_1 \tag{6.51}$$

Thus, the voltmeter reading is proportional to the temperature T_1 at junction J_1.

In some applications, we avoid putting the junction J_3 in an ice bath. Instead, we generate a voltage equal but opposite to the thermal voltage V_3 in Equation (6.48). This makes the voltmeter voltage to be equal to

$$\Delta V = V_1 - V_3 + V_3 = V_1 \tag{6.52}$$

This technique is called cold-junction compensation. There is software and hardware cold-junction compensation techniques that will be described in the following paragraphs.

Consider a generic thermocouple with metals A and B, shown in Figure 6.17. Junction J_2 is formed by metal A and copper and junction J_3 by metal B and copper. We can use the cold-junction compensation to obtain the temperature T_1. We can use an isothermal block that surrounds the junctions J_2 and J_3. The isothermal block keeps the area around the junctions J_2 and J_3 to be of the same temperature. In addition, the block is an electrical isolator and also a good heat conductor. The temperature inside the block is measured using another temperature sensor, such as a thermistor. If a thermistor is used, we measure its resistance to obtain the reference temperature T_{ref} of the block. The equivalent reference junction voltage V_{ref} of the thermocouple is obtained. We add the measured voltage V_M to V_{ref} for obtaining the voltage V_1 at the hot junction J_1. This procedure is termed software compensation. A computer uses the temperature T_{ref} and Equation (6.48) to obtain the junction voltage V_1. In addition, after obtaining the voltage V_1, the computer software uses Equation (6.47) to obtain the temperature at the hot junction T_1.

Another method for cold-junction compensation uses electronic hardware. In hardware compensation, a variable-voltage source is placed in the circuit to cancel the influence of the cold-junction temperature. Figure 6.18 shows a circuit that can be

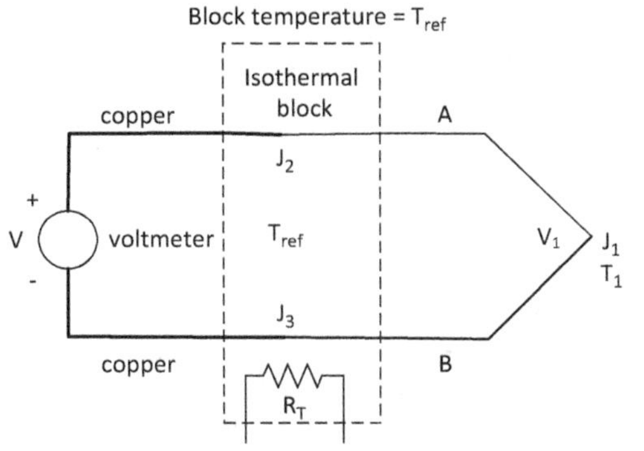

FIGURE 6.17 Thermocouple software compensation.

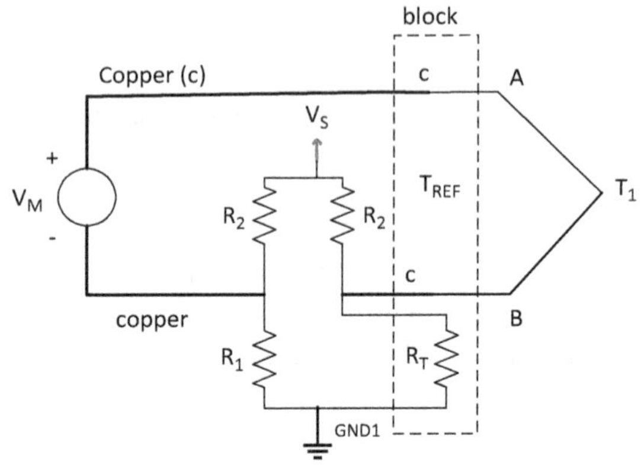

FIGURE 6.18 Thermocouple hardware compensation.

used for hardware compensation. The bridge circuit consists of R_T, R_1, R_2 and voltage source V_S. The thermistor monitors the terminal contact. The thermistor voltage, which depends on the temperature on the junction, is used to compensate for junction voltage. The resistor R_1 is set such that the measured voltage, V_M, is zero when the thermocouple is at 0 °C.

Thermocouples have very low input resistance, and the output voltages may vary from 6 to 90 μV/°C. The thermocouple voltages are normally amplified before being displayed. Inverting or noninverting amplifiers are normally used to increase the gain of the thermocouple thermal emf.

Example 6.6: Type J Thermocouple

The NIST polynomial coefficients for Type J thermocouple are shown in Table 6.8. (i) Plot the voltage generated by the thermocouple for temperatures between 0°C and 700°C. (ii) What are the voltages generated at 25°C and 124°C?

Solution

MATLAB® CODE

```
% Solution to Example 6.6
% From Table 6.8, the coefficients for Type J thermocouple is c
c = [1.56317257e-20 -1.2538395e-16 2.0948090e-13 -1.7052958e-10 ...
    1.322819530e-7 -8.56810657e-5 3.047583693e-2 50.38118782 0];
% Temperatures are generated
T = 0:50:700;
len = length(T);
for i=1:len
    v(i) = polyval(c,T(i))*1.0e-6;
end
plot(T,v,'k',T,v,'k*')
title('Voltage vs. Temperature of Type J Thermocouple')
xlabel('Temperature in degrees Centigrade')
ylabel('Voltage Generated by Thermocouple')
v25 = polyval(c,25)*1.0e-6;
```

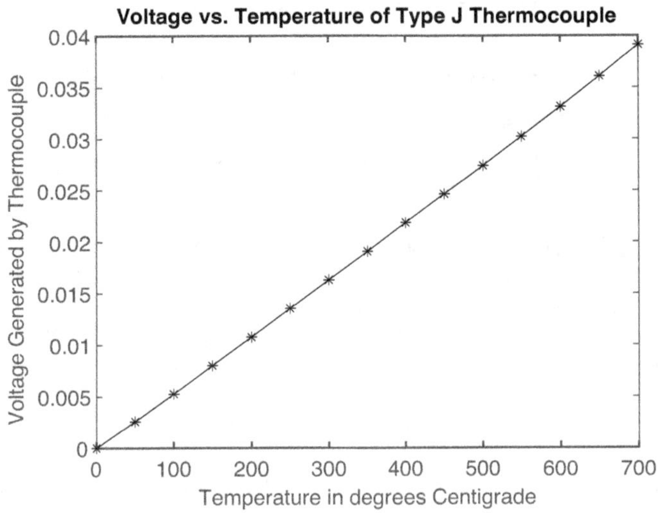

FIGURE 6.19 Voltage versus temperature of type J thermocouple.

```
v124 = polyval(c,124)*1.0e-6;
fprintf('Thermocouple voltage at 25 degrees Cent. is %9.3e volts
\n',v25)
fprintf('Thermocouple voltage at 124 degrees Cent. is %9.3e volts
\n',v124)
```

Figure 6.19 depicts the voltage versus temperature of Type J thermocouple.

Answers

Thermocouple voltage at 25°C is 1.277e-03 V.
Thermocouple voltage at 124°C is 6.579e-03 V.

6.7 COMPARISON OF COMMON TEMPERATURE TRANSDUCERS

Table 6.9 shows the comparison of various temperature sensors from the perspective of temperature range, accuracy, linearity, sensitivity, excitation, form of output, and susceptibility to self-heating. Table 6.10 shows the sensors and their areas of application.

TABLE 6.9
Comparison of Characteristics of Temperature Sensor Types

Criteria	Thermocouple	RTD	Thermistor	Integrated Silicon
Temperature Range	−210°C to 1760°C	−240°C to 650°C	−40°C to 250°C	−55°C to 150°C
Accuracy	Medium	High	Medium	Poor
Linearity	Fair, requires at least a 4th-order polynomial	Good, requires at least a 2nd-order polynomial	Fair, requires at least a 3rd-order polynomial	Good, if the diode current is constant.
Susceptibility to self-heating	No	Yes, but minimal	Yes, highly susceptible	Yes, if the sensor current is large
Sensitivity	Low	Medium	Very high	Medium
Excitation	None required	Current source	Voltage source	Typically, a supply voltage
Form of output	Voltage	Resistance	Resistance	Voltage, current or digital

TABLE 6.10
Advantages and Disadvantages of Temperature Sensor Types

Sensor	Advantages	Disadvantages	Application Areas
Thermocouple	• Fastest response • Rugged; • Wide temperature range • No external power; • Point temperature sensor	• Nonlinear response; • Small sensitivity; • Small output voltage; • Least stable, • Requires CJC	• Furnace monitoring; • Steel mills; • Power generation; • Food processing
RTD	• Most stable; • Good linearity; • Most accurate; • Very resistant to corrosion of an RTD material	• Low sensitivity to small temperature changes; • Slow response time; • Small output resistance; • Has self-heating error; • Externally powered	• Air conditioning servicing; • Furnace servicing; • Textile production
Thermistor	• Fast; • High sensitivity to small temperature changes; • High input	• Limited temperature range; • Nonlinear response; • Has self-heating error; • Externally powered	• Electronic equipment inside automobiles; • Medical devices; • HVAC; • Overcurrent protection
Silicon Semiconductor	• Small and lighter; • Highly accurate; • Easily integrated with electronic components	• Limited temperature range; • Self-heating error if the current is large; • Requires a constant current source	• Hard disk drives; • Personal computers; • Cellular phones; • Process control

PROBLEMS

Problem 6.1 Table P6.1 shows the diode voltage versus temperature of a silicon diode at 10 uA and 100 uA. (i) Determine the line of best fit for the two currents and (ii) calculate the change in voltage versus temperature for the two currents.

Problem 6.2 For the Widlar current source shown in Figure 6.3, $R_C = 40\ \mathrm{k}\Omega$, $V_{CC} = 5\ \mathrm{V}$, $V_{BE1} = 0.7\ \mathrm{V}$, $R_E = 20\ \mathrm{k}\Omega$, and $\beta = 120$. Determine the voltage across the resistance R_E for temperatures from -100°C to 100°C.

Problem 6.3 For the Widlar current source shown in Figure 6.3, determine the output current I_{C1} versus temperature if $R_C = 50\ \mathrm{k}\Omega$, $V_{CC} = 5\ \mathrm{V}$, $V_{BE1} = 0.7\ \mathrm{V}$, $R_E = 25\ \mathrm{k}\Omega$, and $\beta = 50$.

TABLE P6.1
Diode Voltage versus Temperature at Two Diode Currents

Temperature (°C)	Diode Voltage at Current of 10 μA	Diode Voltage at Current of 100 μA
−40	0.61	0.75
−20	0.55	0.71
0	0.49	0.67
20	0.43	0.62
40	0.37	0.58
60	0.31	0.54
80	0.25	0.51
100	0.19	0.47

TABLE P6.2
Resistance versus Temperature of a RTD

T (°C)	*R* (Ω)
−200	240
−100	600
0	1000
100	1400
200	1800
300	2100
400	2500
500	2700
600	3100

TABLE P6.3
Resistance versus Temperature of a RTD

T (°C)	*R* (Ω)
0	1002
10	1038
20	1078
30	1114
40	1155
50	1193
60	1232
70	1271
80	1308
90	1347
100	1386

TABLE P6.4
Resistance versus Temperature of a Thermistor

Temperature (°K)	Resistance (Ω)
343	1000
353	550
363	430
373	350
393	220
413	150
423	90

Problem 6.4 Information obtained from the data sheet of a HEL – 700 series platinum RTD is shown in Table P6.2. If the resistance, R, is related to the temperature, T, by the expression $R = a + bT + cT^2 + dT^3$, use MATLAB to determine the coefficients a, b, c, and d.

Problem 6.5 The following empirical data were obtained from a platinum resistance thermometer (Table P6.5).

a. Plot the temperature versus resistance characteristics.
b. The characteristic equation for the sensor is given as

$$R_T = R_0(1 + \alpha T)$$

where R_T = resistance at temperature T, R_0 = resistance at 0 °C, and α is the temperature coefficient. We are to determine R_0 and α.

TABLE P6.5
Temperature versus Resistance of a Thermistor

Temperature (°K)	Resistance (kΩ)
328	6.92
350	5.36
380	4.26
400	3.52
430	2.89
460	2.24
480	1.87
510	1.45
530	1.13

TABLE P6.6
Temperature versus Generated Emf

Temperature (°C)	Emf (mV)
0	0
30	1.203
120	4.920

TABLE P6.7
Temperature versus Generated Emf

Temperature (°C)	Emf (mV)
0	0
20	1.019
200	10.779

Problem 6.6 To the first order in $1/T$, the relationship between resistance R and temperature T of a thermistor is given by $R(T) = R(T_0)\exp\left(\beta\left(\frac{1}{T} - \frac{1}{T_0}\right)\right)$, where T is expressed in degrees Kelvin; T_0 is the reference temperature in degrees Kelvin. $R(T_0)$ is the resistance at the reference temperature, and β is the temperature coefficient. If $R(T_0)$ is 10,000 Ω at the reference temperature, use MATLAB and the data in the Table P6.4 to obtain the constants β and T_0.

Problem 6.7 The relationship between resistance and temperature of a thermistor is approximated by Equation (6.39). If $R(T_0) = 10.29$ kΩ, use the values in Table P6.5 to obtain the constant β and the reference temperature T_0

Problem 6.8 A chrome–alumel thermocouple has an approximate Seebeck coefficient of 40 μV/°K. If the temperatures at the hot and cold junctions are 500°K and 300°K, respectively, calculate the Seebeck emf.

Problem 6.9 For a type K thermocouple, NIST data for temperature and generated emf are shown in Table P6.6.

The temperatures of the hot and cold junctions of the thermocouple are 120°C and 30°C, respectively. What is the thermoelectric emf generated?

Problem 6.10 For a type J thermocouple, NIST data for temperature and generated emf are shown in Table P6.7.

If the temperatures of hot and cold junctions of the thermocouple are 120°C and 30°C, respectively, what is the thermoelectric emf generated?

Problem 6.11 A type T thermocouple is intended for use in an environment where the temperature is 300°C. If the reference junction temperature is 100°C, (a) calculate the generated emf if the first three terms of the polynomial coefficients, shown in Table 6.8, are used; (b) determine the generated emf if all the terms of the polynomial coefficients, shown in Table 6.8, are used; (c) calculate the error involved if the first three terms of the polynomial coefficients are used.

Problem 6.12 The NIST polynomial coefficients for a type T thermocouple is shown in Table 6.8. (i) Plot the voltage generated by the thermocouple for temperatures between 0°C and 400°C. (ii) Determine the temperature when the thermocouple voltage is 10 mV.

BIBLIOGRAPHY

Alexander, Charles and Matthew N.O. Sadiku, *Fundamentals of Electric Circuit, ISF*, 7th Edition, McGraw-Hill Education, 2020.

Aminu, K.T., A. G. Jumba, A. A. Jimoh, A. Shehu, B. D. Halilu, S. A. Baraza, A. S. Kabiru, and M. A. Sule, "NTC Thermistor Performance and Linearization of its Temperature-Resistance Characteristics Using Electronic Circuit," *International Journal of Advances in Scientific Research and Engineering*, Vol. 6, Issue 8, pp. 35–49, 2020.

Asadi, Farzin, *Essential Circuit Analysis using NI Multisim and MATLAB*, 1st Edition, Springer, 2022.

Attia, John O., *Electronics and Circuit Analysis Using MATLAB*, CRC Press, 2nd Edition, 2004.

Attia, John O., *Circuits and Electronics: Hands-on Learning with Analog Discovery*, CRC Press, 2018.

Bolton, William, *Instrumentation and Control Systems*, 3rd Edition, Elsevier, 2021.

Chapman, Stephen J., *MATLAB Programming for Engineers*, 7th Edition, Cengage Learning, 2024.

Diarah, Reuben S., Christian Osueke, Adefemi Adekunle, Segun Adebayo, Adedayo Banji Aaron, and Olaluyi Olawale Joshua, "Types of Temperature Sensors," in Jaydip Sen, Mingqiang Yi, Fenglei Niu, and Hao Wu (Eds.), *Wireless Sensor Networks – Design, Applications and Challenges*, Intechopen, pp. 1–19, 2023.

Etter, Delores, *Engineering Problem Solving with MATLAB*, 3rd Edition, Pearson, 2021.

Gierth, Paul, Lars Rebenklau, Klaus Augsburg, Eric Bachmann, and Lars Niedermeyer, "Novel Thermocouples for Automotive Applications," *Journal of Sensors and Sensor Systems*, Vol. 7, pp. 43–49, 2018.

Hayes, Thomas C., David Abrams, and Paul Horowitz, *Learning the Art of Electronics: A Hands-On Lab Course*. 2nd Edition, Cambridge University Press, 2025.

Hayt, William, Jack Kemmely, Jamie Phillips, and Steven M. Durbin, *Engineering Circuit Analysis*, 10th Edition, McGraw Hill, 2023.

Horowitz, Paul and Winfield Hill, *The Art of Electronics*, 3rd Edition, Cambridge University Press, 2015.

Ida, Nathan, *Sensors, Actuators, and Their Interfaces: A Multidisciplinary Introduction*, 2nd Edition, Institution of Engineering and Technology, 2020.

Irwin, J. David and R. Mark Helms, *Engineering Circuit Analysis*, 12th Edition, Wiley, 2021.

Ivanov, Valentin Ganchev, "Temperature Transducer," *International Journal of Scientific Research in Science and Technology (IJSRST)*, Vol. 4, Issue 5, pp. 1571–1574, 2018.

Jaeger, Richard C. and Travis Blalock, *Microelectronic Circuit Design (ISE)*, 6th Edition, McGraw Hill, 2022.

Mohammed, Abba, Suleiman Babani, Abdurrashid Ibrahim Sanka, Nura Ahmed Abdullahi, "A Comparative Study between Different Types of Temperature Sensor," *Proceedings of 13th ISERD International Conference*, Dubai, UAE, pp. 22–24, 2015

Morris, Alan S. and Reza Langari, *Measurement and Instrumentation: Theory and Applications*, 3rd Edition, Academic Press, 2020.

Northrop, Robert B., *Introduction to Instrumentation and Measurements*, 3rd Edition, CRC Press, 2014.

Potter, David, "Measuring Temperature with Thermocouples – a Tutorial," National Instruments Application Note 043, 1996.

Rana, Kanwar Pal Singh, Vineet Kumar, Amit Kumar Dagar, Ankit Chandel and Akshay Kataria, "FPGA Implementation of Steinhart–Hart Equation for Accurate Thermistor Linearization," *IEEE Sensors Journal*, vol. 18, no. 6, pp. 2260–2267, 2018.

Razavi, Behzad, *Fundamentals of Microelectronics*, 3rd Edition, Wiley, 2021.

Sedra, Adel S. and Kenneth Smith, *Microelectronic Circuits*, 8th International Edition, Oxford University Press, 2020.

Tougas, Ian M., Matin Amani and Otto J. Gregory, "Metallic and Ceramic Thin Film Thermocouples for Gas Turbine Engines," *Sensors*, Vol. 13, pp. 15324–15347, 2013.

Tutunea, Dragos, George Gherghina, Ilie Dumitru, and Alexandru Dima, "Evaluation of Thermistors Used for Temperature Measurement in Automotive Industry," *Applied Mechanics and Materials*, Vol. 880, pp. 157–162, March 2018.

Valentine, Daniel T. and Brian D. Hahn, *Essential MATLAB for Engineers and Scientists*, 8th Edition, Academic Press, 2022.

Wu, Joseph, "A Basic Guide to RTD Measurements," Texas Instruments Application Note, SBAA275A – June 2018 – Revised March 2023, https://www.ti.com/lit/an/sbaa275a/sbaa275a.pdf?ts=1754003722887 (Accessed August 4, 2025).

Yee, Angela, "NIST ITS-90 Thermocouple Database - SRD 60," National Institute of Standards and Technology, 2000, https://doi.org/10.18434/T4S888 (Accessed August 4, 2025).

7 Displacement Sensors

Displacement transducers can be used to sense linear or angular displacement of an object. Linear displacement transducers sense the displacement along a line, whereas angular displacement transducers sense a rotation about an axis. Resistive, capacitive, inductive, and piezoelectric displacement sensors will be discussed in this chapter.

7.1 RESISTIVE TRANSDUCERS

There are two popular forms of resistive transducers: (i) potentiometers and (ii) strain gages.

7.1.1 POTENTIOMETERS

Potentiometers for measuring displacement consist of a single resistive wire with sliding contact. The sliding contact is connected to an object that might move in the linear or angular direction. Figures 7.1 and 7.2 show potentiometers for measuring translational and angular displacements, respectively.

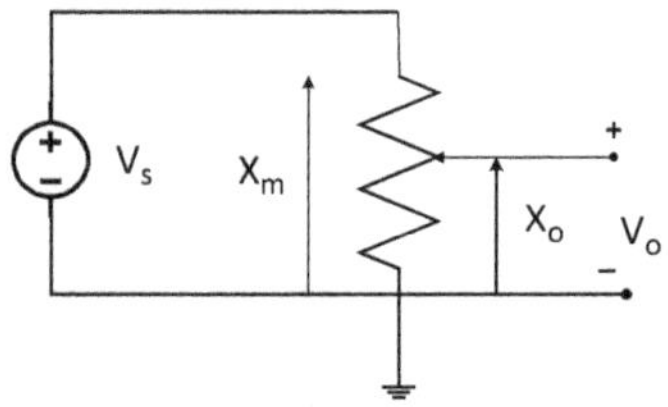

FIGURE 7.1 Potentiometer transducer for translational displacement.

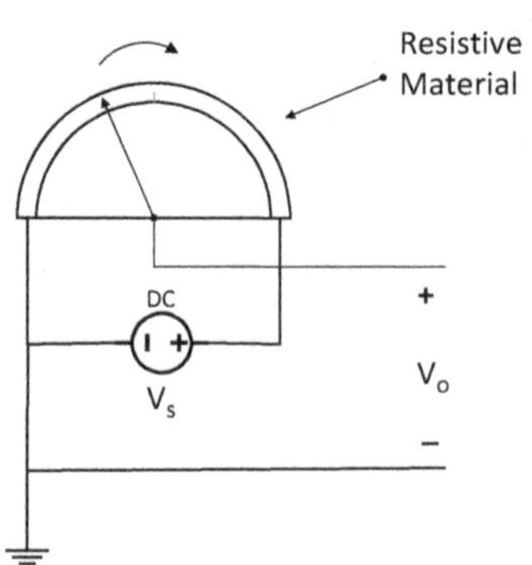

FIGURE 7.2 Potentiometer transducer for angular displacement.

DOI: 10.1201/9781003765554-7

If the potentiometer transducer is not loaded, then from Figure 7.1,

$$V_0 = iR_{X_0} = i\frac{\rho X_0}{A} \tag{7.1}$$

where

ρ = resistive of the potentiometer wire.
A = area of the potentiometer.
i = current flowing through the potentiometer.

Similarly,

$$V_S = iR_{Xm} = i\frac{\rho X_m}{A} \tag{7.2}$$

Thus,

$$V_0 = \frac{X_0}{X_m}V_S \tag{7.3}$$

Equation (7.3) shows that the output voltage is proportional to the linear displacement. Figure 7.3 shows a displacement sensor connected to a load.

If a load is connected to the output port of the potentiometer, some of the current from the voltage source will flow through resistance R_L. The output voltage will be nonlinearly related to the displacement X_0. The nonlinearly related voltage can be reduced if the load is very high compared to the resistance of the potentiometer. This is illustrated through Example 7.1. Figure 7.4 shows one approach of reducing the loading by using a buffer amplifier.

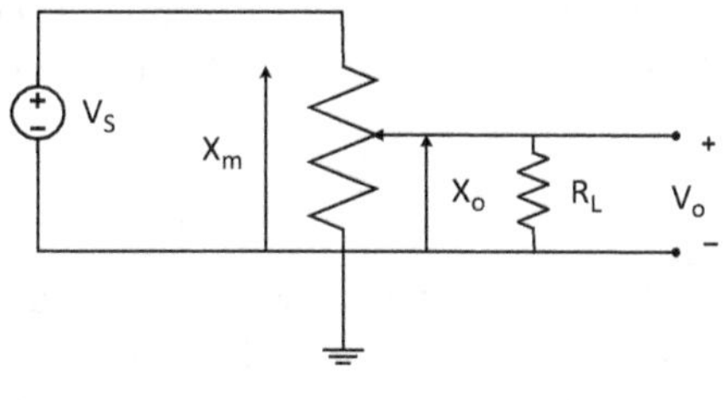

FIGURE 7.3 Resistive sensor with a load.

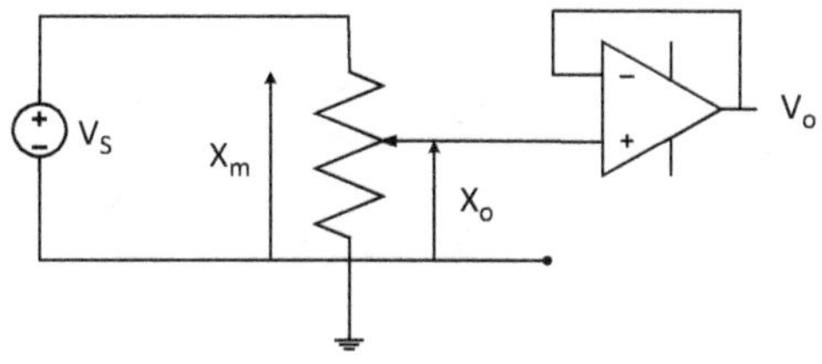

FIGURE 7.4 Buffer amplifier used to reduce the loading effect.

Example 7.1: Loading Effect Error

For the circuit shown in Figure 7.5, the track resistance $R_P = 10\ \text{k}\Omega$, $V_S = 5\ \text{V}$, $\frac{x}{L} = 20\%$, and $L = 5\ \text{cm}$. If $R_L = 50\ \text{k}\Omega$, find the error in reading the voltage across x, if x is 20% of L.

Solution

For $\frac{x}{L} = 20\%$,

$$R_{XL} = R_X \| R_L = (0.2 * 10\ \text{K} \| 50\ \text{K}) = (2\ \text{K} \| 50\ \text{K}) = \frac{(2\ \text{K})(50\ \text{K})}{52\ \text{K}} = 1.923\,\text{k}\Omega$$

$$R_M = R_P - R_X = 10\ \text{K} - 2\ \text{K} = 8\ \text{k}\Omega$$

$$V_0 = \frac{R_{\text{XL}} * V_S}{R_M + R_{\text{XL}}} = \frac{(1.923\ \text{K}) * 5}{8\ \text{K} + 1.923\ \text{K}} = 0.969\ \text{V}$$

If there is no loading, that is, as R_L approaches infinity, then $R_{\text{XL}} = R_X$

$$V_0 = \frac{R_X * V_S}{R_P} = \frac{(2K) * 5}{10K} = 1.0\ \text{V}$$

Error = 1.0 − 0.969 = 0.031 V

7.1.2 Strain Gages

Strain gages are resistive elements, made of wire, metal foil, or semiconductor material. Strain gages can be used to measure displacement and applied force. For a wire of cross-sectional area A, resistivity ρ, and length L, the resistance is given as

$$R = \frac{\rho L}{A} \tag{7.4}$$

When a wire is stretched, the cross-sectional area will be reduced, and the resistivity of the wire may change due to the change in the lattice structure by the strain. The differential change in resistance R is obtained by taking the differential of Equation (7.4).

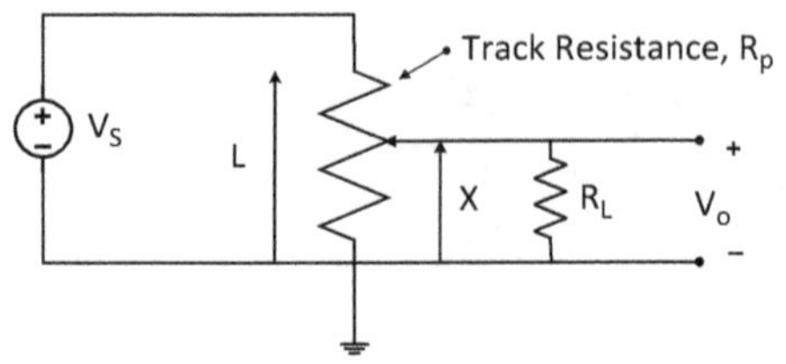

FIGURE 7.5 Resistive sensor with a load.

$$dR = \frac{\rho dL}{A} - \rho A^{-2} L dA + L\frac{d\rho}{A} \tag{7.5}$$

Dividing each item in Equation (7.5) by Equation (7.4), we have

$$\frac{dR}{R} = \frac{dL}{L} - \frac{dA}{A} + \frac{d\rho}{\rho} \tag{7.6}$$

Using incremental values, Equation (7.6) becomes

$$\frac{\Delta R}{R} = \frac{\Delta L}{L} - \frac{\Delta A}{A} + \frac{\Delta \rho}{\rho} \tag{7.7}$$

The cross-section area A of a circular wire with diameter D is given as

$$A = \frac{\pi}{4} D^2 \tag{7.8}$$

Thus,

$$\frac{\Delta A}{A} = 2\frac{\Delta D}{D} \tag{7.9}$$

The Poisson ratio, μ, relates the change in diameter to the change in length, that is

$$\frac{\Delta D}{D} = -\mu \frac{\Delta L}{L} \tag{7.10}$$

When Equation (7.10) is substituted into Equation (7.7), we have

$$\frac{\Delta R}{R} = (1+2\mu)\frac{\Delta L}{L} + \frac{\Delta \rho}{\rho} \tag{7.11}$$

Note that the change in resistance is a function of changes in dimension (length change $\frac{\Delta L}{L}$ and area change $2\mu\frac{\Delta L}{L}$) and change in resistivity ($\frac{\Delta \rho}{\rho}$) with strain.

The gage factor, G, of the strain gage is defined as

$$G = \frac{\Delta R / R}{\Delta L / L} \tag{7.12}$$

Which is used in comparing the performance of various strain gage materials. From Equation (7.11), the gage factor is

$$G = (1+2\mu) + \frac{\Delta \rho / \rho}{\Delta L / L} \tag{7.13}$$

Table 7.1 shows the gage factors of some materials. For most metals, the Poisson's ratio is approximately 0.3; thus, the gate factor is a least 1.6. However, the gage factor of semiconductor materials is approximately 50 to 70 times that of metals. For metals, the gage factor is primarily a function of dimensional effects, whereas for

TABLE 7.1
Gage Factors

Material	Gage Factor
Metal foil	1–5
Semiconductor (crystal)	80–150
Diffused semiconductor	10–200

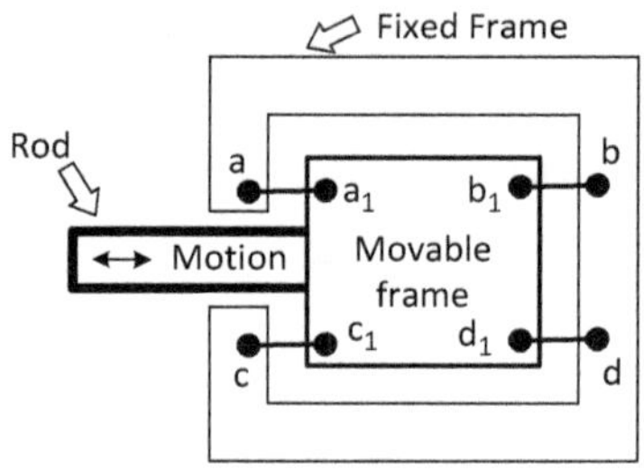

FIGURE 7.6 Unbonded strain gage.

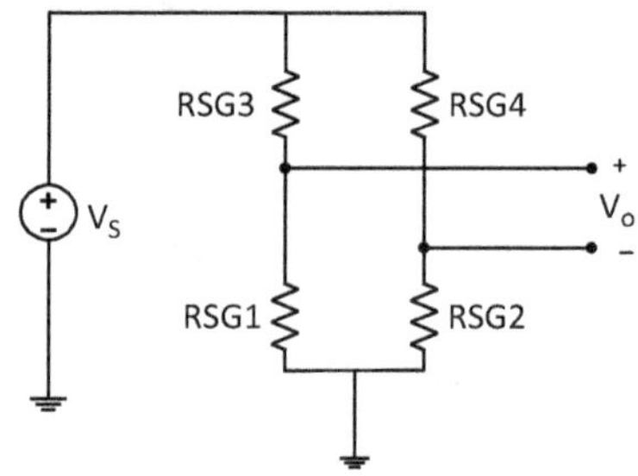

FIGURE 7.7 Wheatstone bridge with four unbound gages.

semiconductors, the change in resistivity with length predominates. The desirable feature for semiconductor strain gages is high gage factor, which offsets their greater resistivity and high temperature coefficient. Thus, it is advisable to incorporate temperature compensation for instruments that use semiconductor gages.

Strain gages can be classified as bound or unbound. The unbounded strain gage consists of a wire wound between a fixed and movable frame. Figure 7.6 shows four sets of strain gages that have been arranged such that two of the gages can be shortened or lengthened by the movement of the movable part relative to the fixed part. As given in Figure 7.6, the strain gages $SG1$ (b–b_1), $SG2$ (a–a_1), $SG3$ (c–c_1), and $SG4$ (d–d_1) will be configured such that SGl and $SG4$ are in tension and $SG2$ and $SG3$ are in compression. The motion of the movable object to the left causes the resistance of $RSG1$ and $RSG4$ to increase and $RSG2$ and $RSG3$ to decrease.

Bridge amplifier configurations are the best readout circuitry for strain gages. Figure 7.7 shows a bridge circuit for an unbound strain gage. The four sets of strain gages are connected to the four arms of a Wheatstone bridge. Thus,

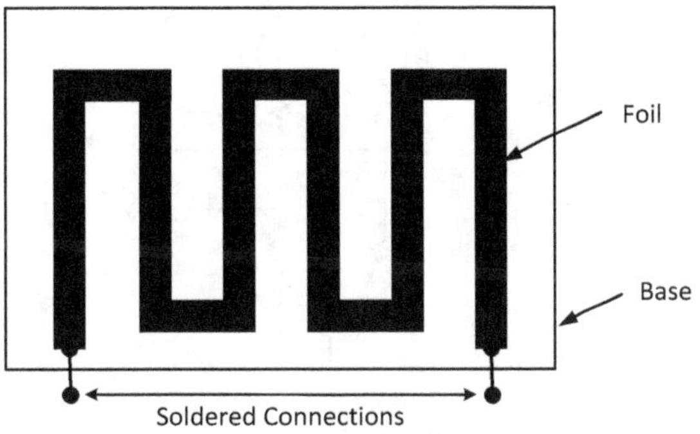

FIGURE 7.8 Bonded strain gage.

$$\begin{aligned} RSG1 = RSG4 = R + \Delta R \\ RSG2 = RSG3 = R - \Delta R \end{aligned} \tag{7.14}$$

The voltage at the output of the bridge circuit is

$$V_0 = \left(\frac{R + \Delta R}{2R} - \frac{R - \Delta R}{2R} \right) V_S = -\frac{2\Delta R}{2R} V_S = \frac{\Delta R}{R} V_S \tag{7.15}$$

The four strain-gage set is less sensitive to temperature since changes in the temperature affect each wire almost to the same extent and leaves the output voltage essentially unchanged with temperature.

The bonded strain gage consists of a metal foil, or etched metal foil, or vacuum deposited metal or a semiconductor bar connected to a surface whose strain is to be measured. Figure 7.8 shows examples of a bonded strain gage.

Temperature changes the gage resistance and produces strain due to different coefficients of expansion of the gage metal and the test structure. Since the fractional change in resistance with respect to strain is fairly small, the effect of temperature on the strain gage is considerable. One way to achieve temperature compensation is to have a dummy gage mounted on the same structure or similar (but unstressed) material and has the same temperature as the stressed strain gage. The output of the dummy strain gage is then used to correct the readings on the main strain gage.

Example 7.2: Temperature Effects on Strain Gage

In Figure 7.9, $R_2 = R_3 = R_U = R_V = 500\,\Omega$ at 0°C. If $R_U = 600\,\Omega$ at 25°C and $R_U = 700\,\Omega$ at 50°C, find the output voltage at those temperatures.

$$I_1 = \left(\frac{V_i}{R_U + R_3} \right); \qquad I_2 = \left(\frac{V_i}{R_V + R_2} \right) \tag{7.16}$$

$$V_{AD} = I_1 R_U = \left(\frac{V_i R_U}{R_U + R_3} \right) \tag{7.17}$$

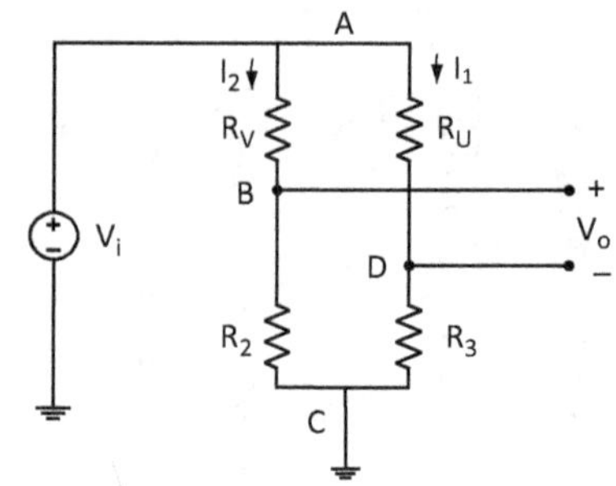

FIGURE 7.9 Bridge circuit with a strain gage.

$$V_{AB} = I_2 R_V = \left(\frac{V_i R_V}{R_V + R_2} \right) \tag{7.18}$$

$$V_0 = V_{BD} = V_{BA} + V_{AD} = -V_{AB} + V_{AD} \tag{7.19}$$

$$V_0 = -\frac{V_i R_V}{R_V + R_2} + \frac{V_i R_U}{R_U + R_3} \tag{7.20}$$

At balance, $V_0 = 0$, and

$$\frac{V_i R_V}{R_V + R_2} = \frac{V_i R_U}{R_U + R_3} \tag{7.21}$$

$$R_2 = R_3 = R_U = R_V = 500\,\Omega \text{ at } 0°\text{C}$$

$$R_U = 600\ \Omega \text{ at } T = 25°\text{C}$$

$$R_U = 700\ \Omega \text{ at } T = 50°\text{C}$$

Using Equation (7.20), we have

$$\text{At } T = 25°\text{C}, V_0 = -\frac{10 * 600}{1100} - \frac{10 * 500}{1000} = 0.459 \text{ V} \tag{7.22}$$

$$\text{At } T = 50°\text{C}, \ V_0 = -\frac{10 * 700}{1200} - \frac{10 * 500}{1000} = 0.833 \text{ V} \tag{7.23}$$

7.2 CAPACITANCE TRANSDUCER

For two parallel plates with cross-sectional area A and separation distance x, the capacitance C is given as

$$C = \frac{\varepsilon_0 \varepsilon_r A}{x} \tag{7.24}$$

where

ε_0 = permittivity of free space = 8.85×10^{-12} F/m.
ε_r = relative dielectric constant of the medium between the plates.

In theory, it is possible to monitor displacement by changing the dielectric constant, or cross-sectional area, or distance between plates. Figure 7.10 shows capacitive sensors with various features of capacitance employed for displacement measurement.

The sensitivity, K, of a capacitance transducer relates the change in capacitance to the change in plate separation. It is defined as

$$K = \frac{\Delta C}{\Delta x} \tag{7.25}$$

From Equations (7.24) and (7.25),

$$K = \frac{\Delta C}{\Delta x} \simeq \frac{dC}{dx} = -\frac{\varepsilon_0 \varepsilon_r A}{x^2} \tag{7.26}$$

From Equation (7.26), it can be seen that the sensitivity increases as the plate separation decreases. From Equation (7.24) and (7.26), we have

$$\frac{dC}{dx} = -\left(\frac{\varepsilon_0 \varepsilon_r A}{x}\right)\frac{1}{x} = -\frac{C}{x}$$

Thus,

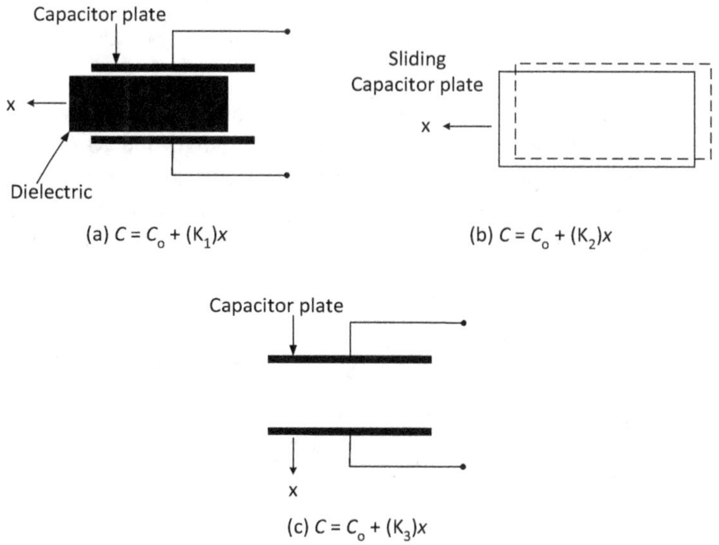

FIGURE 7.10 Capacitive transducers with the following: (a) movable dielectric for line displacement; (b) movable plate sliding; (c) movable plate for linear displacement.

$$\frac{dC}{C} = -\frac{dx}{x} \text{ or}$$
$$\frac{\Delta C}{C} = -\frac{\Delta x}{x} \tag{7.27}$$

From Equation (7.27), the percentage change in capacitance C is equal to the percentage change in plate separation, x, for small displacements.

A bridge amplifier network, shown in Figure 7.11, can be used to measure the change in capacitance with respect to displacement. As shown in Figure 7.11, capacitor C_1 is chosen such that it is equal to C_2 when the displacement is zero. The resistors R_1 and R_2 are chosen to be very large and equal in value. The resistors provide a DC current return path to ground for the op amp. Care should be taken to reduce the stray capacitance in wiring the circuit. Shielding and mounting the bridge and amplifier very close to the capacitive sensor are necessary to reduce the stray capacitance.

The ac output voltage is given as

$$V_0 = G(V_a - V_b) \tag{7.28}$$

where G = gain of the instrumentation amplifier

$$V_0 = GV_S\left(\frac{1}{2} - \frac{1/jwC_2}{1/jwC_1 + 1/jwC_2}\right) = GV_S\left(\frac{C_1 - C_2}{C_1 + C_2}\right) \tag{7.29}$$

If $C_1 = C_0 + kx$ and $C_2 = C_0$, then

$$V_0 = \frac{GV_S}{2}\left(\frac{kx}{2C_0 + kx}\right) \simeq \frac{GV_S}{4C_0}kx \tag{7.30}$$

Thus, the output voltage is proportional to displacement.

Another method of measuring the capacitance change due to displacement is to make the capacitance part of an LC oscillator circuit. As the capacitance value changes, the frequency of the oscillator will change. Through the use of a ratio detector or a discriminator circuit, the change in frequency is converted into an equivalent

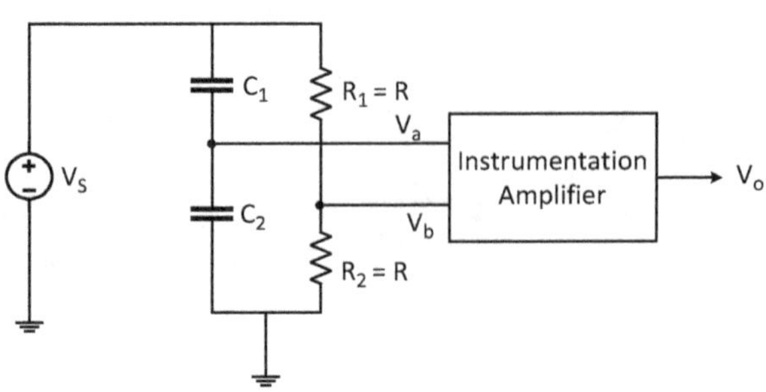

FIGURE 7.11 Bridge circuit for capacitive sensors.

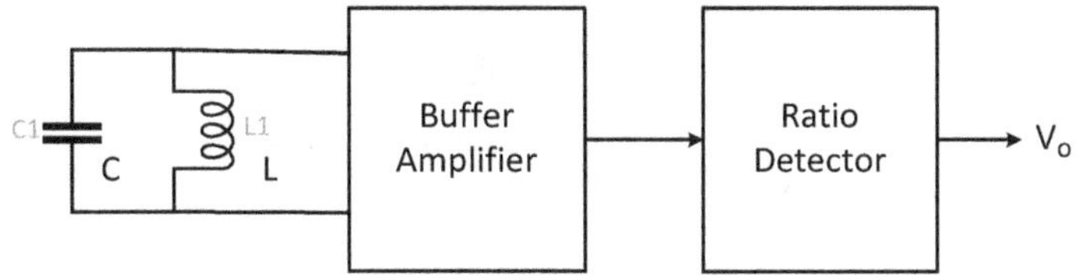

FIGURE 7.12 Use of the LC oscillator and ratio detector for the measurement of the displacement of the capacitor sensor.

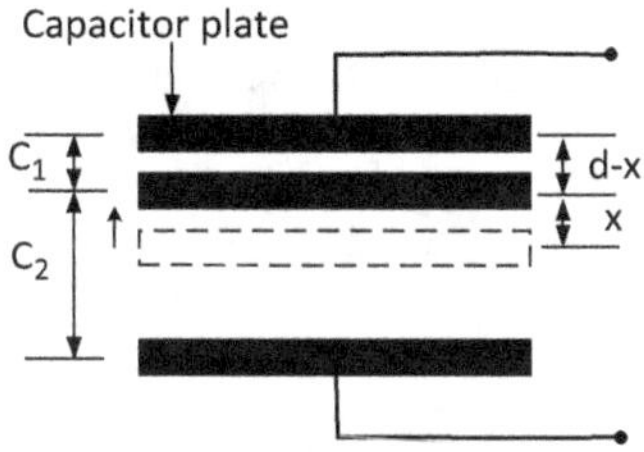

FIGURE 7.13 Displacement measurement through change in plate separation.

output voltage. Figure 7.12 shows a block diagram of the use of the LC oscillator for measuring the change in capacitance.

For more accurate measurement of displacement, a differential capacitor system, shown in Figure 7.13, can be used. Two capacitors C_1 and C_2 are formed from three plates. At equilibrium, the plate separation is d. Assuming the middle plate moves upward by a small distance x, then

$$C_1 = \frac{\varepsilon_0 \varepsilon_r A}{d - x}, \text{ and } C_2 = \frac{\varepsilon_0 \varepsilon_r A}{d + x} \tag{7.31}$$

Thus,

$$\frac{C_1 - C_2}{C_1 + C_2} = \frac{\varepsilon_0 \varepsilon_r \left(\frac{1}{d - x} - \frac{1}{d + x} \right)}{\varepsilon_0 \varepsilon_r \left(\frac{1}{d - x} + \frac{1}{d + x} \right)} = \frac{d + x - d + x}{d + x + d - x} = \frac{x}{d} \tag{7.32}$$

Thus, the displacement x is proportional to fractional difference of the capacitances C_1 and C_2.

A bridge circuit, shown in Figure 7.14, can be used to convert the fractional difference in capacitance into voltage.

When the bridge is balanced, $C_1 = C_2$ and $C_3 = C_4$. When there are slight changes in C_1 and C_2, we have from Equation (7.32).

$$\frac{C_1 - C_2}{C_1 + C_2} = \frac{x}{d}$$

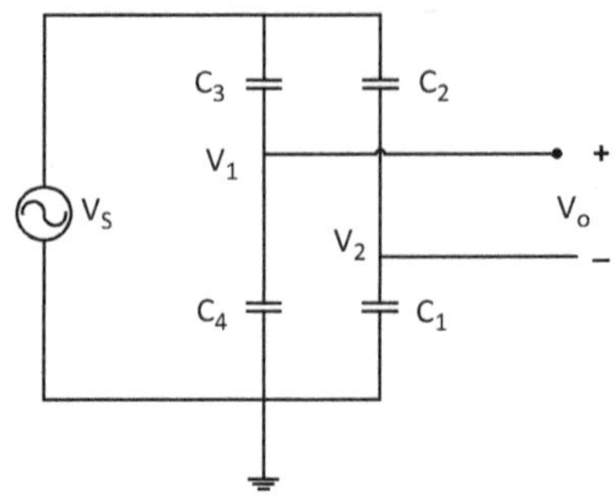

FIGURE 7.14 Bridge circuit for differential capacitance (C_1 and C_2 are differential capacitors).

But

$$V_0 = V_1 - V_2 = \left(\frac{1/jwC_4}{1/jwC_4 + 1/jwC_3} - \frac{1/jwC_1}{1/jwC_2 + 1/jwC_1} \right) V_S \tag{7.33}$$

Assuming $C_3 = C_4$, Equation (7.33) is simplified to

$$V_0 = \left(\frac{C_3}{C_3 + C_4} - \frac{C_2}{C_1 + C_2} \right) V_S = \left(\frac{1}{2} - \frac{C_2}{C_1 + C_2} \right) V_S = \frac{1}{2} \left(\frac{C_1 - C_2}{C_1 + C_2} \right) V_S = \frac{x}{2d} V_S \tag{7.34}$$

Thus, the output voltage is proportional to the displacement and is independent of the frequency of the source and area and permittivity of the capacitors.

Capacitance transducers can be used for liquid level sensors. This is illustrated by Figure 7.15. The inner probe of the capacitance is enclosed in a cylinder with port holes. The liquid in the tank forms a dielectric between the inner probe and the outer cylinder of the capacitor sensor. The liquid must be nonconducting. The change in capacitance ΔC in the tank is proportional to the change in the height of the liquid Δx. The capacitance variation is given as follows:

$$\Delta C = K \Delta x \tag{7.35}$$

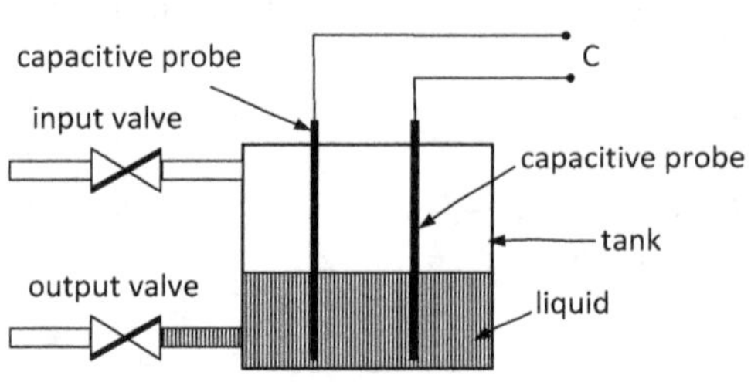

FIGURE 7.15 Capacitance sensor for liquid level measurement.

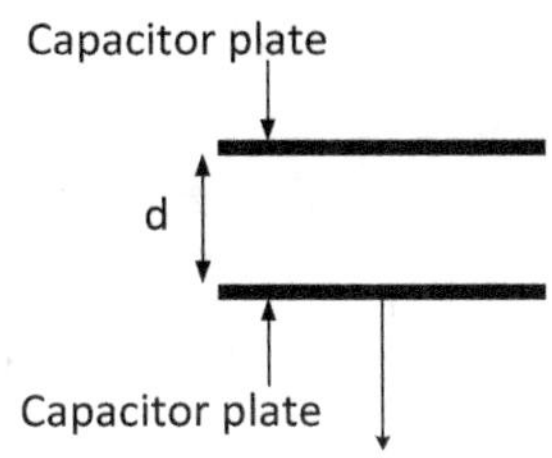

FIGURE 7.16 Capacitive position sensor.

where

K is a constant that depends on the dielectric constant of the liquid and the inner and outer diameters of the capacitance sensor.
x is the height of the liquid.

Example 7.3: Capacitance Movable Plate for Linear Displacement

Figure 7.16 shows two plates of a capacitive sensor for measuring displacement. The area of each plate is 144 mm^2. The plates move toward or away from each other with a minimum displacement of 0.2 mm and maximum displacement of 1.6 mm. (a) Calculate the capacitance with respect to the displacement and (b) determine the sensitivity with respect to the displacement. Assume that the space between the plates is air.

Solution

We use Equation (7.24) to calculate the capacitance and Equation (7.26) to calculate the sensitivity.

MATLAB® CODE

```
% Solution to Example 7.3
% Capacitance and Sensitivity with respect to displacement
% The constants of the capacitor
e0 = 8.85e-12; % permittivity of free space
er = 1.0; % relative permittivity of free space
A = 12e-3*12e-3; % area of the plate
Kons = e0*er*A;
% displacements are generated
d = 0.2e-3:0.1e-3:1.6e-3;
len = length(d);
for i=1:len
 c(i) = Kons/d(i);
 s(i) = -Kons/(d(i)*d(i));
end
figure (1)
plot(d,c,'k',d,c,'k*')
title('Displacement vs. Capacitance')
xlabel('Displacement in meters')
ylabel('Capacitance in Farads')
figure (2)
```

```
plot(d,s,'k',d,s,'k*')
title('Displacement vs. Sensitivity')
xlabel('Displacement in meters')
ylabel('Capacitance Sensitivity in Farads/meters')
```

Figure 7.17 shows the capacitance versus displacement, and Figure 7.18 shows the sensitivity of the capacitance with respect to displacement.

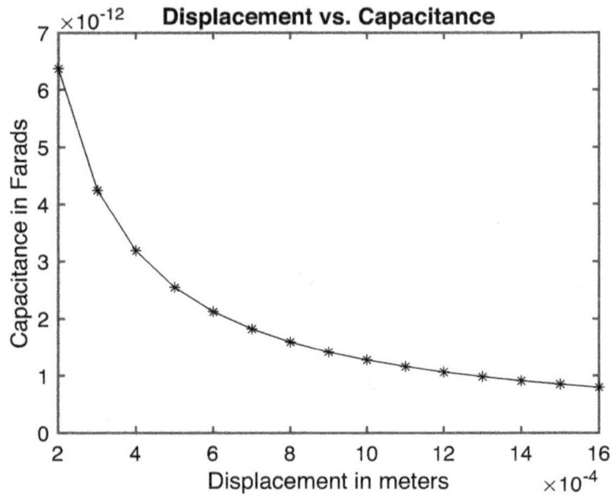

FIGURE 7.17 Capacitance versus displacement.

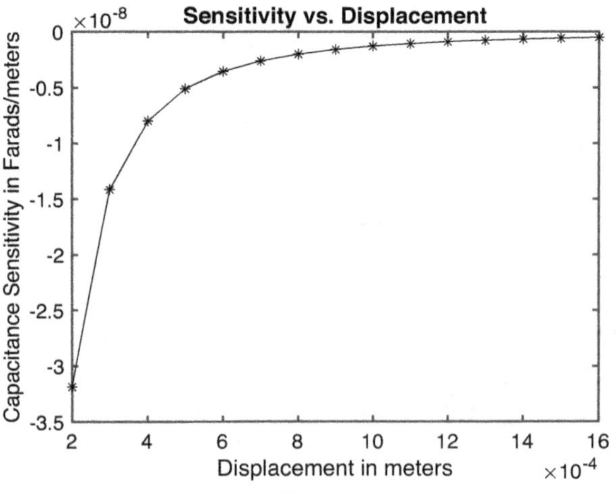

FIGURE 7.18 Capacitance sensitivity versus displacement.

7.3 LINEAR VARIABLE DIFFERENTIAL TRANSFORMER

A wide variety of techniques exist for the measurement of displacement using methods based on the variation of inductance of single coils and the mutual inductance of two coils. The methods include the following: (i) change in coil geometry, (ii) change in path reluctance by change in the air gap, and (iii) change in mutual coupling of two coils by movement of a coil. The most popular displacement measurement by inductance is the Linear Variable Differential Transformer (LVDT).

The LVDT is a sensor that has a wide variety of applications. It can be used to measure displacements, force, and pressure. One advantage of the LVDT is that it provides a very high output voltage for small changes in the displacement of a movable element.

The LVDT consists of three coils: a primary coil and two identical secondary coils. The two secondary coils are connected in series opposition. A movable core (normally a ferrite core) is used to alter the coupling between the primary and secondary coils. A circuit diagram of the LVDT is shown in Figure 7.19. When the core is in the central or null position, the magnetic coupling of the primary coil and the two secondary coils will be identical and the output voltage is zero since $i_1 = i_2$.

When the core moves upward, being displaced more toward secondary coil S1, the current from secondary coil S1 will be higher than that generated by secondary coil S2. The output voltage will be proportional to the core displacement and will be in phase with voltage V_{IN}. When the core moves down, being displaced more toward the secondary coil S2, the current generated by secondary coil S2 will be higher than that generated by secondary coil S1. The output voltage will be proportional to displacement, but the output voltage will be 180° out of phase with primary input voltage. Thus, there is 180° phase shift when the core passes the null position. The output voltage versus displacement of the core is shown in Figure 7.20.

The primary coil of the LVDT is normally excited by a sinusoidal voltage of 1–15 V rms and frequency of 60 Hz to 20,000 Hz. If the displacement is small, the output voltage is linearly proportional to the displacement of the movable core. That is,

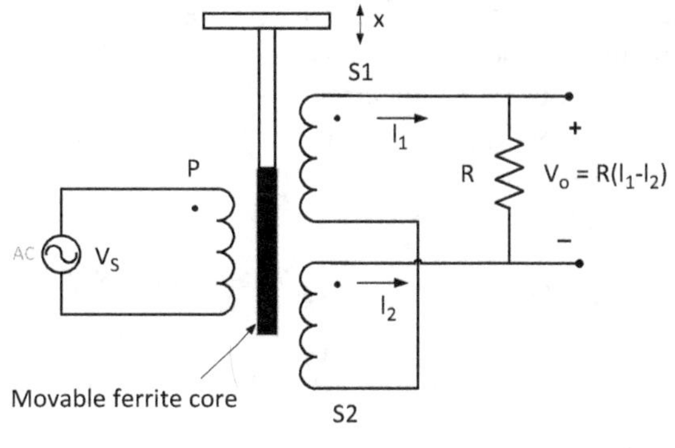

FIGURE 7.19 Circuit diagram of an LVDT.

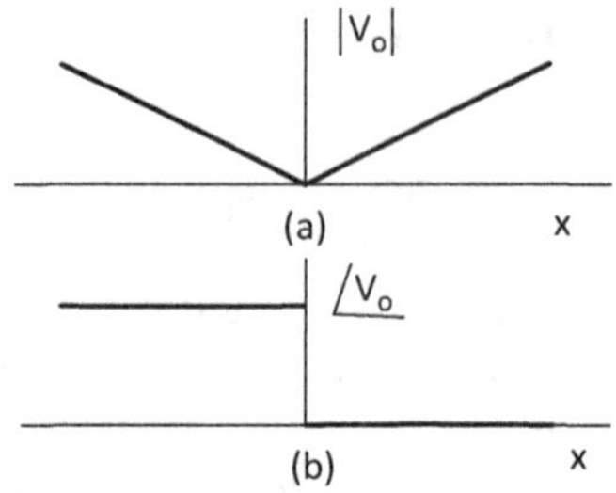

FIGURE 7.20 Output voltage: (a) magnitude versus displacement ; (b) phase versus displacement.

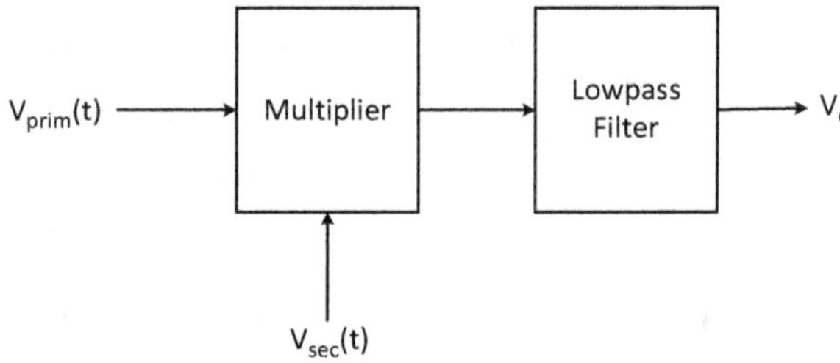

FIGURE 7.21 Phase-sensitive network.

$$|V_0| = kV_{IN}x \tag{7.36}$$

where

k is a constant (sensitivity).
x is the displacement of the core.
V_{IN} is the excitation voltage

LVDTs have the advantage that they have a higher output voltage for small changes in the core position. A phase-sensitive demodulator network is used to determine the direction of the core displacement. A phase-sensitive network is shown in Figure 7.21. It is similar to a synchronous detection circuit.

If the primary voltage is $V_{prim}(t) = A\sin(wt)$, then the secondary voltage is given as $V_{sec}(t) = kV_{prim}(t)x(t) = kAx(t)\sin(wt)$

The voltage at the input of the low-pass filter is

$$V_{Lpf}(t) = V_{prim}(t)V_{sec}(t) = kA^2x(t)\sin^2(wt)$$
$$V_{Lpf}(t) = \frac{1}{2}kA^2x(t) - \frac{1}{2}kA^2x(t)\cos(2wt) \tag{7.37}$$

We can filter out the second term of Equation (7.37) using a low-pass filter. Thus,

$$V_{\text{Lpf}}(t) = \frac{1}{2}kA^2x(t) \tag{7.38}$$

An LVDT with its phase-sensitive network is shown in Figure 7.22.

Example 7.4: Sensitivity of a LVDT Sensor

Table 7.2 shows the data that were taken while calibrating an LVDT. Use a least-squares fit to calculate the following: (i) the calibration constant (V/m) for a linear calibration curve and (ii) the mean deviation of the measurements.

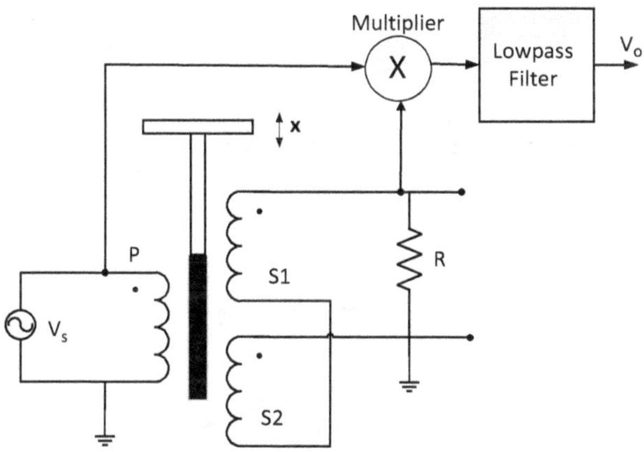

FIGURE 7.22 LVDT with a phase-sensitive demodulator.

TABLE 7.2
Displacement and Output Voltage of LVDT

Core Displacement (in mm)	Signal-Conditioned Output Voltage (mV)
−30	−5.2
−20	−4.7
−10	−2.9
0	0
10	1.8
20	4.8
30	5.3

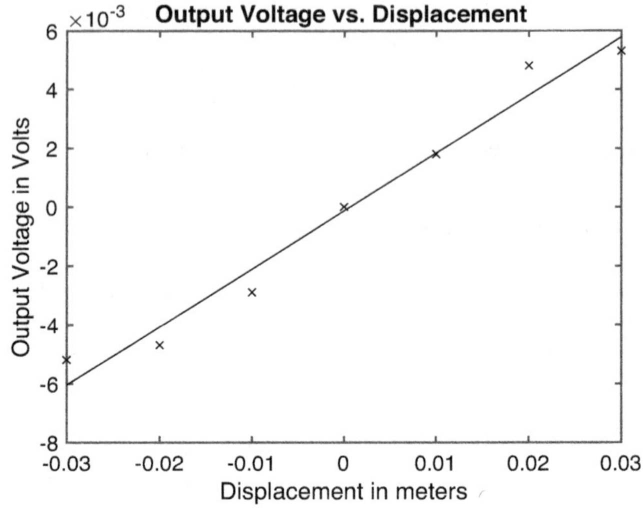

FIGURE 7.23 LVDT output voltage versus core displacement.

Solution

MATLAB® CODE

```
% Solution to Example 7.4
% LVDT
D = -30e-3:10e-3:30e-3;
V = [-5.2e-3 -4.7e-3 -2.9e-3 0 1.8e-3 4.8e-3 5.3e-3];
len = length(D);
%
% Linear equation is y = mx + b
% coefficient
vfit = polyfit (D, V, 1);
b = vfit(2);
m = vfit(1);
%
vfit = m*D + b;
% Plot V versus T and best fit model
plot (D, vfit, 'k', D, V, 'kx')
xlabel ('Displacement in meters')
ylabel('Output Voltage in Volts')
title('Displacement versus Output Voltage of LVDT Sensor')
for i = 1:len
 vx(i) = m*D(i);
 dev(i) = abs(V(i)-vx(i));
end
cal_cons = m
max_dev = max(dev)
```

Figure 7.23 shows the LVDT core displacement versus the signal-conditioned output voltage. From the MATLAB program results, the calibration constant is 0.1971 V/m and the maximum deviation is 9.2857e-04 V

7.4 PIEZOELECTRIC TRANSDUCER

Piezoelectric transducers generate electrical potential when they are mechanically deformed. Piezoelectric materials have an asymmetrical crystal structure. When the crystal structure is distorted through mechanical force on the piezoelectric material, a change in the orientation takes place, causing separation of positive and negative charges across the crystal surface.

Some materials that exhibit piezoelectricity while in their natural form are quartz, Rochelle salt, and ammonium dihydrogen phosphate. Barium titanate and lead zirconate titanate compositions are ceramics that also exhibit piezoelectricity. In addition, polymers, such as polypyridine fluoride, show excellent piezoelectric properties. Table 7.3 shows the properties of some piezoelectric materials.

As mentioned earlier, when piezoelectric materials are mechanically deformed, positive and negative charges are induced across the surface of the material. The total induced charge is directly proportional to the applied force; that is

$$q = D.f \tag{7.39}$$

where

q is total induced charge.
D is the piezoelectric constant (C/N).
f is the applied force (N).

Within elastic limits, the applied force is proportional to the deflection, x.

$$f = k_f x \tag{7.40}$$

where k_f is a constant.

Substituting Equation (7.40) into Equation (7.39), we get

$$q = D.k_f x = k_q x \tag{7.41}$$

where $k_q = D.k_f$, and it is a constant.

TABLE 7.3
Properties of Some Piezoelectric Materials

Material	D-Coefficient, C/N	Relative Dielectric Constant	Resistivity (Ω · m)
Multiply by	10^{-12}	1	10^9
Quartz	2.3	4.5	>1000
Barium titanate ceramic	140	1200	>100
Lead zirconate titanate	105	1600	>30
Lead niobate	200	1500	>100

With the charge separation obtained when piezoelectric materials are strained and high resistivity of the piezoelectric materials, the latter can be modeled as a parallel-plate capacitor, C, with the voltage across the capacitor given as follows:

$$v(t) = \frac{q(t)}{C} = \frac{k_q x(t)}{C} = \frac{k_q x(t)}{\varepsilon_0 \varepsilon_r A / d} = \left(\frac{k_q d}{\varepsilon_0 \varepsilon_r A} \right) x(t) \tag{7.42}$$

where ε_r = relative dielectric constant and d = charge separation distance

Table 7.3 shows that piezoelectric materials have very high resistivity. Thus, if static deformation is applied to the piezoelectric material, the induced charge leaks very slowly through the large resistance of piezoelectric material.

The equivalent circuits of the piezoelectric transducer at low and medium frequencies are shown in Figures 7.24 and 7.25, respectively. In the model, R_t is the piezoelectric leakage resistance and C_t is the capacitance of the sensor. From Equation (7.41), the current $i(t)$) is given as

$$i(t) = \frac{dq(t)}{dt} = k_q \frac{dx}{dt} \tag{7.43}$$

When a piezoelectric transducer is connected to an amplifier, the input impedance of the amplifier should be an order of magnitude higher than that of the piezoelectric transducer.

Piezoelectric transducers have high-pass response. This can be shown by considering the circuit shown in Figure 7.26, wherein a piezoelectric transducer is connected to an amplifier with internal resistance R_a and internal capacitance C_a. The circuit shown in Figure 7.27 is equivalent to that in Figure 7.26.

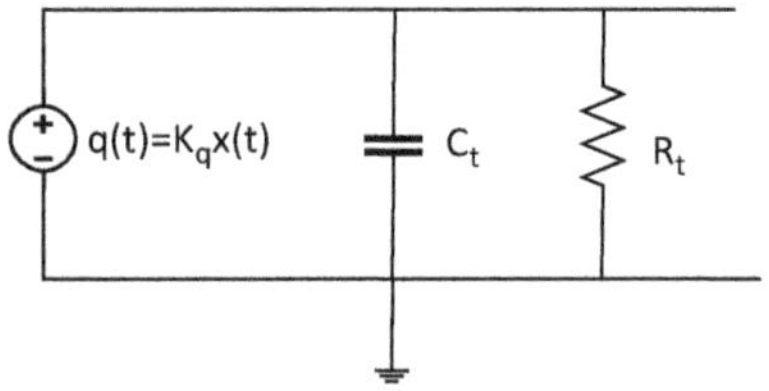

FIGURE 7.24 Low- and medium-frequency equivalent circuit of piezoelectric transducers with a charge generator.

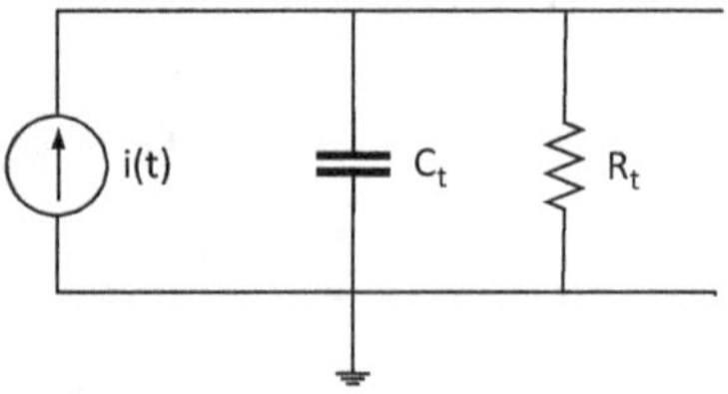

FIGURE 7.25 Low- and medium-frequency equivalent circuit of piezoelectric transducers with a current generator.

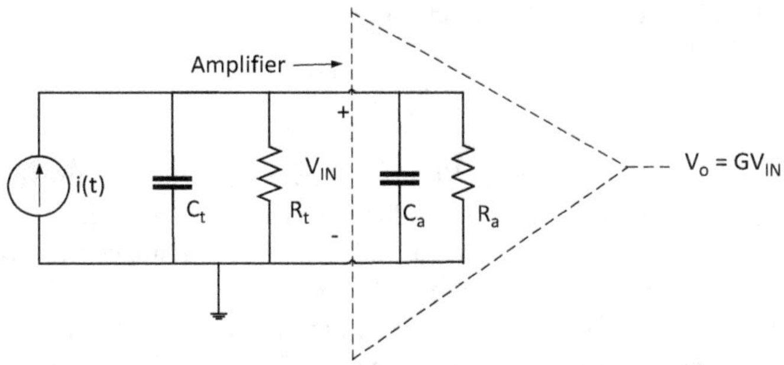

FIGURE 7.26 Circuit diagram of a piezoelectric transducer connected to an amplifier.

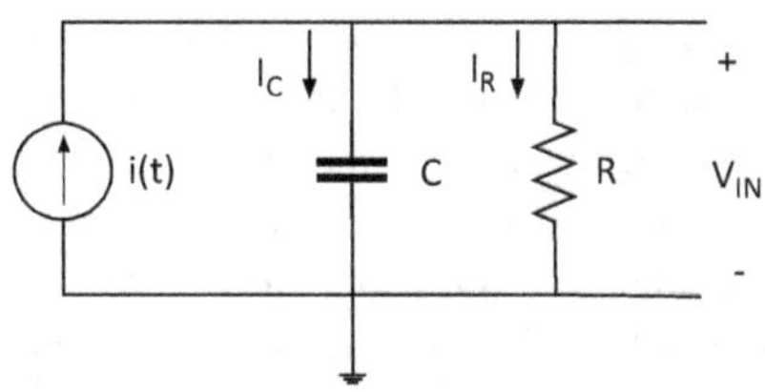

FIGURE 7.27 Equivalent circuit of a piezoelectric transducer connected to an amplifier, where $C = \left(C_t \| C_a\right)$ and $R = \left(R_t \| R_a\right)$.

$$i_C = i(t) - i_R = k_q \frac{dx}{dt} - \frac{V_{IN}(t)}{R} = C\frac{dV_{IN}(t)}{dt} \tag{7.44}$$

Thus,

$$\frac{dV_{IN}(t)}{dt} = \frac{k_q}{C}\frac{dx}{dt} - \frac{V_{IN}(t)}{CR} \tag{7.45}$$

In the frequency domain, Equation (7.45) becomes

$$sV_{IN}(s) = \frac{k_q}{C} sX(s) - \frac{V_{IN}}{CR}(s) \tag{7.46}$$

With $k_x = \dfrac{k_q}{C}$ and $\tau = RC$, Equation (7.46) becomes

$$sV_{\text{IN}}(s) = k_x sX(s) - \frac{V_{\text{IN}}(s)}{\tau}$$

Thus,

$$V_{\text{IN}}(s)\left(s + \frac{1}{\tau}\right) = k_x sX(s)$$

$$\frac{V_{\text{IN}}(s)}{X(s)} = \frac{k_x s}{s + \frac{1}{\tau}} = \frac{k_x s\tau}{1 + s\tau} \tag{7.47}$$

Since $V_0(s) = GV_{IN}(s)$, the transfer function

$$H(s) = \frac{V_0(s)}{X(s)} = \frac{Gk_x s\tau}{1 + s\tau} \tag{7.48}$$

From Equation (7.48), when $s=0$, $H(0)=0$ However, when s approaches infinity, $H(s) = Gk_x$ Thus, the piezoelectric transducer has a high-frequency response. The transducer lacks a DC response. This implies that the displacement or the force should vary in time for an output voltage to be generated.

Piezoelectric materials resonate at high frequencies. Typically, the resonant frequencies lie between 20 and 80kHz. The resonant features of the piezoelectric material can be modeled by adding a series RLC circuit in parallel with the transducer capacitance and leakage resistance. The complete equivalent circuit of piezoelectric materials is shown in Figure 7.28, and its frequency response is shown in Figure 7.29. L_m, C_m, and R_m are used to model the mechanical resonance.

In the resonant mode, piezoelectric transducers emit high-frequency signals. They can be operated in pulse-echo or continuous-wave modes depending on the excitation of the ultrasonic transducer. In addition, in the resonant mode, a piezoelectric transducer behaves as a crystal oscillator. The latter can be used as frequency-conditioning clocks for digital systems.

There are two popular methods for conditioning the signals from piezoelectric transducers: voltage amplification and charge amplification. In voltage amplification, a high input impedance amplifier is connected to the piezoelectric transducer. Since the piezoelectric transducer has a high resistance, the amplifier input impedance should have a far higher input impedance. A MOSFET operational amplifier,

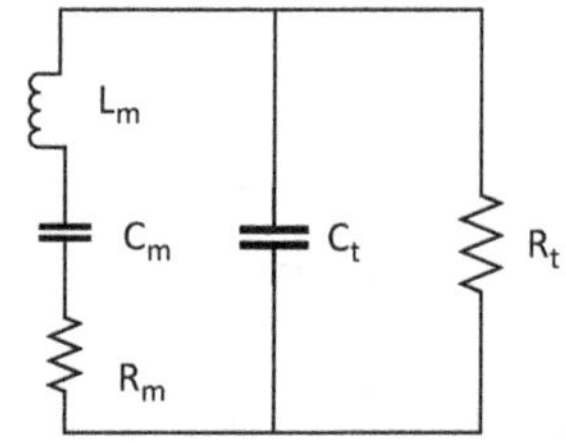

FIGURE 7.28 Complete equivalent circuit of the piezoelectric material.

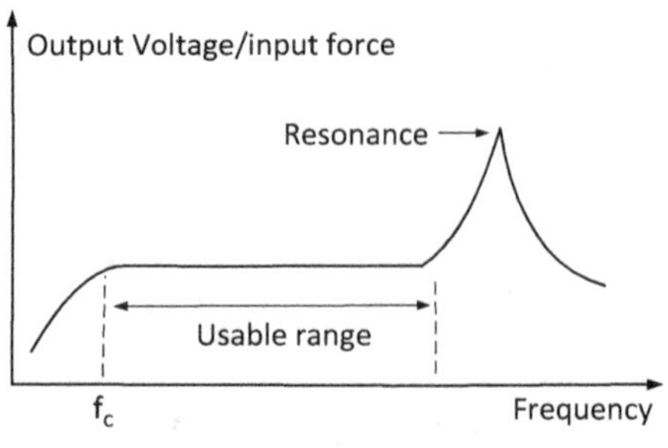

FIGURE 7.29 Frequency response of the piezoelectric transducer.

connected in the non-inverting configuration, is normally used. Figure 7.30 shows a circuit for voltage amplification.

In Figure 7.30, R_L is a high-valued resistor that provides the necessary DC current path from the op amp input to the ground. The cable capacitance, C_C, reduces the transducer voltage sensitivity, given by

$$V_{\text{in}} = \frac{q}{C_a} \tag{7.49}$$

where

C_t = transducer capacitance.
C_c = cable capacitance.
$C_a = C_c + C_t$
q = charged generated by the transducer.

As the cable capacitance increases, the net capacitance at the input of the amplifier increases. This reduces the voltage seen by the amplifier. The output voltage of the amplifier is given as

$$V_0 = V_{\text{in}}\left(1 + \frac{R_2}{R_1}\right) = \frac{q}{C_a}\left(1 + \frac{R_2}{R_1}\right) \tag{7.50}$$

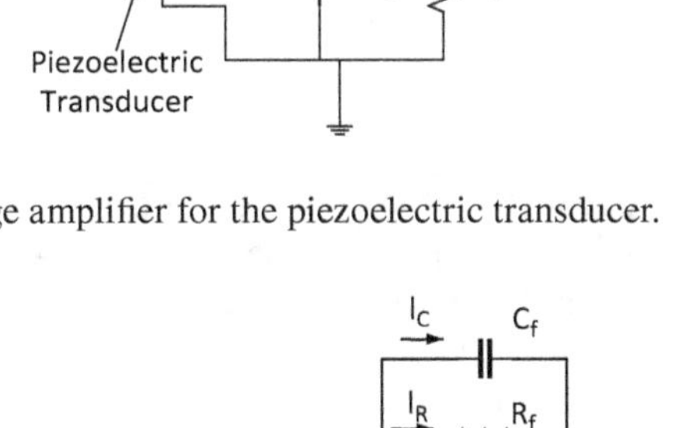

FIGURE 7.30 Voltage amplifier for the piezoelectric transducer.

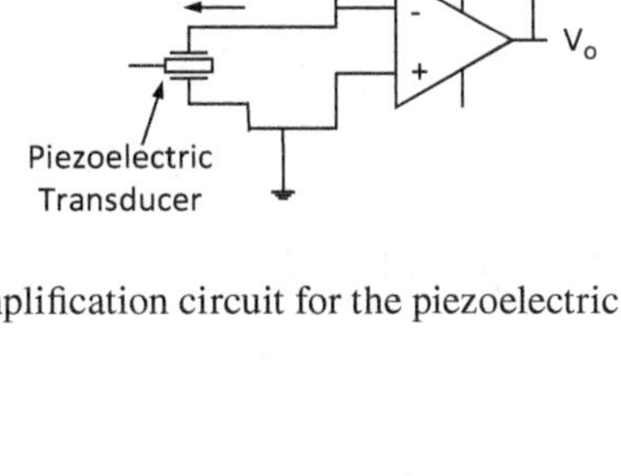

FIGURE 7.31 Charge amplification circuit for the piezoelectric transducer.

For the charge amplification, the charge generated by the transducer is converted to voltage readout. A circuit for the charge amplification is shown in Figure 7.31.

A very large resistance is connected across the capacitor C_f to provide a path for the bias current required by the op amp. Using Kirchoff's current law, we have

$$I_T + I_C + I_R = 0 \tag{7.51}$$

From Equation (7.43), $I_T = k_q \frac{dx}{dt}$. Therefore, Equation (7.51) becomes

$$k_q \frac{dx}{dt} - C_f \frac{dV_0}{dt} - \frac{V_0}{R_f} = 0 \tag{7.52}$$

In the frequency domain, Equation (7.52) becomes

$$\frac{V_0(s)}{X(s)} = \frac{k_q}{C_f} \left(\frac{s}{s + \frac{1}{C_f R}} \right) \tag{7.53}$$

From Equation (7.53), we have

$$V_0(t) = \frac{k_q}{C_f} x - \frac{k_q}{C_f^2 R_f} e^{-x/R_f C_f} \tag{7.54}$$

If we choose R_f to be far greater than C_f, then the second term at the right-hand side of Equation (7.54) can be removed. The output voltage is then proportional to transducer displacement. Equation (7.53) shows that the charge amplifier circuit behaves as a high-pass filter with time constant $R_f C_f$. The lower cutoff frequency is

$$f_{3dB} = \frac{1}{2\pi R_f C_f} \tag{7.55}$$

From Equation (7.55), if we reduce C_f, the lower corner frequency increases. It should be noted from Figure 7.30 that the voltage at the two inputs of the high-impedance MOSFET operational amplifier is a virtual ground. Thus, the voltage across the transducer and input cable is essentially zero.

Example 7.5: Break Frequency of a Piezoelectric Sensor

An equivalent circuit of a piezoelectric sensor is shown in Figure 7.32. $C=600$ pF, and the sensor leakage resistance is 10 GΩ. The amplifier input impedance is 5 MΩ. What is the corner frequency? Note that $i_S(t) = \frac{dq(t)}{dt} = k_q \frac{dx}{dt}$, $C = C_t + C_c + C_a$, and $R = R_t \| R_a$, where C_t, C_c, and C_a are capacitances of the

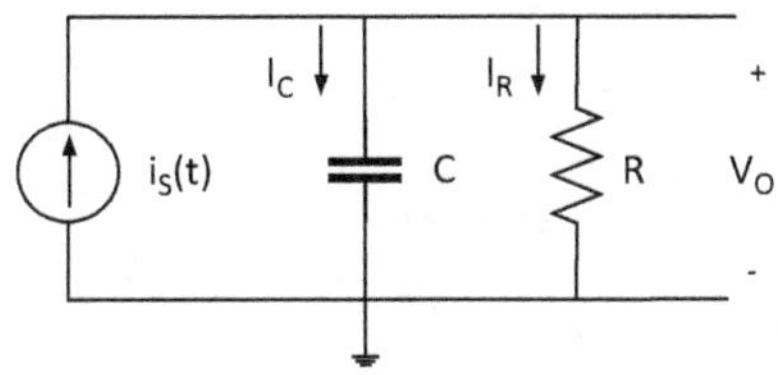

FIGURE 7.32 Equivalent circuit of a piezoelectric sensor.

piezoelectric transducer, cables, and amplifier connected to the circuit, respectively. R_t, and R_a are transducer and amplifier resistances, respectively.

Solution

The output voltage is given as follows:

$$V_o = I_S Z_{CR} = I_S\left(R \middle\| \frac{1}{jwC}\right) \tag{7.56}$$

$$i_s = k_q \frac{dx}{dt}$$

In the frequency domain, we have $I_S = k_q jwX(w)$

$$\frac{V_o}{X}(w) = \frac{jwRk_q}{1 + jwCR} \tag{7.57}$$

The cutoff frequency occurs at $w_C = \dfrac{1}{RC}$

$$fc = \frac{1}{2\pi RC} = \frac{1}{2\pi(600\times10^{-12})(5\times10^{6})} = 53.04 \text{ Hz}$$

PROBLEMS

Problem 7.1 For the circuit shown in Figure 7.5, the track resistance $R_P = 10\ \text{k}\Omega$, $V_S = 5\ \text{V}$, $L = 5$ cm, and $R_L = 50\ \text{k}\Omega$, find the error in reading the voltage across x, if (a) x is 30% of L, (b) x is 40% of L, and (c) x is 50% of L.

Problem 7.2 For the circuit shown in Figure 7.5, the track resistance $R_P = 10\ \text{k}\Omega$, $V_S = 5\ \text{V}$, and $L = 5$ cm. Plot $\dfrac{V_O}{V_S}$ versus $\dfrac{x}{L}$ for (a) $R_L = 10\ \text{k}\Omega$ and (b) $R_L = 100\ \text{k}\Omega$.

Problem 7.3 Figure P7.1 shows a capacitor with a movable dielectric that can be used as a displacement sensor. The distance between the two plates of the capacitor is d. The length of the plate is L. The width of each plate is W. If the dielectric constant of the medium between the plates is K and the relative dielectric constant of air is unity, show that the equivalent capacitance is given as follows:

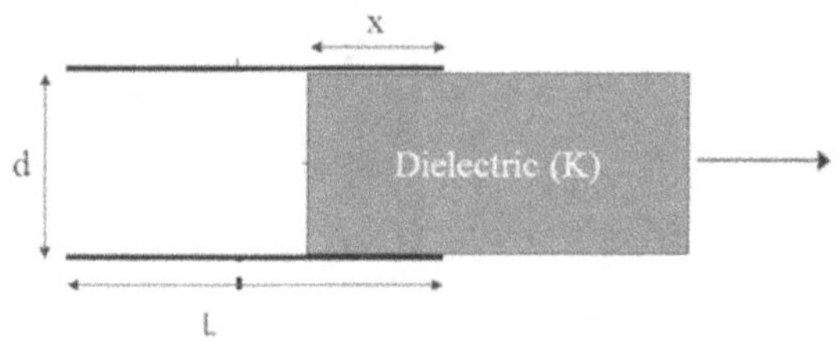

FIGURE P7.1 Capacitor with a movable dielectric.

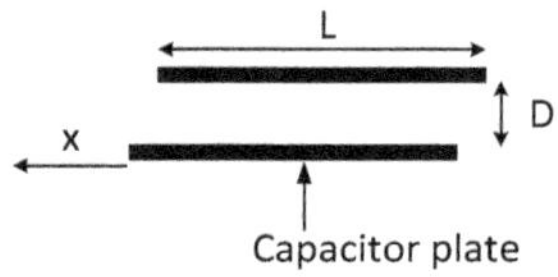

FIGURE P7.2 Capacitor with a sliding plate.

$$C_{eq} = \frac{\varepsilon_o W}{d}\left(Kx + (L - x)\right) \quad 0 \le x \le L \tag{P7.1}$$

Assuming that $L = 20\,\text{mm}$, $W = 20\,\text{mm}$, and $d = 2\,\text{mm}$ and $K = 5$, determine the equivalent capacitance as x is varied.

Problem 7.4 Figure P7.2 shows two parallel plates of a capacitor, with one of the plates sliding to the left. L is the length of a plate, W is the width of a plate, and D is the distance between the plates. If one of the plates moves to the left by X, (a) determine the capacitance between the plates, (b) plot the value of the capacitance as a function of X, assuming that $L = 40\,\text{mm}$, $W = 30\,\text{mm}$, and $D = 3\,\text{mm}$ and $K = 1$, and determine the equivalent capacitance as x is varied.

Problem 7.5 The following data have been taken from the calibration of an LVDT. Using a least-squares fit, determine the sensitivity of the LVDT (V/m) and the mean deviation of the measurements (see Table P7.1 for the data).

Problem 7.6 A circuit with a piezoelectric transducer is shown in Figure P7.3. $i_S(t) = \frac{dq(t)}{dt} = k_q \frac{dx}{dt}$, $C = C_t + C_c + C_a$, and $R = R_t \| R_a$, where C_t, C_c, and C_a are capacitances of the piezoelectric transducer, cables, and amplifier connected to the circuit, respectively. R_t, and R_a are transducer and amplifier resistances, respectively. Assuming that $C = 550$ pF, $R = 4.5\ \text{M}\Omega$, $R_1 = 1000\ \Omega$, and $R_2 = 9000\ \Omega$, (i) find the transfer function of the circuit and (ii) determine the cutoff frequency of the circuit.

Problem 7.7 A circuit with a piezoelectric transducer is shown in Figure P7.4. $i_S(t) = \frac{dq(t)}{dt} = k_q \frac{dx}{dt}$, $C = C_t + C_c + C_a$, and $R = R_t \| R_a$, where C_t, C_c, and C_a are capacitances of the piezoelectric transducer, cables, and amplifier connected to the circuit, respectively. R_t, and R_a are transducer and amplifier resistances, respectively. Assuming that $C = 500$ pF, $R = 5\ \text{M}\Omega$, $C_f = 0.1\ \mu\text{F}$,

TABLE P7.1
Output Voltage Versus Displacement of an LVDT

Core Displacement (in mm)	Signal-Conditioned Output Voltage (mV)
0	0.6
1	8.2
2	16.8
3	23.4
4	34.3
5	42.9
6	52.3
7	62.6
8	71.6
9	80.6
10	90.3

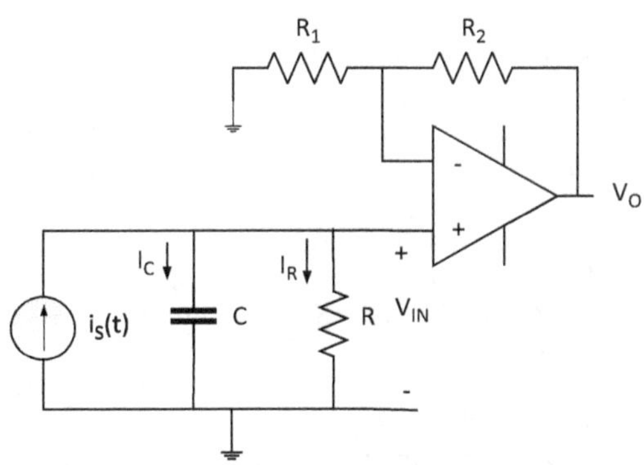

FIGURE P7.3 A piezoelectric transducer with a buffer amplifier.

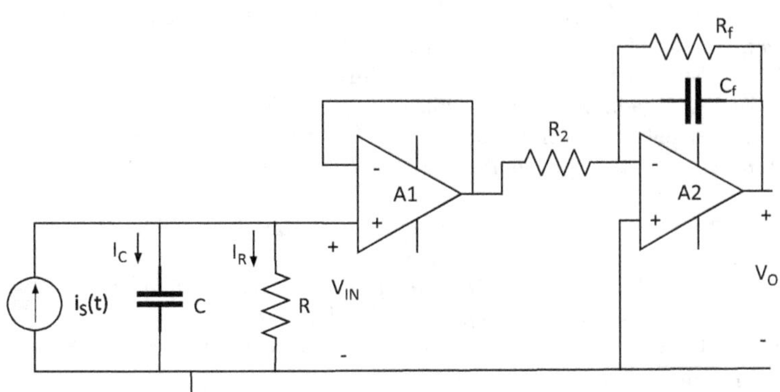

FIGURE P7.4 A piezoelectric transducer circuit.

FIGURE P7.5 Piezoelectric sensor system.

R_1 = 1000 Ω, and R_2 = 5000 Ω, (i) find the transfer function of the circuit, (ii) determine the low cutoff frequency of the circuit, (iii) calculate high cutoff frequency, and (iv) find the bandwidth of the system.

Problem 7.8 A piezoelectric sensor is connected to a microcontroller through an amplifier, as shown in Figure P7.5. The output of the piezoelectric sensor ranges from -10 to +10 mV. If the input voltage of the microcontroller is −5 V to +5 V, design an amplifier that will allow the piezoelectric sensor to provide the above-mentioned range of voltages to the input of the microcontroller.

BIBLIOGRAPHY

Alexander, Charles Alexander and Matthew Sadiku, *Fundamentals of Electric Circuit, ISF*, 7th Edition, McGraw-Hill Education, 2020.

Attia, John O., *Electronics and Circuit Analysis Using MATLAB*, CRC Press, 2nd Edition, 2004.

Attia, John O., *Circuits and Electronics: Hands-on Learning with Analog Discovery*, CRC Press, 2018.

Behl, Akshay, Akash Bhatia and Aviral Puri, "Piezoelectric Sensors," *International Journal of Innovative Research in Technology*, Vol. 1, Issue 7, pp. 477–480, 2014.

Bolton, William, *Instrumentation and Control Systems*, 3rd Edition, Elsevier, 2021.

Cajuhi, Aguinaldo Jose, and Iuri Muni Pepe, "Strain Gage as a Displacement Transducer," *Proceedings of COBEM 2009, 20th International Congress of Mechanical Engineering*, Gramado, RS, Brazil, November 15–20, 2009.

Chapman, Stephen J., *MATLAB Programming for Engineers*, 7th Edition, Cengage Learning, 2024.

Etter, Delores, *Engineering Problem Solving with MATLAB*, 3rd Edition, Pearson, 2021.

Graxinha, Artur M. F. and J. M. Dias Pereira, "A LVDT Measurement System for Industrial Applications with Conditioned-Based Maintenance Capabilities," *Sensors & Trans ducers*, Vol. 245, Issue 6, pp. 11–19, 2020.

Gruber, Gabriel, Markus Neumayer, Bernhard Schweighofer, T. Leitner, M. Berger, G. Klosch, and, Hannes Wegleiter, "Linear Variable Differential Transformer in Harsh Environments - A Displacement and Thermal Study," *2023 IEEE International Instrumentation and Measurement Technology Conference (I2MTC)*, Kuala Lumpur, Malaysia, pp. 01–06, 2023.

Hayt, William, Kemmely, Jack, Phillips, Jamie and Durbin, Steven M., *Engineering Circuit Analysis*, 10th Edition, McGraw Hill, 2023.

Ida, Nathan, *Sensors, Actuators, and Their Interfaces: A Multidisciplinary Introduction*, 2nd Edition, Institution of Engineering and Technology, 2020.

Jaeger, Richard C. and Travis Blalock, *Microelectronic Circuit Design (ISE)*, 6th Edition, McGraw Hill, 2022.

Kehileche, Brahim, Abdelhalim Telmçani, Noureddine Henini and Younes Chiba, "Piezoelectric Energy Conversion and Its Applications: Sensors and Actuators," *2nd International Conference on Electronics, Energy and Measurement (IC2EM)*, Medea, Algeria, pp. 1–5, 2023.

Maundy, Brent and Stephan J. G. Gift, "Strain Gauge Amplifier Circuits," *IEEE Transactions on Instrumentation and Measurement*, Vol. 62, pp. 1–8, 2013.

Morris, Alan S. and Reza Langari, *Measurement and Instrumentation: Theory and Applications*, 3rd Edition, Academic Press, 2020

Northrop, Robert B., *Introduction to Instrumentation and Measurements*, 3rd Edition, CRC Press, 2014.

Pelegrin, Juliano de, Bruno Monte de Carvalho, Fábio Luiz Bertotti, Jean-Marc Stephane Lafay, Jessé de Pelegrin, "Development and Evaluation of a Linear Variable Differential Sensor," *2017 2nd International Symposium on Instrumentation Systems, Circuits and Transducers (INSCIT)*, Fortaleza, Brazil, pp. 1–6, 2017.

Rana, Srinivas, Boby George, and Jagadeesh Kumar V, "Signal Conditioning of a Resistive Potentiometric Displacement Sensor with a Floating Slide," *2015 IEEE International Instrumentation and Measurement Technology Conference (I2MTC) Proceedings*, Pisa, Italy, pp. 1769–1773, 2015.

Razavi, Behzad, *Fundamentals of Microelectronics*, 3rd Edition, Wiley, 2021.

Sedra, Adel S. and Kenneth Smith, *Microelectronic Circuits*, 8th International Edition, Oxford University Press, 2020.

Sekhar, Beera Chandra, B. Dhanalakshmi, B. Srinivasa Rao, Singampalli, Ramesh, K. Venkata Prasad, P.S.V. Subba Rao and B. Parvatheeswara Rao, "Piezoelectricity and Its Applications," in Dipti Ranjan Sahu (Ed.), *Multifunctional Ferroelectric Materials*, IntechOpen, 2021.

Shadwani, Mayank, Shivani Sachan, Palak Sachan, "Capacitive Sensing & Its Applications," *International Journal of Engineering Research and General Science*, Vol. 4, Issue 3, pp. 185–193, 2016.

Ştefănescu, Dan, "Strain gauges and Wheatstone bridges - Basic Instrumentation and New Applications for Electrical Measurement of Non-electrical Quantities," *Proceedings of International Multi-Conference on Systems, Signals and Devices, SSD'11 - Summary Proceedings*, Sousse, Tunisia, pp. 1–5, 2011.

Valentine, Daniel T. and Brian D. Hahn, *Essential MATLAB for Engineers and Scientists*, 8th Edition, Academic Press, 2022.

Yoo, Yongsang and Byong-Deok Choi, "Readout Circuits for Capacitive Sensors," *Micromachines*, Vol. 12, p. 960, 2021. https://doi.org/10.3390/mi12080960 (Accessed August 8, 2025).

8 Pressure and Force Sensors

8.1 PRESSURE TYPES

Pressure, a scalar quantity, is defined as the force per unit area. At sea level, the atmospheric pressure is about $1.013 \times 10^5 \text{N/m}^2$, and it is termed as one atmosphere (atm). Another unit of pressure is the bar: ($1 \text{ bar} = 1.00 \times 10^5 \text{N/m}^2$). Table 8.1 shows the units of pressure and their conversion factors.

For fluids, the pressure P at a depth h below the surface of a liquid, shown in Figure 8.1, is due to the weight of the liquid above it. The pressure P is given as follows:

$$P = \frac{\text{Force}}{\text{Area}} = \frac{\rho Ahg}{A} = \rho hg \tag{8.1}$$

where

ρ is the liquid density,

TABLE 8.1
Units of Pressure and Their Conversion Factors

	Pascal	Atmosphere	Bar	psi
Pascal	1 Pa	9.869	10^{-5}	1.45×10^{-4} psi
Atmosphere	101.325 kPa	1 atm	1.01325 bar	14.7 psi
Bar	100 kPa	0.986923 atm	1 bar	0.01935 psi
psi	6.89 kPa	0.068 atm	0.0689 bar	1 psi

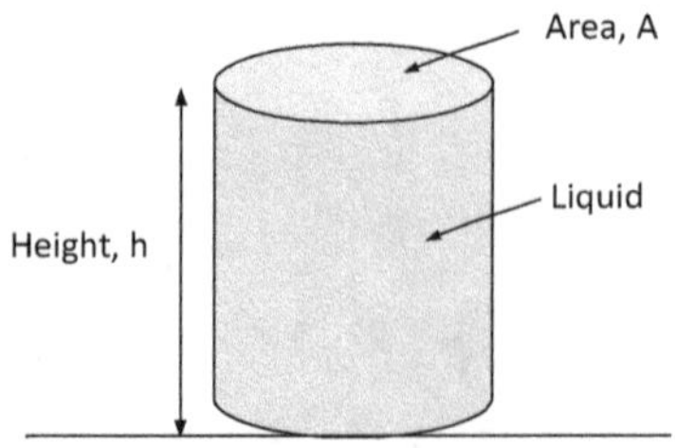

FIGURE 8.1 Liquid pressure.

DOI: 10.1201/9781003765554-8

A is area, h is the liquid height, and
g is the acceleration due to gravity.

Equation (8.1) is valid for any liquid whose density does not change with depth. Most pressure gages measure the pressure above the atmospheric pressure—this is called the **gage pressure**. The absolute pressure P_{abs} is the sum of the atmospheric pressure P_a and the gage pressure P_g. Mathematically, we have

$$P_{\text{abs}} = P_a + P_g \tag{8.2}$$

where

C is the atmospheric pressure.
P_g is the gage pressure.

Figure 8.2 shows an open-tube manometer for measuring pressure. The pressure in the open end is at atmospheric pressure. The difference between the pressure being measured P_{measd} and the atmospheric pressure P_a will be proportional to the change in liquid height, h_{diff}. Thus,

$$P_{\text{diff}} = P_{\text{measd}} - P_a = \rho g(h_{\text{diff}}) \tag{8.3}$$

where

ρ is the liquid density,
h_{diff} is the change in liquid height, and
g is the acceleration due to gravity.

Elastic elements, diaphragms, and bellows can be configured to measure the following: (i) absolute pressure; (ii) gage pressure; and (iii) differential pressure. Figures 8.3–8.5 show configurations of the diaphragm for measuring the above-mentioned types of pressures.

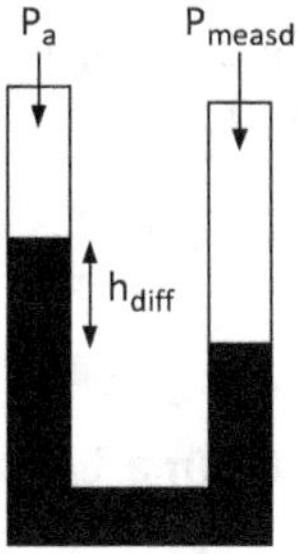

FIGURE 8.2 Open-tube manometer.

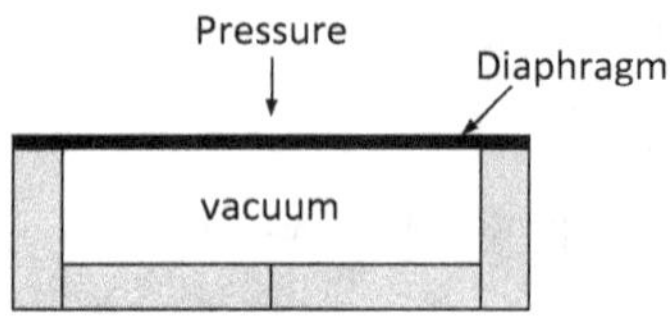

FIGURE 8.3 Absolute pressure measurement.

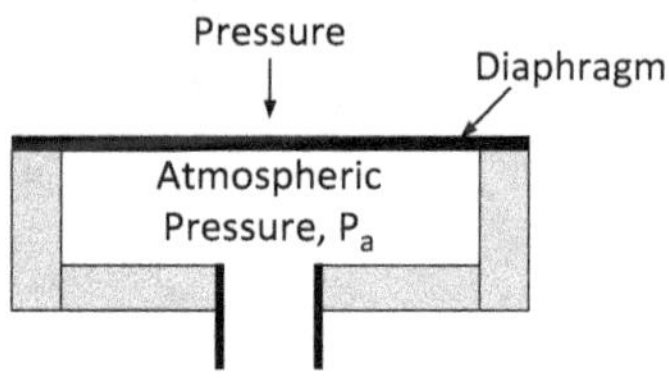

FIGURE 8.4 Gage pressure measurement

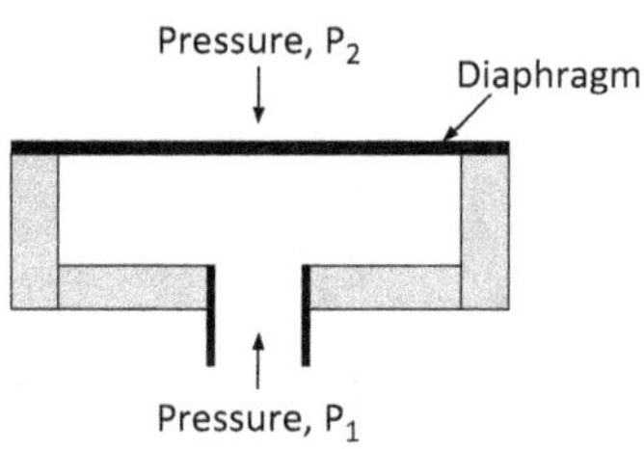

FIGURE 8.5 Differential pressure measurement.

For absolute pressure measurement, one side of the chamber is connected to a vacuum, as shown in Figure 8.3. The pressure measured is relative to vacuum. The deflection of the diaphragm is proportional to the absolute pressure.

When one side of the chamber is connected to the atmosphere and the other side of the chamber is connected to the pressure to be measured, as shown in Figure 8.4, the pressure measured is gage pressure.

In the case of differential pressure measurement, the two chambers are connected to pressures P_2 and P_1, as shown in Figure 8.5. The displacement of the diaphragm is proportional to the differential pressure,

$$\Delta P = P_2 - P_1 \tag{8.4}$$

Example 8.1: Pressure of a Liquid in a U-Tube Manometer

As shown in Figure 8.2, the liquid in the manometer is mercury with a density of 13,593 kg/m³. The change in height $\Delta h = h_{diff} = 45$ cm. If the atmospheric pressure is

101,325 Pa, what is the pressure being measured, assuming one side of the U-tube manometer is opened to the atmosphere?

Solution

From Equation (8.3),

$$P_{\text{measd}} = \rho g(\Delta h) + P_a$$

$$P_{\text{measd}} = (13593)(9.81)(0.45) + 101325 = 60006 + 101325 = 161331 = 161.33 \text{ kPa}$$

8.2 CAPACITIVE PRESSURE SENSOR

The equations related to capacitance transducers were given in Section 7.2. In this section, the application of capacitance transducers to measure pressure is described. Most common forms of pressure sensors use a diaphragm as a flexible element. In its simplest form, a diaphragm consists of a thin metallic plate that is rigidly clamped at its edge. The plate is deflected in the center when pressure is applied. The pressure sensor readout is given in terms of pressure rather than capacitance.

Two capacitive pressure sensors are shown in Figures 8.6 and 8.7. In Figure 8.6, there is one static plate. A diaphragm acts as the other plate. The diaphragm deflects under pressure, thus reducing the distance between the plates and causing a change in capacitance. Bridge circuits are normally used to measure the changes in capacitance.

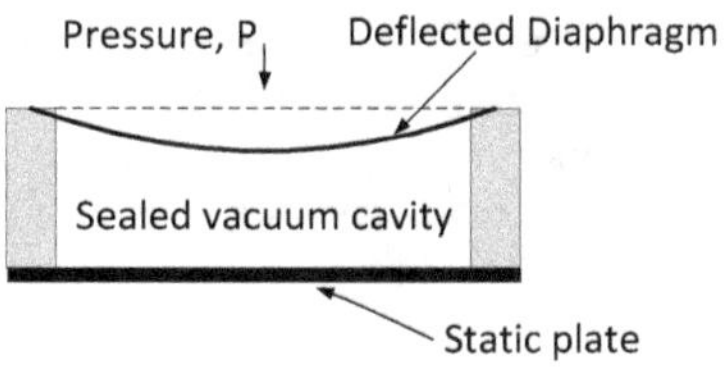

FIGURE 8.6 Capacitive pressure measurement.

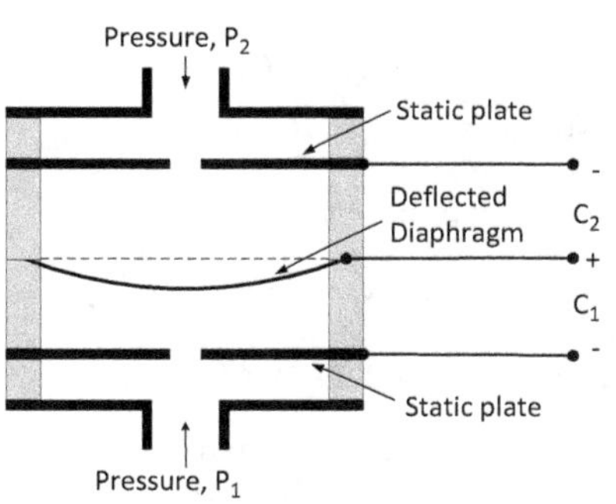

FIGURE 8.7 Three-plate capacitive system for pressure measurement.

Figure 8.7 shows a three-plate capacitive system that may be used to measure differential pressure. Two capacitors C_1 and C_2 are formed from the three plates. It was shown in Section 7.2 that

$$\frac{C_1 - C_2}{C_1 + C_2} = \frac{x}{d} \tag{7.32}$$

where

d is the plate separation when the differential pressure is zero.
x is the movement of a plate due to non-zero differential pressure.

The fractional difference of the capacitances C_1 and C_2 will be proportional to the difference between pressures P_1 and P_2. A bridge circuit, similar to the one shown in Figure 7.14, can be used to convert the fractional difference in capacitance into voltage and then from voltage to pressure.

Example 8.2: Capacitance versus Pressure of a Capacitive Sensor

A capacitive pressure sensor is made from two parallel plates to measure pressure, similar to Figure 8.6. Both parallel plates of the capacitive sensor have a diameter of 40 mm and are spaced 2.2 mm apart when there is no pressure. One of the plates act acts as a flexible diaphragm, and the other is fixed. Air fills the plates (instead of a vacuum shown in Figure 8.6). (i) Determine the capacitance when no pressure is applied. (ii) Assuming the effective area between the plates remains unchanged, calculate the new capacitances when (i) an applied pressure of 240 psi and the effective distance between the plates is 1.9 mm, (ii) an applied pressure of 450 psi and the effective distance between the plates is 1.7 mm, and (iii) an applied pressure of 720 psi and the effective gap between the plates is 1.5 mm. (iii) Plot the relationship between the capacitance versus pressure of the capacitive sensor.

Solution

We use Equation (7.24) to obtain the capacitance at various applied pressures. The MATLAB® program is shown below.

MATLAB® CODE

```
% Solution to Example 8.2
% Capacitance versus Pressure
% The constants of the capacitor
e0 = 8.85e-12; % permittivity of free space
er = 1.0; % relative permittivity of free space
d = 0.04; % diameter of the plate
A = 3.142*d*d/4; % area of the plate
Kons = e0*er*A;
p = [0 240 450 720]; % pressure values
d = [2.2e-3  1.9e-3  1.7e-3  1.5e-3]; % distances between plates
len = length(d);
for i=1:len
```

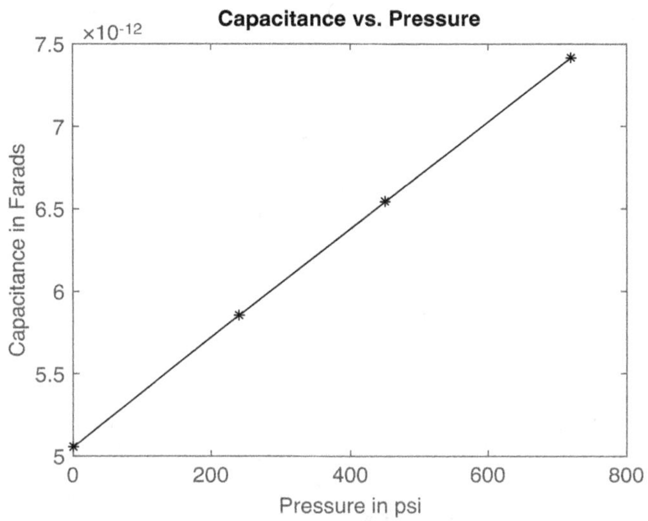

FIGURE 8.8 Capacitance versus pressure of a capacitive sensor measurement.

```
     c(i) = Kons/d(i);
end

plot(p,c,'k',p,c,'k*')
title('Capacitance vs. Pressure')
xlabel('Pressure in psi')
ylabel('Capacitance in Farads')
```

Answers are as follows:

Pressure values are as follows: 0 240 450 720 psi psi
Capacitance values are as follows: $1.0e\text{-}11 * \left[0.5056 \quad 0.5854 \quad 0.6543 \quad 0.7415\right]$ F.

Figure 8.8 shows the relationship between capacitance and pressure of the capacitive sensor.

8.3 STRAIN GAGE PRESSURE SENSOR

As discussed in Section 7.1.2, strain gages are resistive elements made of wire, metal foil, or semiconductor material. It was shown in Section 7.1.2 that for a wire of cross-sectional area A, resistivity ρ, and length L, the resistance is given as

$$R = \frac{\rho L}{A} \tag{7.4}$$

and,

$$\frac{\Delta R}{R} = (1 + 2\mu)\frac{\Delta L}{L} + \frac{\Delta \rho}{\rho} \tag{7.11}$$

where μ is the Poisson ratio.

The gage factor, G, of the strain gage was defined as

$$G = \frac{\Delta R / R}{\Delta L / L} \tag{7.12}$$

The gage factor for the resistive element is shown to be

$$G = (1 + 2\mu) + \frac{\Delta \rho / \rho}{\Delta L / L} \tag{7.13}$$

For metals, the gage factor is primarily a function of dimensional effects, whereas for semiconductors, the change in resistivity with length predominates. This leads to two types of strain gages: (i) metal strain gages and (ii) semiconductor strain gages. Table 8.2 shows the comparison between metal and semiconductor strain gages. It should be noted that the gage factors for semiconductor strain gages is far higher than those of metal strain gages. In addition, since the semiconductor strain gages have poor temperature stability, they require temperature compensation.

Figure 8.9 shows a block diagram of the use of a strain gage as a pressure sensor. A strain gage pressure sensor consists of a diaphragm, which is a thin, flexible membrane that deflects under the influence of an applied pressure. On the surface of the diaphragm, strain gage or gages are bonded or etched. When pressure is applied to the diaphragm, a strain gage deforms, causing it to experience mechanical stress. Thus, stress leads to a change in its electrical resistance. The change in resistance is detected and converted into an electrical signal, normally through a Wheatstone bridge.

TABLE 8.2
Comparison between Metal and Semiconductor Strain Gages

Parameter	Metal Strain Gage	Semiconductor Strain Gage
Gage Factor	2.0–4.4	50–200
Resistance in Ohms	10–5000	1000–5000
Resistance Tolerance	0.1%–0.2%	1%–2%
Size in mm	0.4–150	1–5
Linearity	Very good	Poor
Temperature Stability	Excellent	Poor

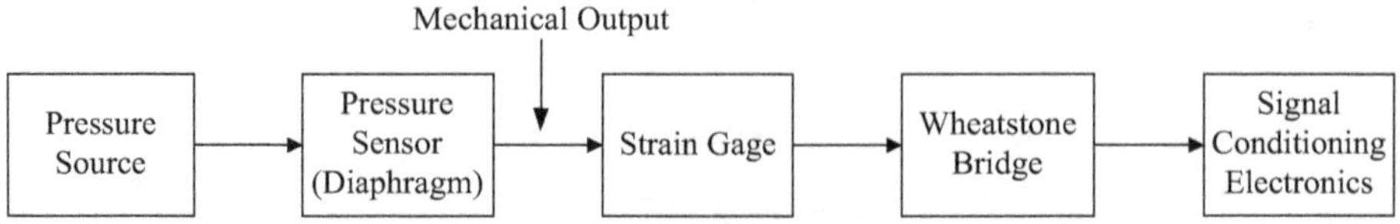

FIGURE 8.9 Block diagram of using strain gages as a pressure sensor.

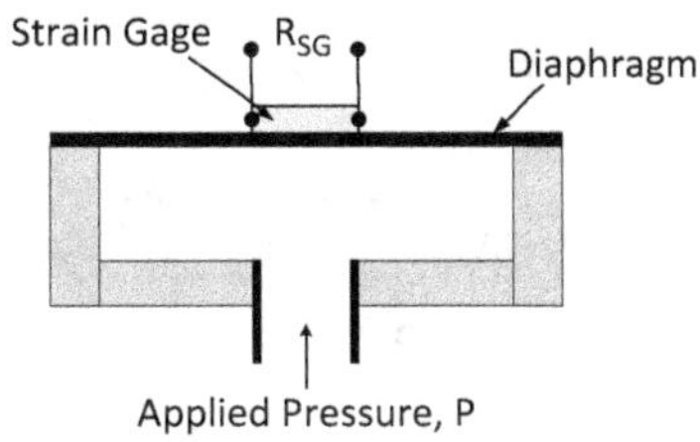

FIGURE 8.10 Strain gage for measuring pressure.

The output voltage of the bridge will be proportional to the pressure being sensed. The bridge output voltage may be conditioned and further processed.

An example of the use of a strain gage to measure pressure is shown in Figure 8.10. The gage is bonded onto the diaphragm. As the diaphragm deflects due to pressure, the resistance of the gage changes, which can be detected using a bridge circuit.

8.3.1 Metal Strain Gage

The metal strain gages change their resistance due to changes in geometry. It is the most commonly used strain gage. They are normally made by etching a metal wire on a polyimide film. A copper–nickel alloy is one of the most commonly used materials. When the wire is under a tensile strain, it increases in length and its cross-sectional area reduces, which increase the resistance of the wire. Under compressive strain, the length reduces and the cross-sectional area increases, which reduces the resistance. In Equation (7.13), the change in length and cross-sectional area predominate for metal strain gages. Thus, for metal strain gages, the gage factor can be approximated as follows:

$$G \simeq (1+2\mu) \tag{8.5}$$

8.3.2 Semiconductor Strain Gage

The basic principle behind semiconductor strain gages is the piezoresistive effect, which refers to the change in the electrical resistivity of a material when subjected to mechanical stress. Semiconductor materials, such as silicon and germanium, exhibit piezoresistive effects. For materials that exhibit piezoresistive effects, the resistivity increases when stretched. In Equation (7.13), the change in the resistivity predominates for semiconductor strain gages. Thus, for semiconductor strain gages, the gage factor can be approximated as follows:

$$G \simeq \frac{\Delta\rho/\rho}{\Delta L/L} \tag{8.6}$$

Semiconductor pressure sensors are popular for measuring pressure in various industries such as automotive, medical, and aerospace.

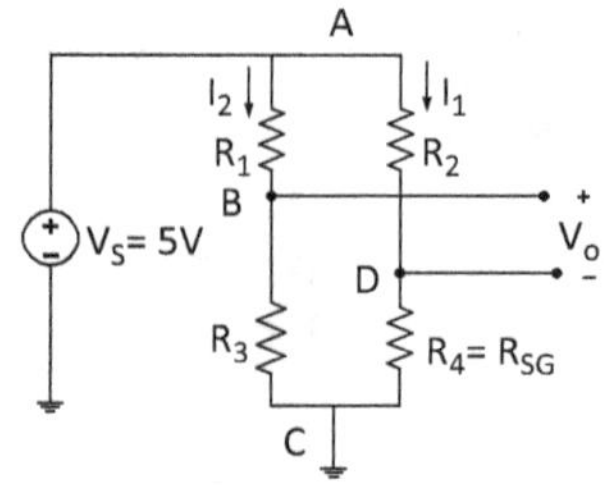

FIGURE 8.11 Wheatstone bridge with a strain gage with resistance R_{SG}.

Example 8.3: Strain Gage bonded to a Cantilever Beam

A strain gage is bonded to a bottom of a cantilever beam. The gage has an initial resistance of 150 Ω and a gage factor of 2.5. The gage is connected as one leg in a Wheatstone bridge, shown in Figure 8.11, and is powered by a 5.0-V battery. The bridge is balanced when there is no force on the beam. When a force is placed on the beam, the output voltage is 40 mV. What is the measured strain?

Solution

When there is no force applied,
$R_1 = R_2 = R_3 = R_{SG} = R_0 = 150\ \Omega$, and the output voltage is zero
When the force is non-zero, then,

$$R_1 = R_2 = R_3 = R_0 = 150\ \Omega$$

$$R_4 = R_{SG} = R_0 + \Delta R$$

Using KVL, the output voltage is given as follows:

$$V_0 = V_S\left(\frac{R_1}{R_1 + R_3} - \frac{R_2}{R_2 + R_4}\right) \tag{8.7}$$

$$V_S = 5\ \text{V},\ V_0 = 0.04\ \text{V}$$

$$0.04 = 5\left(\frac{150}{300} - \frac{150}{300 + \Delta R}\right) \tag{8.8}$$

Solving for ΔR, we get

$$\Delta R = 4.87\ \Omega$$

For a strain gage, an input strain $\left(\frac{\Delta L}{L}\right)$ produces a change in the resistance $\left(\frac{\Delta R}{R}\right)$. From Equation (7.12), the gage factor is given as

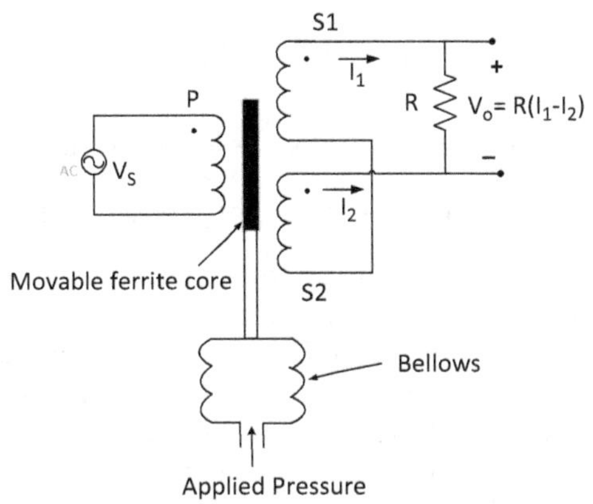

FIGURE 8.12 Bellow and LVDT combination for pressure measurement.

$$G = \left(\frac{\Delta R/R}{\Delta L/L} \right) = 2.5 = \left(\frac{4.87/150}{\Delta L/L} \right) = \left(\frac{0.0325}{\Delta L/L} \right) \tag{8.9}$$

Thus, the strain,

$$\varepsilon = \left(\frac{\Delta L}{L} \right) = \frac{0.0325}{2.5} = 0.013 \tag{8.10}$$

8.4 DISPLACEMENT PRESSURE SENSOR

Some pressure transducers convert pressure to displacement, and the latter is converted to an electrical output voltage using displacement transducers. Figure 8.12 shows bellows connected to an LVDT, whose operation is discussed in Section 7.3. An increase in pressure makes the bellows to expand. The motion of the bellows causes the LVDT ferrite core to move. This creates an output voltage of the LVDT which is proportional to the pressure applied. Bellow-type pressure transducers produce large displacements and hence large output voltages for the LVDT.

8.5 LOAD CELLS

Because of the close relationship between pressure and force (see Equation (8.1)), pressure sensors can be used to measure force. Force transducers are also called *load cells*. A load cell is a transducer which converts force into a measurable electrical output. Although there are many varieties of force sensors, strain gage load cells are the most common.

In practice, loads are applied to strain gages that cause deflections on the material that the strain gages are attached to and also change resistances in them. A strain gage is calibrated by applying weights that cause known deflections ΔL on a measuring system, such as cantilever beam, and measuring the corresponding change in gage resistance ΔR. The calibrated curve is of the form:

$$\frac{\Delta L}{L} = b_0 + b_1\left(\frac{\Delta R}{R}\right) \tag{8.11}$$

where b_0 and b_1 are constants. The slope b_1 is a local gage factor. However, the engineering gage factor $G = \frac{\Delta R/R}{\Delta L/L}$ is the slope over the entire region of the strain.

Normally, a Wheatstone bridge circuit with four resistors in the circuit is used. Any of the resistors may be replaced by strain gage(s). If one resistor is replaced by a strain gage, it is known as a quarter bridge; if two are replaced, it is called a half bridge; and if all four resistors are strain gages, it is called a full bridge.

Figure 8.13 shows a bending beam force transducer. A force is applied to the cantilever beam through a rectangular block of steel that is mounted on the beam. Strain gages, mounted on the top and bottom of the beam, are used to measure the applied force. The strain gages measure the bending stresses of the beam which are proportional to the applied force.

In Figure 8.13, the force transducer consists of four strain gages mounted on the opposite sides of a cantilever beam. The gages $GA1$ and $GA2$ are in tension, and the gages $GB1$ and $GB2$ are in compression. Four strain gages connected in a bridge configuration, shown in Figure 8.14, are normally used to increase the output voltage and also to reduce the effect of temperature on the measurements. Known pressures are needed for calibration. Calculations associated with load cells are discussed in Example 8.4.

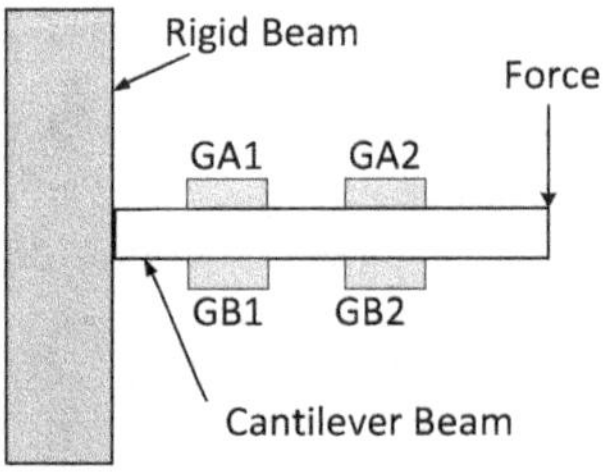

FIGURE 8.13 Force transducer with four strain gages.

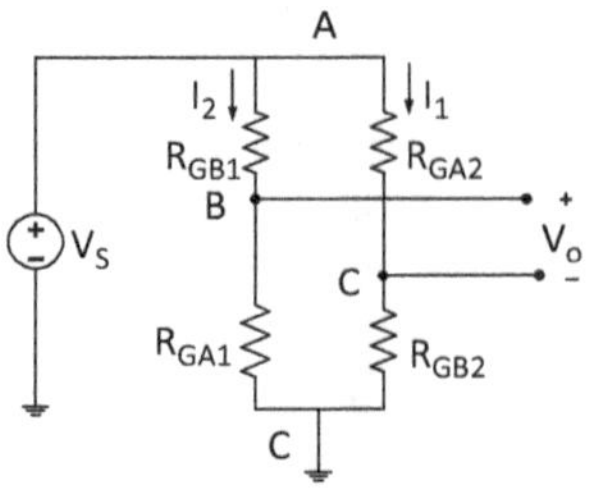

FIGURE 8.14 Bridge circuit configuration.

Example 8.4: Four Strain Gages for Force Measurement

Four strain gages are bonded to the top and bottom of a cantilever beam, shown in Figure 8.13. The gages are placed in two legs of a Wheatstone Bridge, shown in Figure 8.14, and are powered by a V_S battery. The impedances are given as

$$R_{GA1} = R_0 + \Delta R$$

$$R_{GA2} = R_0 + \Delta R$$

$$R_{GB1} = R_0 - \Delta R$$

$$R_{GB2} = R_0 - \Delta R$$

where

R_0 is the nominal resistance for the strain gages when there is no force on the beam.
ΔR is the magnitude change in resistance due to strain.

a. Derive an expression for the bridge circuit voltage output V_0.
b. If the gages have a resistance of 120 Ω when there is no force on the beam and the voltage source is 5 V, determine the output voltage V_0 versus ΔR for $0 \le \Delta R \le 3\ \Omega$
c. If the gage factor of a strain gage is 3.5, determine the strain versus ΔR for $0 \le \Delta R \le 3\ \Omega$.

Solution

a. Derivation for output voltage V_0

$$I_1 = \frac{V_S}{(R - \Delta R) + (R + \Delta R)} = \frac{V_S}{2R} \tag{8.12}$$

$$I_2 = \frac{V_S}{(R + \Delta R) + (R - \Delta R)} = \frac{V_S}{2R} \tag{8.13}$$

$$V_0 = V_{BC} = V_{BA} + V_{AC} = -V_{AB} + V_{AC} \tag{8.14}$$

$$V_0 = V_{BC} = -\frac{V_S}{2R}(R - \Delta R) + \frac{V_S}{2R}(R + \Delta R) = \frac{V_S}{R}\Delta R \tag{8.15}$$

b. Given that $V_S = 5\text{ V}$, $R = 120\ \Omega$, and $0 \le \Delta R \le 3\ \Omega$, we use Equation (8.15) to calculate the output voltages for the various values of ΔR.
c. From Equation (7.12), the strain is given as

$$\text{Strain} = \Delta L/L = \frac{\Delta R/R}{G} \tag{8.16}$$

The gage factor, G, is 3.5, and resistance R is 120 Ω. We use Equation (8.15) to obtain the strain as ΔR changes. A MATLAB program was written for the calculations involved in parts (b) and (c). Figure 8.15 shows the effect of the change in resistance on the output voltage of the Wheatstone bridge circuit, and Figure 8.16 shows the change in resistance with strain.

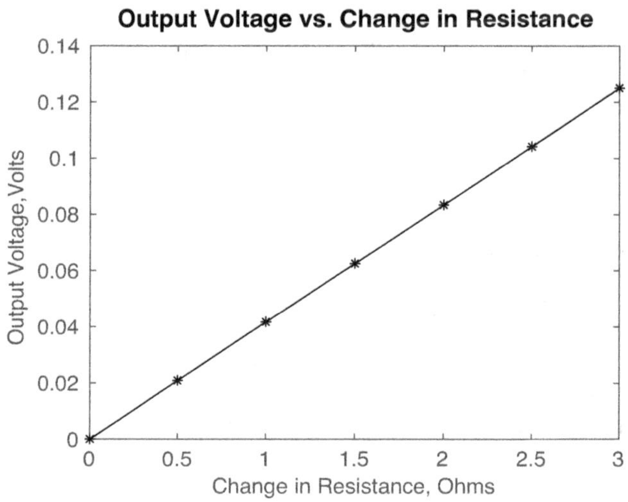

FIGURE 8.15 Change in resistance due to strain versus bridge output voltage.

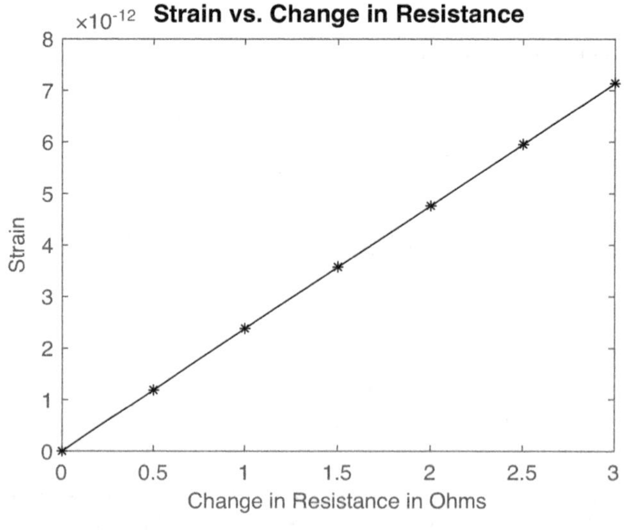

FIGURE 8.16 Change in resistance versus strain.

MATLAB® CODE

```
% Solution to Example 8.4
% Bridge voltage and Strain versus change in resistance
%
G = 3.5; % gage factor
vs = 5.0; % voltage source
R = 120; % resistance with zero force
% dr is change in resistance due to strain
% vo is the output voltage of the bridge
% st is the strain
dr = 0:0.5:3.0;
len = length(dr);
for i=1:len
    vo(i) = vs*dr(i)/R;
    st(i) = dr(i)/(G*R);
end
dr
vo
st
figure (1)
plot(dr,vo,'k', dr,vo,'k*')
title('Output Voltage vs. Change in Resistance')
xlabel('Change in Resistance in Ohms')
ylabel('Output Voltage in Volts')
figure (2)
plot(dr,st,'k', dr,st,'k*')
title('Strain vs. Change in Resistance')
xlabel('Change in Resistance in Ohms')
ylabel('Strain')
```

ANSWERS

Change in resistance, dr = 0 0.5000 1.0000 1.5000 2.0000 2.5000 3.0000

Output Voltage, vo = 0 0.0208 0.0417 0.0625 0.0833 0.1042 0.1250

Strain, st = 0 0.0012 0.0024 0.0036 0.0048 0.0060 0.0071

Example 8.5: Cantilever Beam as a Force Transducer

A cantilever beam with a single load at the end is shown in Figure 8.17.

The strain gage is connected as one of the arms of a Wheatstone bridge circuit, shown in Figure 8.18. The gage resistance is $R_{SG,}$ and the voltage source $V_S = 5\,\text{V}$. Information about the beam is shown in Table 8.3. Determine the output voltage of the bridge if the applied load at the tip of the cantilever beam varies from 50 N to 80 N.

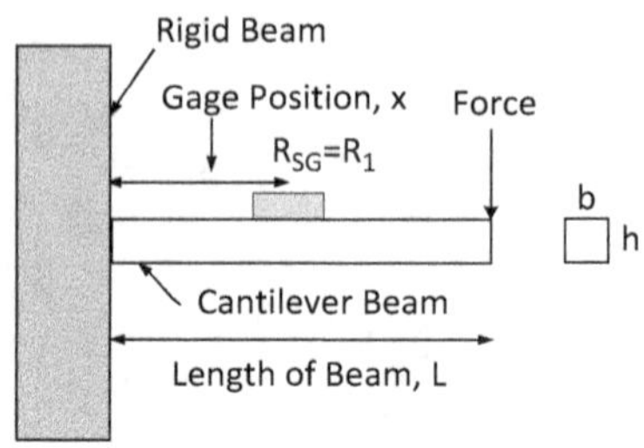

FIGURE 8.17 Cantilever beam with a strain gage.

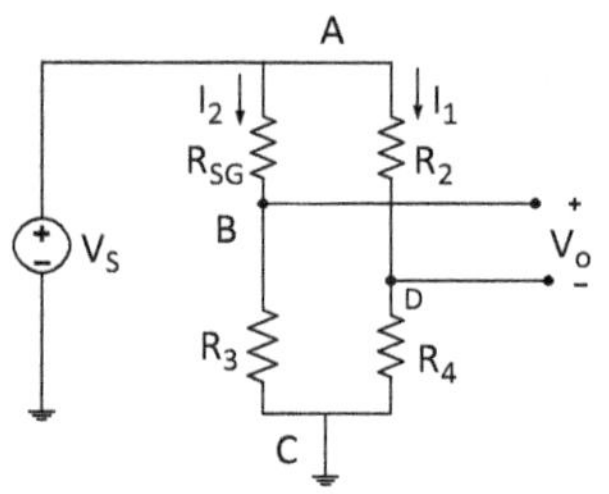

FIGURE 8.18 Wheatstone bridge with a strain gage with resistance R_{SG}.

TABLE 8.3
Information about the Cantilever Beam

Cantilever Beam Data	Symbol	Value
Length	L	60 cm
Width	B	5 cm
Thickness	H	5 cm
Young's Modulus	E	205 GPa
Gage Factor	G	2.4
Resistance	R_0	150 Ω
Gage Position	x	30 cm

Solution

For a cantilever beam of length L with a single load F_L at the end of the beam and with the strain gage placed at the position x, the deflection d at a position x is given as

$$d = \frac{F_L x^2}{6EI}(3L - x) \tag{8.17}$$

where

F_L is the force due to the weight at the tip of the beam.
x is the distance of the gage from the fixed end.
E is the Young modulus of the beam.
I is the moment of inertia.

The moment of inertia of a rectangular beam is given as

$$I = \frac{bh^3}{12} \tag{8.18}$$

where

b is the base of the beam.
h is the height of the beam.

As given from Figure 8.18, if there is no force applied, the resistance values are the same, and the output voltage is zero. However, when force is applied, we have

$$R_2 = R_3 = R_4 = R_0 = 150\ \Omega$$

$$R_{SG} = R_1 = R_0 + \Delta R$$

$$V_S = 5\ \text{V}$$

Using KVL, it can be shown that (see Equation 8.7) the output voltage is given as

$$V_0 = V_S\left(\frac{R_1}{R_1 + R_3} - \frac{R_2}{R_2 + R_4}\right) \tag{8.19}$$

Thus,

$$V_0 = 5\left(\frac{150 + \Delta R}{150 + \Delta R + 150} - \frac{150}{150 + 150}\right) = 5\left(\frac{150 + \Delta R}{300 + \Delta R} - \frac{150}{300}\right) \tag{8.20}$$

To obtain the deflection d of the beam, we use the information in Table 8.3 and Equations (8.17) and (8.18). The gage factor, G, given by Equation (7.12)

$$G = \left(\frac{\Delta R/R}{\Delta L/L}\right)$$

We can obtain the change in resistance ΔR knowing R, G, L, and ΔL. It is assumed that the deflection is small, so the Pythagoras theorem can be used to calculate $L + \Delta L$ from d and L. Knowing ΔR, we can use Equation (8.20) to obtain the output voltage V_0. The MATLAB program is shown below. Figure 8.19 shows the output voltage versus the applied load on the beam.

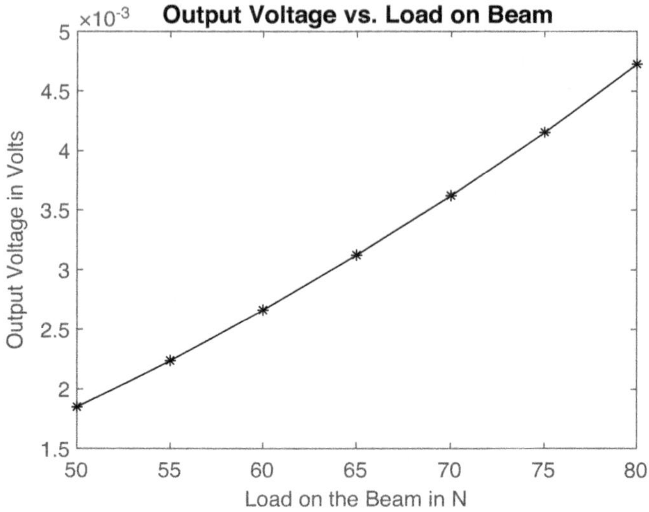

FIGURE 8.19 Output voltage versus applied load on the beam.

MATLAB® CODE

```
% Solution to Example 8.5
% Bridge voltage and applied force on a beame
%
G = 2.4; % gage factor
vs = 5.0; % voltage source
R = 150; % resistance with zero force
L = 0.6; % length of the beam (60 cm)
B = 0.05; % width of the beam (5 cm)
H = 0.05; % height of the beam (5 cm)
X = 0.3; % gage position on the beam
E = 205e6; %Young's Modulus
% I is the moment of inertia
I= (B)*(H.^3)/12;  % moment of inertia
% fl is the force on the beam
% db is the deflection of the beam
% dl is change in length of the beam
% Ltotal is the total length of beam under strain
% dr is change in resistance of the strain gage
% vo is the output voltage of the bridge
% st is the strain
% dr = 0:0.5:3.0;
fl = 50:5:80;
len = length(fl);
for i=1:len
    db(i) = fl(i)*(X*X)*(3*L-X)/(6*I*E);
    Ltotal(i) = sqrt(db(i)*db(i) +X*X);
    dl(i) = Ltotal(i) - X;
    dr(i) = (R*G*dl(i)/X);
    vo(i) = vs*(150+dr(i))/(300+dr(i)) -vs*0.5;
 end
```

```
plot(fl,vo,'k', fl,vo,'k*')
title('Output Voltage vs. Load on Beam')
xlabel('Load on the Beam in N')
ylabel('Output Voltage in Volts')
```

PROBLEMS

Problem 8.1 A capacitive pressure sensor has one fixed plate, and the other plate is an elastic diaphragm. The area between the plates is filled with air. The area of each plate is 200 μm by 200 μm. The distance between the diaphragm and the fixed plate is 4 μm. The diaphragm has a stiffness of 180 μN/m. A capacitive bridge circuit is used to measure the capacitive change due to applied pressure with the capacitive sensor connected to one arm of the bridge circuit. When pressure was applied to the capacitive sensor, there was a corresponding increase in the capacitance compared to the capacitance without applied pressure. For the following increase in the capacitance of the capacitive pressure sensor: 0.1, 0.2, 0.3, 0.4, 0.5, and 0.6 pF, determine the corresponding applied pressure.

Problem 8.2 Table P8.1 shows the pressure versus capacitance of a capacitive pressure sensor. (a) Determine the equation of best fit between capacitance and pressure and (b) what will be the capacitance if the pressure is 350 psi?

Problem 8.3 A strain gage is bonded to a bottom of a cantilever beam. The gage has an initial resistance of 100 Ω and a gage factor of 2.1. The gage is connected as one leg in a Wheatstone bridge circuit, shown in Figure 8.11, and is powered by a 5.0-V battery. The bridge is balanced when there is no force on the beam. When various forces were placed on the beam and the corresponding output voltages were 10, 20, 30, 40, and 50 mV, determine the corresponding change in resistance.

Problem 8.4 Two strain gages are bonded to the top and bottom of a cantilever beam, shown in Figure P8.1a. The gages are placed in two legs in a Wheatstone bridge, shown in Figure P8.1b, and is powered by a V_S V battery.

The impedances are given as

TABLE P8.1
Pressure versus Capacitance of a Capacitive Pressure Sensor

Pressure (psi)	Capacitance (pF)
0	5.056
240	5.854
450	6.543
720	7.415

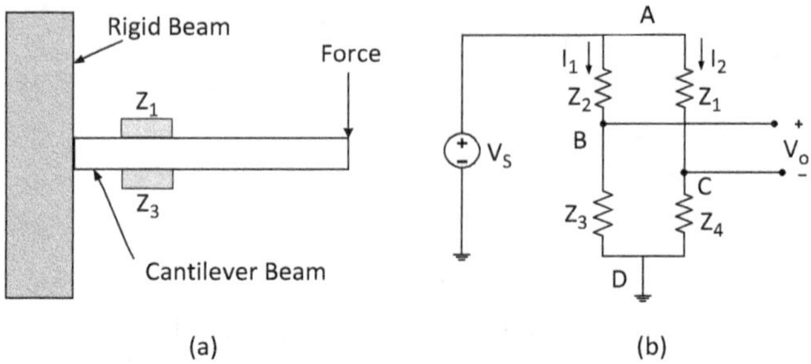

FIGURE P8.1 (a): Cantilever beam with two strain gages. (b): Wheatstone bridge circuit with two strain gages.

$$Z_1 = R_0 + \Delta R$$

$$Z_2 = R_0$$

$$Z_3 = R_0 - \Delta R$$

$$Z_4 = R_0$$

where

R_0 is the nominal resistance for both the strain gages and resistors when there is no force on the beam. ΔR is the magnitude change in resistance due to strain.

a. Derive an expression for the bridge voltage output V_0.
b. The gages have an initial resistance of 120 Ω when there is no force on the beam, and the voltage source is 5 V. Determine the output voltage V_0 versus ΔR for $0 \leq \Delta R \leq 4\ \Omega$.
c. If the gage factor of the strain gage is 4.0, determine the strain versus ΔR for $0 \leq \Delta R \leq 4\ \Omega$.

Problem 8.5 A load cell with a range up to 120 N has a nominal full-scale output of 60 mV. The load cell was calibrated, and the results are found in Table P8.2.

a. Determine the curve of best fit.
b. If the output voltage is 36 mV, what is the corresponding force?

Problem 8.6 A load cell is connected to an 8-bit analog-to-digital converter which has a full-scale voltage of 5 V. (a) What is the resolution of the converter? (b) If the sensitivity of the load cell is 20 mV/KN, what should be the amplifier gain if the load cell is used to measure forces in the range of 0.15 to 5 KN?

Problem 8.7 A strain gage is cemented to a metal bar that increased its length when subjected to a strain from 0.5 to 0.51 m. The gage factor of the gage is 2.2. (a) What is the percentage change in the resistance of the gage? (b) If the unstrained value of the resistance of the gage is 150 Ω, what is the resistance value after the strain gage has been strained?

TABLE P8.2
Load Cell Data

Input Force (N)	Measured Output Voltage (mV)
20	10.5
40	19.5
60	29.5
80	40.0
100	51.5
120	59.5

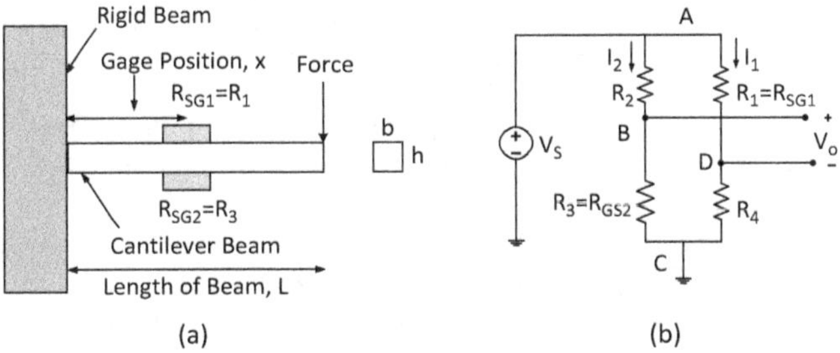

FIGURE P8.2 (a): Cantilever beam with two strain gages R_{SG1} and R_{SG2}. (b): Wheatstone bridge with strain gages with resistance R_{SG1} and R_{SG2}.

TABLE P8.3
Information about the Cantilever Beam (Change the Values)

Cantilever Beam Data	Symbol	Value
Length	L	80 cm
Width	B	4.5 cm
Thickness	H	4.5 cm
Young's Modulus	E	210 GPa
Gage Factor	G	2.1
Resistance	R_0	100 Ω
Gage Position	x	40 cm

Problem 8.8 A cantilever beam with a single load at the end is shown in Figure P8.2.

Strain gages are connected as two of the arms of a Wheatstone bridge, shown in Figure P8.8b. The gage resistances are R_{SG1} and R_{SG2}. The voltage source V_S is 5 V. Information about the beam is shown in Table P8.3. Determine the output voltage of the bridge if the applied load at the tip of the cantilever beam varies from 90 to 120 N.

BIBLIOGRAPHY

Alexander, Charles Alexander and Matthew Sadiku, *Fundamentals of Electric Circuit, ISF*, 7th Edition, McGraw-Hill Education, 2020.

Arshad, Atika, Kushsairy A. Kadir, Sheroz Khan, A. H. M. Zahirul Alam and Rumana Tasnim, "A Pressure Sensor based Inductive Transducer Designed for Biomedical Applications," *2015 IEEE 3rd International Conference on Smart Instrumentation, Measurement and Applications (ICSIMA)*, Kuala Lumpur, Malaysia, pp. 1–6, 2015.

Attia, John O., Circuits and Electronics: Hands-on Learning with Analog Discovery, CRC Press, 2018.

Attia, John O., *Electronics and Circuit Analysis Using MATLAB*, CRC Press, 2nd Edition, 2004.

Behl, Akshay, Akash Bhatia and Aviral Puri, "Piezoelectric Sensors," *International Journal of Innovative Research in Technology*, Vol. 1, Issue 7, pp. 477–480, 2014.

Bolton, William, *Instrumentation and Control Systems*, 3rd Edition, Elsevier, 2021.

Chapman, Stephen J., *MATLAB Programming for Engineers*, 7th Edition, Cengage Learning, May 16, 2024.

Etter, Delores, *Engineering Problem Solving with MATLAB*, 3rd Edition, Pearson, 2021.

Hayt, William, Jack Kemmely, Jamie Phillips, and Steven M. Durbin, *Engineering Circuit Analysis*, 10th Edition, McGraw Hill, 2023.

Ida, Nathan, *Sensors, Actuators, and Their Interfaces: A Multidisciplinary Introduction*, 2nd Edition, Institution of Engineering and Technology, 2020.

Jaeger, Richard C. and Travis Blalock, *Microelectronic Circuit Design (ISE)*, 6th Edition, McGraw Hill, 2022.

Kehileche, Brahim, Abdelhalim Telmçani, Noureddine Henini, and Younes Chiba, "Piezoelectric Energy Conversion and Its Applications: Sensors and Actuators," *2nd International Conference on Electronics, Energy and Measurement (IC2EM)*, Medea, Algeria, pp. 1–5, 2023.

Kim, Kyungrim, Jinwook Kim, Xiaoning Jiang, and Taeyang Kim, "Static Force Measurement Using Piezoelectric Sensors," *Journal of Sensors*, Vol. 2021, Issue 1, pp. 1–8, 2021.

Morris, Alan S. and Langari, Reza, *Measurement and Instrumentation: Theory and Applications*, 3rd Edition, Academic Press, 2020.

Northrop, Robert B., *Introduction to Instrumentation and Measurements*, 3rd Edition, CRC Press, 2014.

Razavi, Behzad, *Fundamentals of Microelectronics*, 3rd Edition, Wiley, 2021.

Sedra, Adel S., and Kenneth Smith, Microelectronic Circuits, 8th International Edition, Oxford University Press, 2020.

Sekhar, Beera Chandra, B. Dhanalakshmi, B. Srinivasa Rao, Singampalli Ramesh, K. Venkata Prasad, P.S.V. Subba Rao and B. Parvatheeswara Rao, "Piezoelectricity and Its Applications," in Dipti Ranjan Sahu (Ed.), *Multifunctional Ferroelectric Materials*, IntechOpen, 2021.

Stavros, Chatzandroulis, Dimitris Tsoukalas, and Peter A. Neukomm, "A Miniature Pressure System with a Capacitive Sensor and a Passive Telemetry Link for Use in Implantable Applications," *Journal of Microelectromechanical Systems*, vol. 9, no. 1, pp. 18–23, 2000.

Suryana, Anang and Melania Suweni Muntini, "Strain Gage for Mass Sensor using Cantilever Beam," *2017 International Conference on Computing, Engineering, and Design (ICCED)*, Kuala Lumpur, Malaysia, pp. 1–4, 2017.

Valentine, Daniel T. and Brian D. Hahn, *Essential MATLAB for Engineers and Scientists*, 8th Edition, Academic Press, Sept. 26, 2022.

9 Flow Sensors

9.1 FLOW, FLOW RATE, AND FLOW VOLUME

Flow is the rate of a fluid movement through a given cross-section of a vessel. Flowmeters are used to measure flow. Fluid flow can either be laminar or turbulent. Laminar flow is orderly, smooth, and each volume cell of a fluid flows in parallel to each other and to the vessel wall. On the other hand, in turbulent flow, eddies, vortexes, and whorls arise and disappear in a chaotic manner, making the flow irregular. Turbulent flow can be discriminated from laminar flow through the use of Reynolds number (Re) of the flow, which is defined as

$$\text{Re} = \frac{\rho v D}{\mu} \tag{9.1}$$

where

ρ is the density of fluid (Kg/m^3).
v is the velocity of the fluid (m/s).
D is the internal diameter of vessel (m).
μ is the dynamic viscosity of the fluid (Ns/m).

In general, Reynolds numbers below 2000 indicate laminar flow, and those above 4000 indicate turbulent flows. The flow rate can be described by the following: (i) mass flow rate and (ii) volumetric flow rate. The mass flow rate is the mass of a fluid passing through a cross-section area in unit time. The volume flow rate is the volume of a fluid passing through a cross-section area of a vessel in unit time.

It has been observed that the velocity of a fluid is never uniform over a cross-section. If v_{avg} is the average velocity across a cross-section of fluid movement, then the mass flow of the fluid is given as

$$m = \rho v_{\text{avg}} A \tag{9.2}$$

where

m is the mass of the fluid (kg).
ρ is the density of the fluid (kg/m^3).
v_{avg} is the average velocity of the fluid (m/s).
A is the area of cross-section (m^2).

The mass flow rate and volume flow rate are related by the following expression:

$$\dot{m} = \rho \dot{V} \tag{9.3}$$

DOI: 10.1201/9781003765554-9

where

$\dot{m}$ is the mass flow rate (kg/s).
$\dot{V}$ is the volume flow rate (m^3/s).

The volume flow rate is also given as the product of the average velocity and the cross-sectional area. That is:

$$\dot{V} = v_{avg} A \tag{9.4}$$

where

v_{avg} is the average velocity (m/s).
A is the cross-section area (m^2).

The following example illustrates the relationship between volume and mass flow rates.

Example 9.1: Volume and Mass Flow Rates of Air in a Duct

Air at a temperature of 24°C and pressure of 40 kPa flows through a circular duct of diameter 600 mm with an average velocity of 2.8 m/s. Determine (i) the volume flow rate and (ii) mass flow rate of the air. Assume that the air flowing at slow subsonic speed is incompressible.

Solution

The density of the air, ρ_{air}, is obtained by using the ideal gas law.

$$\rho_{air} = \frac{P}{RT} \tag{9.5}$$

where

P is pressure (Pa).
T is temperature (K).
R is the specific gas constant of dry air (287.05 J/kg/K).

Using Equation (9.5), the density of air is calculated as follows:

$$\rho_{air} = \frac{P}{RT} = \frac{(40)(1000)}{(287)(273+24)} = \frac{(40)(1000)}{(287)(297)} = 0.469 \text{ kg / m}^3$$

Area of the duct, A, is given as

$$A = \frac{\pi}{4}d^2 = \frac{\pi}{4}(0.6)^2 = 0.283 \text{ m}^2$$

Using Equation (9.4), the volume flow rate is calculated as follows:

$$\dot{V} = v_{avg}A = (2.8)(0.283) = 0.792\ \text{m}^3/\text{s}$$

From Equation (9.3), the mass flow rate is determined as follows:

$$\dot{m} = \rho\dot{V} = (0.469)(0.792) = 0.371\ \text{kg/s}$$

9.2 THERMAL FLOWMETERS

Thermal flowmeters measure the speed of a fluid by using the heat loss from a heated element in the path of a fluid flow. The operation of thermal flowmeters is based upon King's Law, which is given as follows:

$$P = A(\Delta T)\left(\alpha + \beta v_{avg}^{1/2}\right) \tag{9.6}$$

where

P is the heat transfer rate (W).
A is the effective area of the heated element (m^2).
ΔT is the difference between the heated element temperature and the fluid temperature (°C).
v_{avg} is the average fluid velocity (m/s).
α and β are constants dependent on the structure of the heated element and the specific heat of the fluid, respectively.

Figure 9.1 shows a simplified thermal flowmeter. It consists of a heated element that may be any resistive material such as thermistor, tungsten wire, or a metal film on a substrate. There are two temperature sensors, one upstream and the other downstream.

A resistive material is heated to a temperature above the ambient temperature of the fluid by passing a current through the heating element. The average power of the heating element can be expressed as follows:

$$P = I^2R \tag{9.7}$$

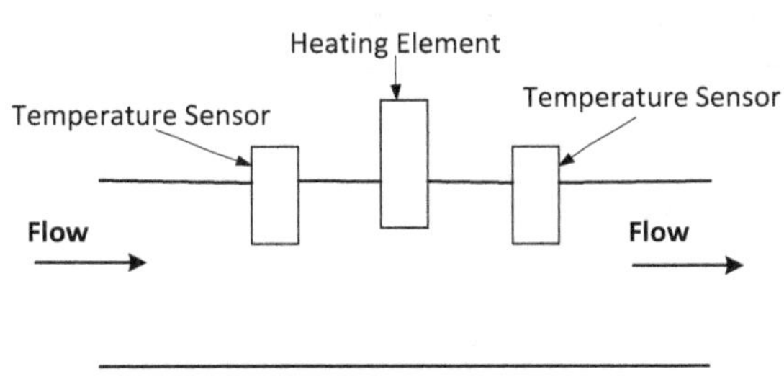

FIGURE 9.1 Simplified thermal flowmeter.

where

P is the average power dissipated by the heating element.
I is the current flowing through the heating element.
R is the resistance of the heating element.

If the heating element and the fluid are in thermal equilibrium, Equations (9.6) and (9.7) can be equated to yield

$$I^2 = \frac{A(\Delta T)\left(\alpha + \beta v_{avg}^{1/2}\right)}{R} \tag{9.8}$$

Rearranging Equation (9.8) and solving for the fluid velocity v_{avg}, we get

$$\begin{aligned} v_{avg} &= \left[I^2 R(\beta A \Delta T)^{-1} - \left(\alpha/\beta\right)\right]^2 \\ &= \left[C_1 I^2 R\left(T_e - T_f\right)^{-1} - C_2 \right]^2 \end{aligned} \tag{9.9}$$

where

C_1, C_2 are constants, $C_1 = (\beta A)^{-1}$ and $C_1 = \dfrac{\alpha}{\beta}$

$$\Delta T = T_e - T_f$$

T_e is the temperature of the heating element.
T_f is the temperature of the fluid.

In practice, the resistance R is held constant, thereby holding T_e constant. Then, from Equation (9.9), we observe that the fluid velocity is proportional to the current I^2. Since R and T_e tend to change with the fluid flow, a feedback system is needed to maintain the resistance R constant.

There are two operations of the thermal flowmeter: (i) temperature difference measurement and (ii) power consumption measurement. In the temperature difference method, the flow is determined by the temperature difference between the upstream and downstream points. The method is used in applications where low fluid flow rates are to be measured. In the power consumption method, the heater is controlled such that there is a fixed temperature difference between the upstream and downstream temperatures. The flow rate is calculated based on the amount of electric power required to maintain the fixed temperature difference.

The heat energy or thermal energy Q is given as follows:

$$Q = mc_p\Delta T \tag{9.10}$$

where

Q is thermal or heat energy in J.

m is mass of a substance in kg.
c_p is the specific heat capacity in J/kg°C.
ΔT is the temperature difference in °C.

The thermal power P can be obtained from Equation (9.10). It is given as follows:

$$P = \dot{m} c_p \Delta T \tag{9.11}$$

where

$P = \frac{dQ}{dt}$ is power in J/s or W.
$\dot{m}$ is the mass flow rate of a substance in kg/s.

Thermal flowmeters have the following advantages: (i) they are able to detect the flow of gases, (ii) there is no pressure loss in the flowmeter, and (iii) they can be used to measure the mass flow rate. The limitations of the thermal flowmeters include the following: (i) when the temperature of the fluid changes, errors occur in the flowmeter measurement, and (ii) errors may also occur in the flow rate calculations if there are deposits on the temperature sensors or the heating element.

Example 9.2: Thermal Flowmeter for Air Mass Flow Rate Measurement

Air flows through a 4-cm-diameter pipe. The mass flow rate is measured by a thermal flowmeter. If 15 W of electric power is delivered to the heating element to maintain a temperature difference of 2°C between the upstream and downstream temperature sensors, calculate the mass flow rate of the air. The specific heat of air is 1006 J/kg°C.

Solution

Specific heat of air, $c_p = 1006$ J/kg°C.
Thermal power, $P = 15$ W.
Temperature difference, $\Delta T = 2$°C.

Using Equation (9.11), we have

$$15 = \dot{m}(1006)(2)$$

The mass flow rate is $\dot{m} = \frac{15}{(1006)(2)} = 0.0075 \text{ kg / s}$

9.3 ELECTROMAGNETIC FLOWMETER

The electromagnetic flowmeter is based on Faraday's Law, which states that when a conductive medium moves through and cuts the flux lines of a magnetic field, an emf which is proportional to the instantaneous velocity of the moving conductive medium is generated. Figure 9.2 shows parts of an electromagnetic flowmeter. There is a magnetic field coil that generates a magnetic field, and the electrodes are attached to a vessel wherein a conducting fluid flows. The sensing electrodes detect the electromotive force generated by the conducting fluid flowing through the magnetic field. It is necessary that the liquid whose velocity is to be measured should be conductive.

If B is the magnetic flux density and $v(t)$ is the instantaneous velocity of the moving fluid, then from Faraday's Law, the induced emf E_{ind} between electrode positions a and b is given as

$$E_{\text{ind}} = \int_a^b v(t).B.dL \tag{9.12}$$

If B is uniform and $v(t)$, B, and E_{ind} are orthogonal, then Equation (9.12) simplifies to the following:

$$E_{\text{ind}} = B.L.v(t) \tag{9.13}$$

where

L is the length between the electrode positions.
B is magnetic flux density.
$v(t)$ is instantaneous velocity.

The electrodes for measuring the induced emf should be in contact with the fluid. In addition, the electrode should be made of a relatively nonreactive metal such as gold or platinum.

There are a few sources of errors that need to be considered while using the electromagnetic flowmeter. For situations where the fluid velocity is nonuniform within the cross-sectional area but remains symmetrical along the axis of the tube (axisymmetric), Equations (9.12) and (9.13) are applicable provided the average flow velocity is used. For known asymmetric flow profiles, a correction factor needs to be applied to the flowmeter reading. In addition, Equations (9.12) and (9.13) are applicable to magnetic fields produced by either DC or AC sources.

Some of the advantages of the electromagnetic flowmeter include the following: (i) there are no moving parts for the flowmeter to break down, (i) the meter can easily measure both low and high flow rates, and (iii) the meter can easily be installed in

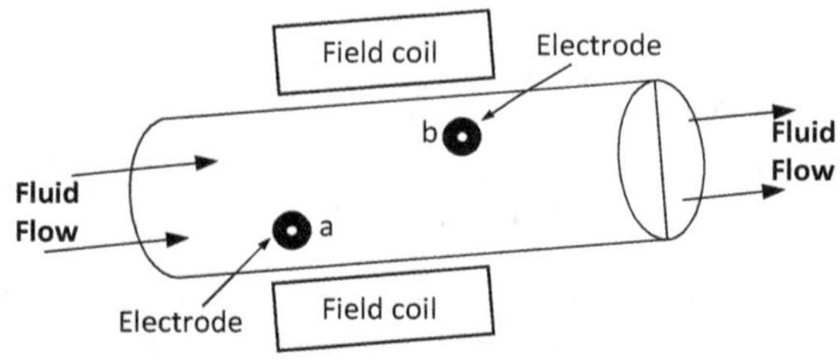

FIGURE 9.2 Electromagnetic flowmeter.

tight spaces. On the other hand, it cannot be used to measure the velocity of non-conductive fluids, such as hydrocarbons or gases, since the flowmeter is limited to conductive fluids.

Example 9.3: Velocity Measurement of a Fluid with Electromagnetic Flowmeter

A fluid flowing through a pipe with a diameter of 15 cm cuts the flux lines of magnetic field with a flux density of 0.12 T. The output of the electromagnetic flowmeter, which is used to measure the average flow rate of a fluid, is connected to an amplifier with a gain of 500 with an input impedance of 1 MΩ. The resistivity of the moving fluid is 200 Ω-m. The area of the electrode is $0.6\,\text{cm}^2$. Determine the average velocity of the fluid when the voltage at the amplifier output is between 0.1 and 0.3 V.

Solution

The input impedance of the amplifier is 1 MΩ.
The resistivity of the fluid is 200 Ω-m.
Area of the electrode $= 0.6\,\text{cm}^2 = 0.6 \times [0.01]^2\,\text{m}^2 = 0.6 \times 10^{-4}\,\text{m}^2$.
For a wire of cross-sectional area A, resistivity ρ, and length L, the resistance R is given by Equation (7.4):

$$R = \frac{\rho L}{A} \tag{7.4}$$

Thus, the length L is given as follows:

$$L = \frac{(R)(A)}{\rho} \tag{9.14}$$

The Gain, G, of the amplifier is given as follows:

$$G = \frac{E_o}{E_{\text{in}}} \tag{9.15}$$

where E_o is the output voltage and E_{in} is the input voltage of the amplifier. From Equation (9.15), we have

$$E_{\text{in}} = \frac{E_o}{G} \tag{9.16}$$

To obtain the average velocity of the fluid, we use Equation (9.13). That is:

$$E_{\text{in}} = B.L.v_{avg}$$

Thus, the average velocity can be obtained from the equation:

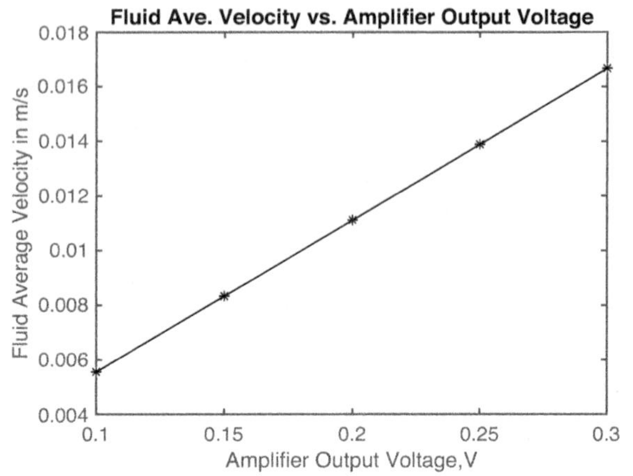

FIGURE 9.3 Fluid average velocity vs. amplifier output voltage.

$$v_{avg} = \frac{E_{in}}{(B)(L)} \tag{9.17}$$

Substituting Equations (9.14) and (9.16) into Equation (9.17), we have

$$v_{avg} = \frac{(E_0)(\rho)}{(G)(B)(R)(A)} \tag{9.18}$$

Given that $\rho = 200\ \Omega$-m., $G = 500$, B = 0.12 T, $R = 1\ \text{M}\Omega$, and A = 0.6 x $(10)^{-4}\text{m}^2$, MATLAB® is used to plot v_{avg} versus E_0, where $0.1 \le E_0 \le 0.3$ V.

MATLAB® CODE

```
% Solution to Example 9.3
% Average velocity of fluid measured with electromagnetic
flowmeter
%
R = 1e6; % resistance
res = 200; % resistivity value
A = 0.6e-4; % area
G = 500; % gain of amplifier
B = 0.12; % magnetic flux density
% Eo is te output voltage values
Eo = 0.1:0.05:0.3;
len = length(Eo);
for i=1:len
    V(i) = Eo(i)*res/(G*B*R*A);
 end

plot(Eo,V,'k', Eo,V,'k*')
title('Fluid Average Velocity vs. Amplifier Output Voltage')
xlabel('Amplifier Output Voltage, V')
ylabel('Fluid Average Velocity in m/s')
```

Figure 9.3 shows a plot of the average velocity of the fluid versus the amplifier output voltage.

9.4 ULTRASONIC FLOW SENSORS

Ultrasonic flow sensors use ultrasonic transducers to measure flow in biological and industrial systems. The operation of ultrasonic flowmeters is based on two physical principles: transit time and Doppler shift. They are described below.

9.4.1 Transit Time Flowmeters

The velocity of fluids, such as water, milk, oil, treated sewage, and blood, can be measured using transit time ultrasonic flowmeters. The velocity of a fluid is obtained by finding the difference between the up- and downstream ultrasonic pulse transit times measured between two ultrasonic transducers. Figure 9.4 shows an ultrasonic transit time flowmeter. Ultrasonic transducers A and B are clamped to the outside of a pipe or blood vessel. The system first sends a pulse signal downstream from A to B and measures the transit time. Then, an upstream pulse is sent from B to A.

The downstream pulse transit time, t_d, is

$$t_d = \frac{L}{c + v_{\text{ave}} \cos\theta} \tag{9.19}$$

The upstream pulse transit time, t_u, is

$$t_u = \frac{L}{c - v_{\text{ave}} \cos\theta} \tag{9.20}$$

where

L is the distance between the transducers.
c is the ultrasonic velocity in the medium.
v_{ave} is the average velocity of the flow of a fluid in the medium.
θ is the angle between the axis of sound propagation and flow axis.

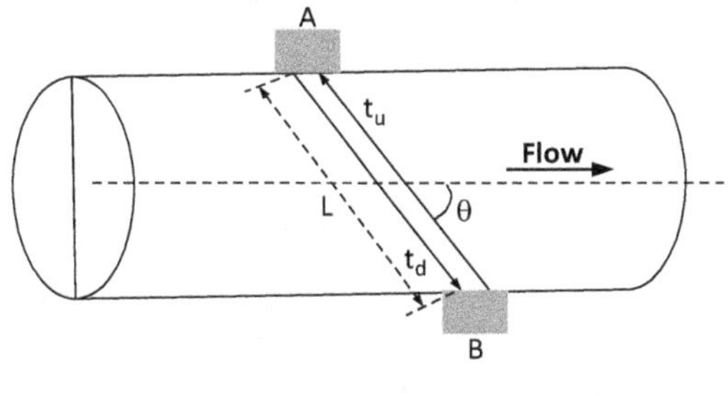

FIGURE 9.4 Ultrasonic transit time flow sensor.

The difference between the upstream and downstream transit times, Δt, is given as follows:

$$\Delta t = t_u - t_d \tag{9.21}$$

Using Equations (9.19)–(9.21), we have

$$\Delta t = \frac{L}{c - v_{\text{ave}} \cos\theta} - \frac{L}{c + v_{\text{ave}} \cos\theta} \tag{9.22}$$

Simplifying Equation (9.22), we have

$$\Delta t = \frac{2L(v_{\text{ave}} \cos\theta)}{c^2 - v_{\text{ave}}^2 \cos^2\theta} \tag{9.23}$$

If the ultrasonic velocity in the medium is far greater than the average flow velocity, Equation (9.23) becomes

$$\Delta t \simeq \frac{2Lv_{\text{ave}} \cos\theta}{c^2} \tag{9.24}$$

and

$$v_{\text{ave}} \simeq \frac{(\Delta t)c^2}{2L\cos\theta} \tag{9.25}$$

From Equations (9.19) and (9.20), we have

$$c + v_{\text{ave}} \cos\theta = \frac{L}{t_d} \tag{9.26}$$

$$c - v_{\text{ave}} \cos\theta = \frac{L}{t_u} \tag{9.27}$$

Subtracting Equation (9.27) from Equation (9.26), we have

$$2v_{\text{ave}} \cos\theta = L\left(\frac{1}{t_d} - \frac{1}{t_u}\right) = L\left(\frac{t_u - t_d}{t_d t_u}\right) = \frac{(\Delta t)(L)}{t_u t_d} \tag{9.28}$$

From Equation (9.28), we have

$$v_{\text{ave}} = \frac{(\Delta t)(L)}{2t_u t_d \cos\theta} \tag{9.29}$$

One drawback of the transit time flowmeter is that the Δt of Equation (9.24) is very small compared to the upstream or downstream transit time. Measurement of a very small-time difference, Δt, presents problems in instrument design. Complex electronic circuitry is required for the measurement. Another problem with this flowmeter is that ultrasonic velocity in the medium may vary with temperature, and this might cause significant errors since the ultrasonic velocity in the medium c appears as c^2 in Equation (9.24). In addition, air bubbles can attenuate the signal in the flowmeters to an extent that the transit signal is no longer detectable.

The advantages of transit time flowmeters are as follows: (i) the method can be used to measure the flow of many liquids and gases because scattering is not required, (ii) it is not restricted to conductive fluids (a requirement of electromagnetic flow meters), and (iii) the method can be used to measure the flow of corrosive fluids and flurries.

Example 9.4: Average Velocity Measurement with an Ultrasonic Transit Time Flowmeter

An ultrasonic flowmeter is used to measure the flow rate of ethanol at 20°C that flows through a pipe of diameter of 15 cm. The speed of sound in ethanol is 1159 m/s. The upward and downward transit times are 66010 and 65950 ns, respectively. If the angle of the transducer is 30 degrees with respect to the longitudinal axis of the pipe, what are (i) the average speed of the ethanol, (ii) volume flow rate, and (iii) mass flow rate? The density of ethanol at 20°C is 789.45 kg/m³.

Solution

Given that:

$c = 1159\ \text{m/s}$ $\quad t_u = 66010\ \text{ns}$ $\quad t_d = 65950\ \text{ns}$

$\rho = 789.45\ \text{kg/m}^3$ $\quad \theta = 30°\text{C}$ $\quad D = 15\ \text{cm}$

The distance between the transducers $L = \dfrac{D}{\sin\theta} = \dfrac{0.15}{\sin 30°} = 0.3\ \text{m}$

i. Using Equation (9.25), we have

The average speed $v_{\text{ave}} \simeq \dfrac{(\Delta t)c^2}{2L\cos\theta} = \dfrac{(60\times 10^{-9})(1159^2)}{2(0.3)\cos 30°} = 0.155\ \text{m / s}$

ii. From Equation (9.4),

The volume flow rate $= \dot{V} = (v_{\text{ave}})\ (\text{area}) = (0.155)\left(\dfrac{\pi D^2}{4}\right) = 0.155$ $\left(\dfrac{\pi 0.15^2}{4}\right) = 0.00274\ \text{m}^3/\text{s}$

iii. From Equation (9.3),

The mass flow rate, $\dot{m} = \rho\dot{V} =$ (density) (volume flow rate) $=$ $(789.45)(0.00274) = 2.163\ \text{kg/s}$

9.4.2 Ultrasonic Doppler Flowmeters

The ultrasonic Doppler flowmeter is based on the change in the frequency of sound waves reflected from particulate matter in a moving fluid. Figure 9.5 shows a transducer arrangement for an ultrasonic Doppler flowmeter.

The transmitter and receiver are piezoelectric crystals that are excited to produce continuous waves. If f_s is the frequency of the transmitter and f_r is the frequency received at the receiver, then it can be shown that the Doppler frequency shift, f_d is given as follows:

$$f_d = f_s - f_r = f_s - f_s\left(\frac{c - v_{\text{ave}}\cos\beta}{c + v_{\text{ave}}\cos\alpha}\right) \tag{9.30}$$

where

c is the ultrasonic velocity in the medium.
v_{ave} is the average velocity of the fluid.
α and β are angles shown in Figure 9.5.

Assuming that $c \gg v_{\text{ave}}$, then Equation (9.30) simplifies to the following:

$$f_d = \pm f_s(\cos\alpha + \cos\beta)\left(\frac{v_{\text{ave}}}{c}\right) \tag{9.31}$$

Where *minus and plus signs* account for flow toward and away from the transducer faces, respectively. It should be noted, from Equation (9.31), that the Doppler frequency shift is directly proportional to the average velocity of the flow.

For the Doppler flowmeter to work, the transmitted sound has to be scattered by small bubbles, solids, or particulate matter. When the particulate matter is a particle in the moving fluid that causes scattering, the particles need only be a few microns in diameter. Thus, the Doppler flowmeter can be used to measure flow velocities of sludge, blood, and slurries or any other fluid that contain air bubbles.

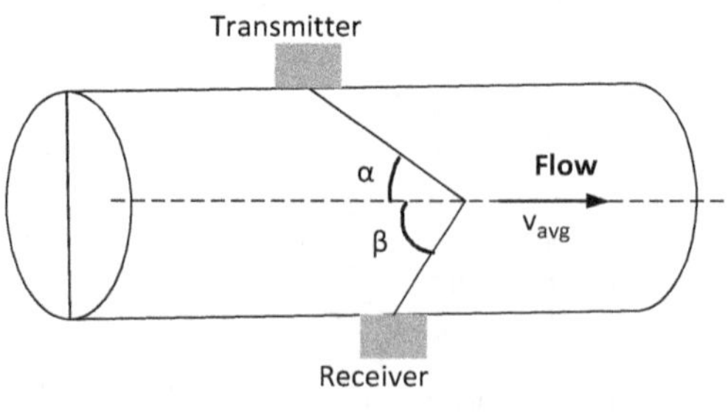

FIGURE 9.5 Ultrasonic Doppler flowmeter with the transmitter and receiver on opposite sides of the vessel with fluid flow.

Example 9.5: Doppler Flowmeter for Measuring the Average Velocity of Blood Flow

A Doppler flowmeter can be used to measure the average velocity of blood flow by reflecting ultrasound waves from red blood cells moving in the blood vessel. If the speed of sound in tissue is 1540 m/s, the Doppler flowmeter uses a sound frequency of 4.8 MHz, and a frequency shift observed in a constricted vein increased from 400 to 450 Hz, determine the corresponding speed of the blood flow. Assume that the angles the transmitter and the receiver make with the horizontal flow axis, α and β, are equal to 45° and the flow is moving away from the transducer face receiver.

Solution

Given that
c = ultrasonic velocity in tissue = 1540 m/s
f_s = sound frequency = 4.8 MHz

$$\alpha = \beta = 45^\circ$$

$$400 \text{ Hz} \le f_d \le 450 \text{ Hz}$$

From Equation (9.31), we have

$$v_{\text{ave}} = \frac{(c)(f_d)}{f_s(\cos\alpha + \cos\beta)} \tag{9.32}$$

With Equation (9.32), we use MATLAB to plot v_{ave} versus f_d.

MATLAB® CODE

```
% Solution to Example 9.5
% Ultrasonic flow meter
%
fs = 4.8e6; %s sound frequency
c = 1450; % ultrasonic velocity in tissue
ang = 45; % angle
ang_r = (ang/180)*pi; % convert degrees to radians
% fd Doppler frequency shift
fd = 400:5:450;
len = length(fd);
for i=1:len
    vave(i)  = c*fd(i)/(fs*cos(ang_r));
 end

plot(fd,vave,'k', fd,vave,'k*')
title('Blood Flow Velocity vs. Doppler Frequency Shift')
xlabel('Doppler Frequency Shift, Hz')
ylabel('Blood Flow Velocity, m/s')
```

Figure 9.6 shows the plot of blood flow velocity versus Doppler frequency shift.

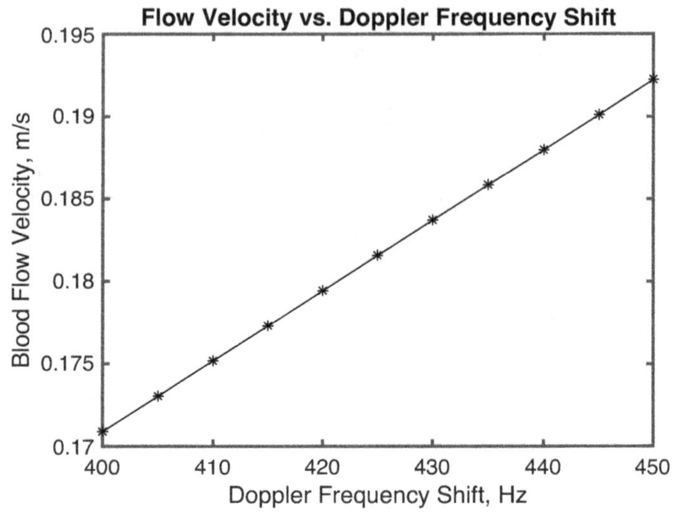

FIGURE 9.6 Blood flow velocity vs. Doppler frequency shift.

9.5 DIFFERENTIAL PRESSURE FLOW SENSORS

Differential pressure flow sensors have devices placed in fluid-carrying vessels that cause obstruction. If an obstruction is placed in a pipe, the velocity of the fluid passing over the obstruction increases and the pressure decreases. Thus, a differential pressure flow sensor normally has a constriction element that introduces pressure drop and sensors to measure the differential pressure. The Bernoulli equation is used to determine the volumetric flow rate based on the differential pressure measurement. The constriction element can be as follows: (i) Venturi tube, (ii) orifice plate, and (iii) Pitot tubes.

Figure 9.7 shows a pipe with a Venturi tube. The tube has a gradual tapering from a full diameter to a constricted diameter. The inlet taper is normally $10.5° \pm 1°$, and the exit taper is between 5° and 15°. The differential pressure for before the constriction and at the constriction may be measured with a U-tube manometer or differential diaphragm pressure cell.

For Venturi tubes, it can be shown that the volumetric flow $\dot{V}$ is proportional to the square root of the pressure difference. That is,

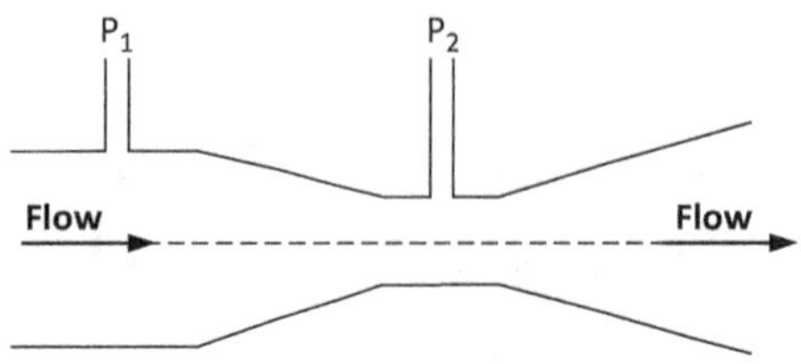

FIGURE 9.7 Venturi tube differential pressure sensor.

$$\dot{V} = k\sqrt{\Delta P} \tag{9.33}$$

where

k is a constant.
ΔP is a pressure difference.

Example 9.6: Flow Through a Venturi Tube

Figure 9.8 shows a flow sensor using a Venturi tube. The cross-section areas at points 1 and 2 are A_1 and A_2, respectively. The fluid velocities at points 1 and 2 are v_1 and v_2, respectively. In addition, the pressures at points 1 and 2 are P_1 and P_2, respectively. Assuming that the fluid is incompressible and the fluid density is ρ,

i. Use the Bernoulli equation to show that $v_1 = \sqrt{\dfrac{2g\Delta h}{\left(\left(\dfrac{A_1}{A_2}\right) - 1\right)}}$, where,

   $\Delta h = h_1 - h_2$

ii. If $A_1 = 600\,\text{cm}^2$ $A_2 = 100\,\text{cm}^2$, and the fluid density is $\rho = 997.77\,\text{kg/m}^3$, determine the fluid volumetric flow rate when Δh various from 1 cm to 6 cm.

Solution

The Bernoulli equation is

$$\frac{1}{2}\rho v_1^2 + \rho g y_1 + P_1 = \frac{1}{2}\rho v_2^2 + \rho g y_2 + P_2 \tag{9.34}$$

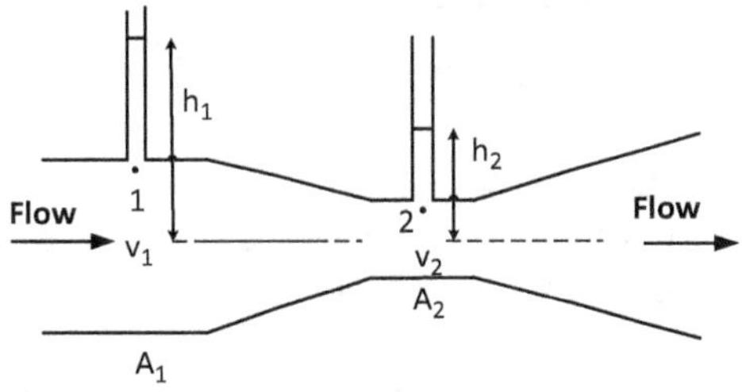

FIGURE 9.8 Flow sensor using a Venturi tube.

where

g is the acceleration due to gravity (9.8 m/s^2).
y_1 and y_2 are heights at points 1 and 2, respectively.

Since y_1 and y_2 are at the same height at the center of the pipe where the fluid velocities are measured, Equation (9.34) becomes

$$P_1 - P_2 = \frac{1}{2}\rho\left(v_2^2 - v_1^2\right) \tag{9.35}$$

The continuity equation allows to relate the speed to the areas as follows:

$$v_1 A_1 = v_2 A_2 \tag{9.36}$$

Substituting Equation (9.36) into (9.35), we have

$$v_1 = \sqrt{\frac{2(P_1 - P_2)}{\rho\left(\left(\frac{A_1}{A_2}\right) - 1\right)}} \tag{9.37}$$

But,

$$P_1 - P_2 = \Delta P = \rho g h \tag{9.38}$$

Substituting Equation (9.38) into (9.37), we have

$$v_1 = \sqrt{\frac{2gh}{\left(\left(\frac{A_1}{A_2}\right) - 1\right)}} \tag{9.39}$$

From Equation (9.4), the volume flow rate $\dot{V}$ is

$$\dot{V} = v_1 A_1 = A_1\sqrt{\frac{2gh}{\left(\left(\frac{A_1}{A_2}\right) - 1\right)}} \tag{9.40}$$

ii. $A_1 = 600\,\text{cm}^2 = 0.06\,\text{m}^2$,$A_2 = 100\,\text{cm}^2 = 0.01\,\text{m}^2$, and $\rho = 997.77\,\text{kg/m}^3$. Using Equation (9.40) and MATLAB, we can find the volumetric flow rate versus the height h for the values of 1 to 6 cm.

MATLAB® CODE

```
% Solution to Example 9.6
% Venturi Tube Flow Detector
%
A1= 0.06; % Area at point 1
```

```
A2= 0.01; % Area at point 2
g = 9.8; % acceleration due to gravity
% h is the height of the fluid
h = 0.001:0.001:0.006;
len = length(h);
den = (A1/A2)-1;
for i=1:len
    vflow(i) = A1*sqrt(2*9.8*h(i)/den);
 end
plot(h,vflow,'k', h,vflow,'k*')
title('Volumetric Flow Rate vs. Fluid height')
xlabel('Fluid Height, m')
ylabel('Volumetric Flow Rate, m/s')
```

Figure 9.9 shows a plot of the volumetric flow rate versus the height of the fluid.

Another obstruction device for differential pressure flow measurement is the orifice plate. It consists of metal disk with a hole in it. An orifice metal disk is shown in Figure 9.10. Figure 9.11 shows a pipe with an orifice plate.

The pressure difference is measured at a point equal to the diameter of the pipe upstream of the orifice and a point equal to half of the diameter downstream. The volume flow rate is proportional to the square root of the pressure difference. The orifice plate differential pressure flow sensor has the advantage of being simple to implement. The orifice plate has the following drawbacks: (i) the edges of the orifice hole gradually get worn out, which affects the relationship between the volumetric flow rate and the differential pressure, and (ii) particles in a fluid may stick behind the orifice hole and reduce the diameter of the orifice.

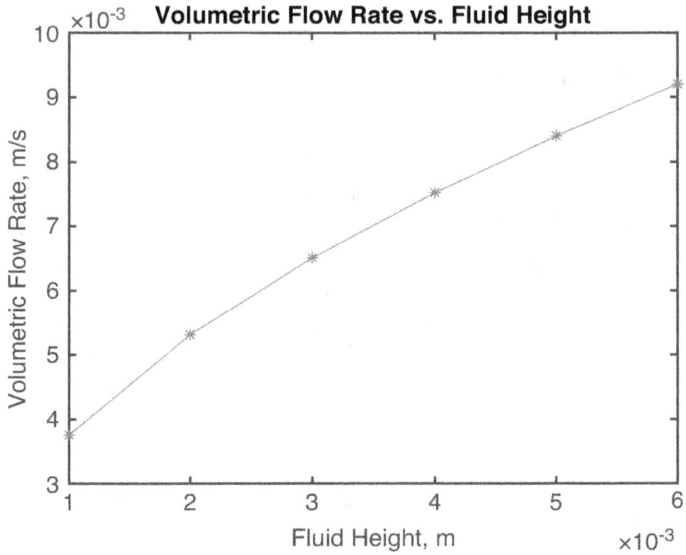

FIGURE 9.9 Volumetric flow rate versus the height of fluid.

FIGURE 9.10 An orifice plate—front view.

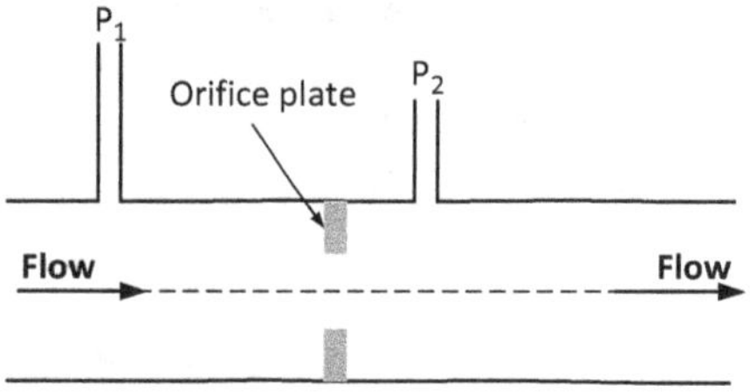

FIGURE 9.11 Orifice plate in a pipe.

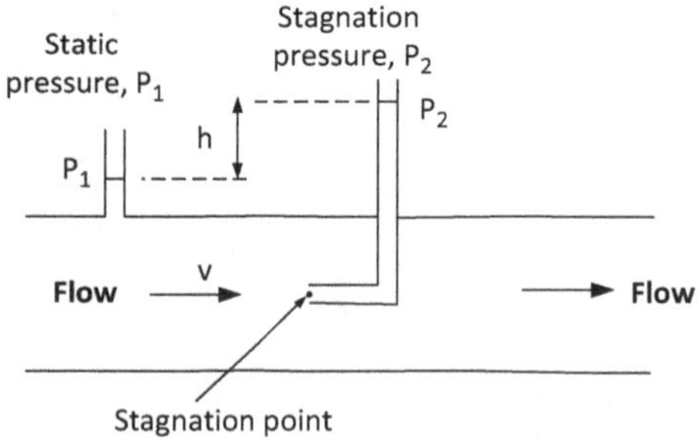

FIGURE 9.12 Volumetric flow rate measurement with a Pitot tube.

Figure 9.12 shows the arrangement with a Pitot tube. In the Pitot tube, as the velocity of the fluid increases, the total pressure in the tube increases, and the head of the tube rises, indicating fluid speed if properly calibrated.

PROBLEMS

Problem 9.1 Air enters a pipe with a diameter of 30 cm with a pressure of 220 kPa and temperature of 25°C. The average velocity of the air at the inlet of the pipe is 4 m/s. The air is heated as it flows through the pipe and leaves the pipe at a pressure of 170 kPa and temperature of 42°C. Calculate (a) the volume flow rate of the air at the inlet of the pipe, (b) the air mass flow rate, (c) the average velocity of the air as it leaves the pipe, and (d) the volume flow rate at the exit of the pipe.

Problem 9.2 A liquid with a specific heat of 2450 J/kg°C flows through a pipe. A power of 25 W was delivered to a heating element that maintained a temperature difference of 3.0°C between the upstream and downstream thermometers. (a) Calculate the mass flow rate of the liquid. Assuming the mass flow rate determined in part (a), determine the power that will maintain the temperature difference from 1°C to 4°C.

Problem 9.3 An electromagnetic flowmeter has a flow tube diameter of 500 mm and gives an output voltage of 10 mV for a magnetic flux density of 0.2 T. (i) Determine (a) the velocity of the fluid and (b) the rate of discharge of liquid through the flowmeter. (ii) If the velocity of the fluid is maintained constant, as in part (a), and the magnetic flux density varies from 0.1 to 0.5 T, plot the induced output voltage versus the magnetic flux density.

Problem 9.4 An ultrasonic flowmeter is used to measure the flow rate of castor oil at 25°C that flows through a pipe with a diameter of 10 cm. The speed of sound in castor oil is 1490 m/s. The upward and downward transit times are 67045 and 66960 ns, respectively. If the angle of transducer is 40° with respect to the longitudinal axis of the pipe, what are (i) the average speed of the ethanol, (ii) volume flow rate, and (iii) mass flow rate? Assume that the density of the oil is about 945 kg/m^3.

Problem 9.5 Figure 9.4 shows a transit time ultrasonic flowmeter for measuring the flow rate of water in a pipe. The pipe has a diameter of 0.8 m, and the transducers are opposite sides of the pipe with an angle of 45°. If the speed of sound in water 1,482 m/s at a 25°C and the flow rate is 2.4 m/s, determine the (i) downstream transit time, (ii) upstream transit time, and (iii) volumetric flow rate. Assume that the density of water at 25°C is 997.05 kg/m^3.

Problem 9.6 A Doppler flowmeter, for blood velocity measurement, emits sound at a frequency of 4.2 MHz. When measuring the blood flow in a vein, a beat frequency of 65 Hz was detected. What is the velocity of the blood cells from where the sound is reflecting from? The speed of sound in human tissue is 1540 m/s. Assume that the angles the transmitter and the receiver make with the horizonal flow axis, α and β, are equal to 45° and the flow is moving away from the transducer face receiver.

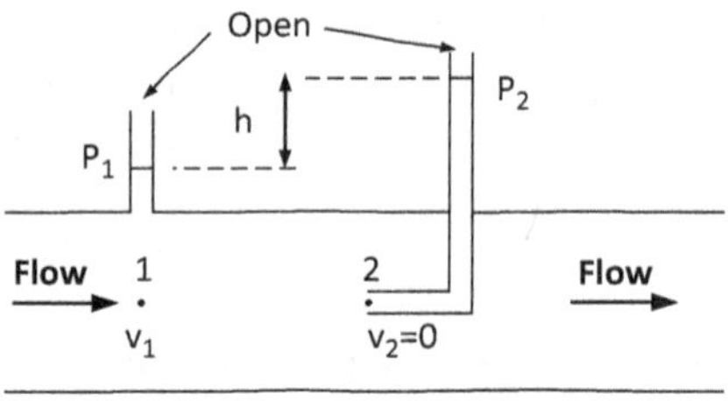

FIGURE P9.1 Pitot tube arrangement for flow measurement.

Problem 9.7 A Pitot tube arrangement, shown in Figure P9.1, is used to measure the speed of water and the volumetric flow rate through a pipe.

The cross-section areas of the pipe at points 1 and 2 are A_1 and A_2, which are equal. The fluid velocities at points 1 and 2 are v_1 and v_2, respectively. However, $v_2 = 0$ since point 2 is a stagnation point. In addition, the pressures at points 1 and 2 are P_1 and P_2, respectively. Assuming that the fluid is incompressible and the fluid density is ρ,

i. Use the Bernoulli equation to show that $v_1 = \sqrt{2g(h)}$,
ii. If $A_1 = 500\,\text{cm}^2$ and the fluid density is $\rho = 997.77\,\text{kg/m}^3$, determine the fluid velocity and volumetric flow rate when h varies from 20 to 100 cm.
iii. When h is 50 cm, what is the mass flow rate?

BIBLIOGRAPHY

Alexander, Charles Alexander and Matthew Sadiku, *Fundamentals of Electric Circuit, ISF*, 7th Edition, McGraw-Hill Education, 2020.

Attia, John O., *Electronics and Circuit Analysis Using MATLAB*, CRC Press, 2nd Edition, 2004.

Baker, Roger C., *Flow Measurement Handbook: Industrial Designs, Operating Principles, Performance, and Applications*, Second Edition, Cambridge University Press, 2016.

Bekraoui, Amina, "An Overview of Thermal Mass Flowmeters Applicability in Oil and Gas Industry," *Energy Procedia*. Vol.141, pp. 299–303, 2017.

Bekraoui, Amina and Ahmed Hadjadj, "Thermal Flow Sensor used for Thermal Mass Flowmeter," *Microelectronics Journal*, Vol. 103, pp. 1–6, 2020.

Bolton, William, *Instrumentation and Control Systems*, 3rd Edition, Elsevier, 2021.

Chapman, Stephen J., *MATLAB Programming for Engineers*, 7th Edition, Cengage Learning, May 16, 2024.

Chen, Hang, Lipin Li, Pei Yang and Hongzhen Han, "Design of an Ultrasonic Doppler Low Flow Measurement System," *10th International Conference on Intelligent Computing and Signal Processing (ICSP)*, Xi'an, China, pp. 153–156, 2025.

Etter, Delores, *Engineering Problem Solving with MATLAB*, 3rd Edition, Pearson, 2021.

Fedoryshyn, Roman, Fedir Matiko, Ihor Kostyk, Yevhen Pistun, "Investigation of Differential Pressure Flowmeters Accuracy in Pulsating Flows," *Energy Engineering and Control Systems*, Vol. 2, Issue 1, pp. 9–18, 2016.

Fukushima, Yukito, Manami Taie, Tomoaki Kageyama, Takahiro Tani, Yuki Kajikawa, and Masamitsu Kuroiwa, "Underground Water Flow Speed and Direction Measurement Using Differential Pressure Sensors," *Proceedings of the 15th Annual IEEE International Conference on Nano/Micro Engineered and Molecular Systems*, San Diego, CA, pp. 14–17, 2020.

Ida, Nathan, *Sensors, Actuators, and Their Interfaces: A Multidisciplinary Introduction*, 2nd Edition, Institution of Engineering and Technology, 2020.

Jaeger, Richard C. and Travis Blalock, *Microelectronic Circuit Design (ISE)*, 6th Edition, McGraw Hill, 2022.

Le, Amy, "Ultrasonic Sensing Basics for Liquid Level Sensing, Flow Sensing, and Fluid Identification Applications," Texas Instruments Application Note SNAA220A, March 2015, Revised June 2015.

Li, Bin, Xinghang Fan, Jie Chen and Junfeng Xu, "Study on the Mechanism of Excitation Switching Process in Electromagnetic Flowmeter to Overcome Slurry Noise," *IEEE Sensors Journal*, Vol. 21, Issue 7, pp. 9023–9037, 2021.

Li, Xuejing, Xinhong Yao, Chongyuan Wang, and Liang Zhu, "An Improved Electromagnetic Flowmeter," *Journal of Physics*, Vol. 1584, pp. 1–6, 2020.

Morris, Alan S. and Reza Langari, *Measurement and Instrumentation: Theory and Applications*, 3rd Edition, Academic Press, 2020.

Nazare, Ganesh, Swapnil Hatte, Amit Gavali, and Santosh Rajgade, "Design and Fabrication of Electromagnetic Flow Meter," *International Journal of Science Technology & Engineering*, Vol. 2, Issue 10, pp. 896–898, 2016.

Northrop, Robert B., *Introduction to Instrumentation and Measurements*, 3rd Edition, CRC Press, 2014.

Rajita, Gurindapalli and Nirupama Mandal, "Review on Transit Time Ultrasonic Flowmeter," *2nd International Conference on Control, Instrumentation, Energy & Communication (CIEC)*, Kolkata, India, pp. 88–92, 2016.

Sârbu, Gabriel Constantin, "Modern Water Flowmeters: Differential Pressure Flowmeters," *2016 International Conference and Exposition on Electrical and Power Engineering (EPE)*, Iasi, Romania, pp. 609–616, 2016.

Sedra, Adel S. and Kenneth Smith, *Microelectronic Circuits*, 8th International Edition, Oxford University Press, 2020.

Suryana, Anang and Muchtar Ali Setyo Yudono, "Ultrasonic Sensor for Measurement of Water Flow Rate in Horizontal Pipes Using Segment Area," *Fidelity: Journal Teknik Elektro*, Vol. 5, Issue 1, pp. 60–68, January 2023.

Tanaka, Hiroaki, Minako Terao and Yoshiaki Tanaka, "Thermal Micro Flow Sensor," 2012 *Proceedings of SICE Annual Conference*, Akita, Japan, pp. 10–15. 2012.

Valentine, Daniel T. and Brian D. Hahn, *Essential MATLAB for Engineers and Scientists*, 8th Edition, Academic Press, Sept. 26, 2022.

Velmurugan Ramasamy and P. Rajalakshmy, "Ultrasonic Flowmeter using Cross-Correlation Technique," *International Journal of Computer Applications*, Vol. 66, Issue 10, pp. 19–22, 2013.

Wang, Sijin, Liang Ge, Guiyun Tian, Guohui Wei, Xiaoting Xiao and Mingwei Zou, "Research Progress on Optimization Techniques for Electromagnetic Flowmeters: A Review," *IEEE Sensors Journal*, Vol. 25, Issue 9, pp. 14557–14574, 2025.

10 Electro-Optical Sensors

10.1 INTRODUCTION

Electro-optical sensors (EOS) are devices that are sensitive to the electromagnetic radiation in the infrared (IR), visible, and ultraviolet regions of the spectrum. EOS convert visible light or IR or ultraviolet radiation into electronic signals. The EOS have many applications. Examples include flash detection, lights that automatically switch on in response to darkness, and position sensors that get activated from the light beam that gets interrupted by an object. Optical sensing is based on two principles: (i) quantum effects of radiation and (ii) thermal effects of radiation.

In quantum effects of radiation, detectors depend on a photon of light interacting with atoms of materials to release electrons, provided that the photon has enough energy to free electrons from covalent bonds of the material. The quantum theory of energy says that energy exists in packets of specific energy levels and sufficient energy is needed to free electrons from covalent bonds. The energy of each photon, E_p, is given by Planck's equation:

$$E_p = hf \tag{10.1}$$

and

$$f = \left(\frac{c}{\lambda}\right) \tag{10.2}$$

where

E_p is the energy in Joules.
E_p can also be expressed as electron-volt (eV), which is equal to 1.602×10^{-19} J.
c is the velocity of light (3×10^8 m/s).
λ is the wavelength in meters.
h is Planck's constant (6.62×10^{-34} Js).
f is the frequency of light in Hertz.

Nearby all optical sensors require photons to have enough energy to free electrons from their chemical bonds. Most chemical bond energies range from 0.2 to 4 eV, thus reflecting the minimum photon energy needed to break chemical bonds in that range. Materials in which the electrons are too tightly bound to the atom to prevent light photon from releasing electrons from the atoms will not work well as EOS. Photon sensors include such detectors as the following: (i) photon-conductive cells, (ii) photovoltaic cells, (iii) photo-emissive cells, (iv) photodiodes, and (v) phototransistors

DOI: 10.1201/9781003765554-10

In the thermal effects of radiation, the absorbed electromagnetic radiation is converted into heat. Thermal detectors exploit the fact that when matter absorbs electromagnetic radiation, it gets heated. The measured increase in temperature is indicative of the absorbed radiation power. Thermal detectors include bolometer, thermopile, and pyroelectric detectors.

10.2 PHOTOCONDUCTIVE CELLS

Photoconductive detectors are devices whose electrical conductance increases with light intensity. Photoconductive detectors are also called photo-resistors. Photoconductive detectors can be built using silicon, germanium, selenium, and cadmium sulfide (CdS). When photons hit a light-sensitive semiconductor, photon energy breaks covalent bonds and electron–hole pairs are produced. Increased illumination leads to the generation of more free electrons and hence decreases in the resistance of the photoconductive cell. This increases the amount of current flowing through the semiconductor, which is then measured. The photoconductive cell acts like a light-sensitive device in which thc internal resistance changes with light intensity. Figure 10. 1 shows the electronic circuit symbol of a photoconductive detector.

The resistance of photoconductive detectors might change from over 1 MΩ in a dark environment to under 10 Ω in bright light. Photoconductive devices are linear at low levels of illumination but become nonlinear at high levels of illumination.

Photoconductive detectors are made from semiconductor materials, normally Pb and Cd sulfides or selenides. CdS and CdSe are the most popular photoconductive detectors. Photoconductive cells are used in photographic exposure meters, densitometers, colorimeters, and other applications where we want a resistance to decrease with an increase in the light intensity. The change in resistance with light is converted to voltage variation with light using a Wheatstone bridge circuit, shown in Figure 10.2.

Example 10.1: Characteristics of a Photoconductive Cell

Figure 10.3 shows a circuit used to obtain the characteristics of a photoconductor cell. The resistance of the photoconductive cell decreases with increase in

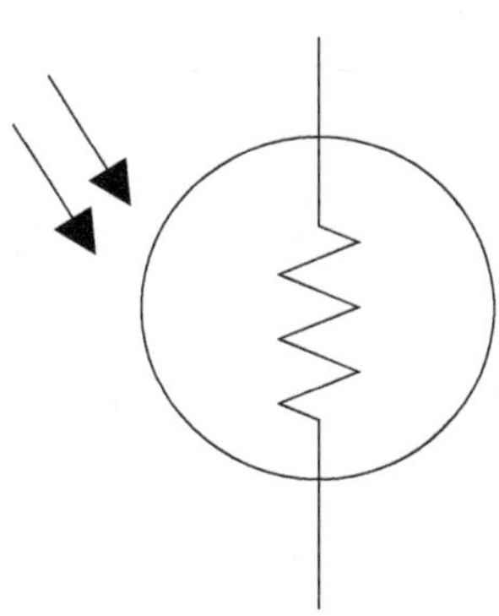

FIGURE 10.1 Electronic circuit symbol of the photoconductive cell.

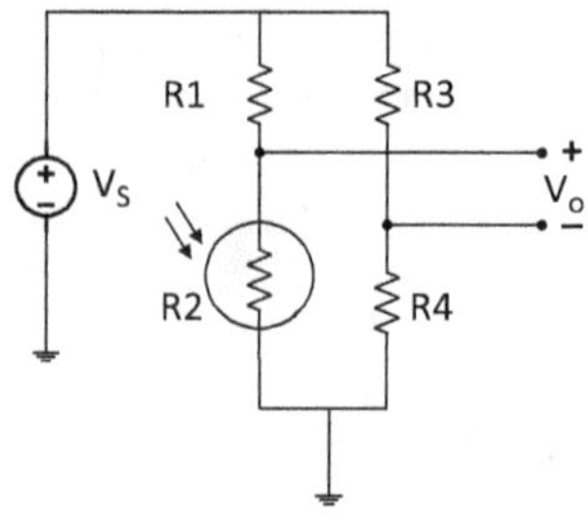

FIGURE 10.2 Wheatstone bridge circuit with the photoconductive cell.

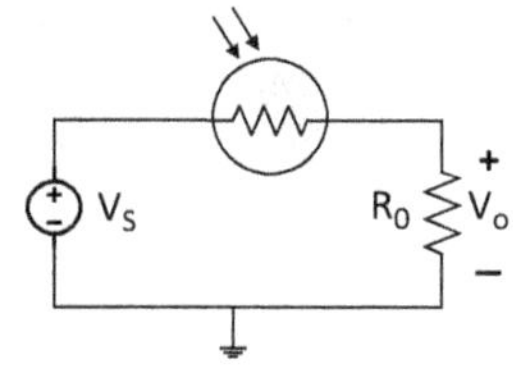

FIGURE 10.3 Photoconductive sensor circuit.

TABLE 10.1
Resistance of a Photoconductive Sensor versus Light Intensity

Light Intensity (μW/cm²)	Resistance of the Photoconductive Cell, R_C (Ω)
1	2700
10	1500
100	600
500	150
1000	50

illumination, as shown in Table 10.1. Use MATLAB to plot the voltage across the 10-K resistor as a function of illumination. Assume that $R_o = 1000\ \Omega$ and $V_S = 5\ \text{V}$

Solution

Using voltage division, the output voltage is given as follows:

$$V_o = \frac{R_o}{R_0 + R_c} V_S$$

With $R_o = 1000\ \Omega$ and $V_S = 5\ \text{V}$, we use MATLAB to plot output voltage versus light intensity.

MATLAB® CODE

```
% Solution to Example 10.1
% Output voltage versus light intensity
%
vs = 5; % source voltage
ro = 1000; % output resistance
li = [1e-6 10e-6 100e-6 500e-6 1000e-6]; % li is light
intensities
rc = [2700 1500 600 150 50]; % resistance of photoconductive
detector
len = length(li);
% vo is the output voltage
for i=1:len
    vo(i) = ro*vs/(ro+rc(i));
    end
figure (1)
semilogx(li,rc,li,rc,'*')
title('Photoconductor Resistance vs Light Intensity')
xlabel('Light Intensity, W/sq. cm')
ylabel('Resistance, Ohms')
figure (2)
semilogx(li,vo,li,vo,'*')
title('Output Voltage versus Light Intensity')
xlabel('Light Intensity, W/sq. cm')
ylabel('Output Voltage, V')
```

Figure 10.4 shows a graph of the photoconductive cell resistance versus light intensity, and Figure 10.5 shows the output voltage versus light intensity.

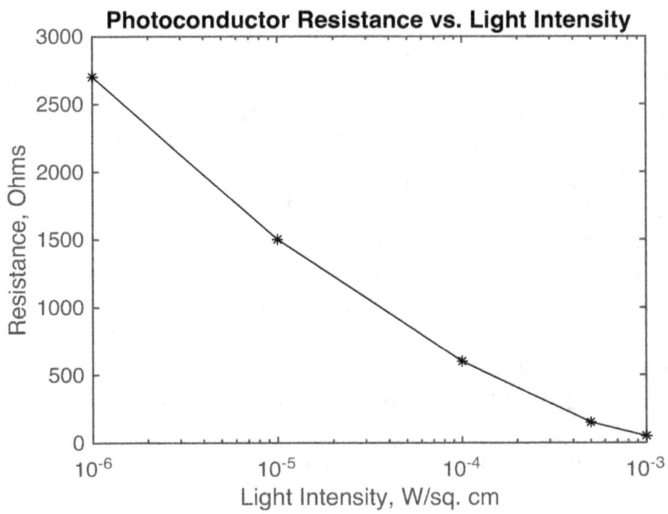

FIGURE 10.4 Photoconductor resistance versus light intensity.

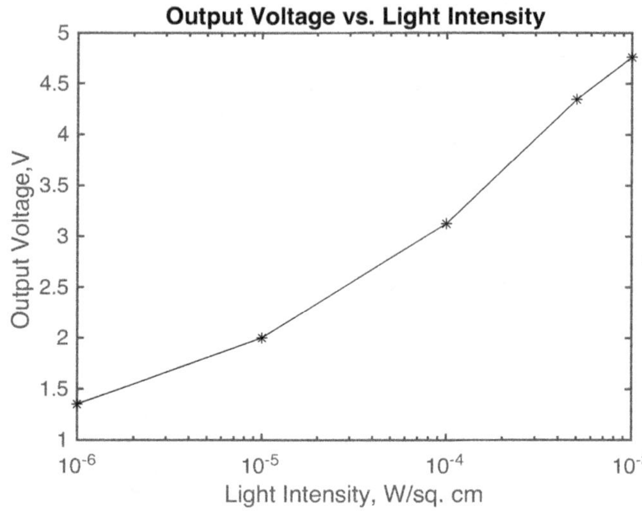

FIGURE 10.5 Output voltage versus light intensity.

10.3 PHOTOVOLTAIC CELLS

A photovoltaic cell consists a thin layer of transparent metal which is normally placed on top of the light-sensitive material. When the light-sensitive material is exposed to light, free electrons are removed from the material, and the electrons may move to the dissimilar material. The most popular photovoltaic cells are made of silicon. Figure 10.6 shows the structure of a silicon photovoltaic cell. The silicon photovoltaic cell consists of a PN junction of P- and N-type silicon. When light illuminates the cell, as a result of photon interactions, hole–electron pairs are generated in the P-device. The holes are attracted to the N-type silicon. The electrode, connected to the annular ring deposited on the exposed surface of the P-type silicon, becomes positive. The electrode at the metal base plate becomes negative. The solar cell is an example of a photovoltaic cell. However, solar cells have a large area.

A simplified equivalent circuit of a photovoltaic cell is shown in Figure 10.7. I_{ph} is the photocurrent generated that is proportional to light intensity. There is a pn junction that produces the current I_{diode}. R_p is parallel resistance owing to the leakage current passing through the p-n junction. R_s is a series resistance due to the combined resistance of contacts, metal grids, and p and n layers. When no resistance is connected to the solar cell, the voltage at the output is termed the open-circuit voltage, V_{oc}.

The current I_L, shown in Figure 10.7, can be written as follows:

$$I_L = I_{\text{ph}} - I_{\text{diode}} - I_p = I_{\text{ph}} - I_0\left(e^{q(V+I_L R_s)/(nKT)} - 1\right) - \frac{V + I_L R_s}{R_p} \tag{10.3}$$

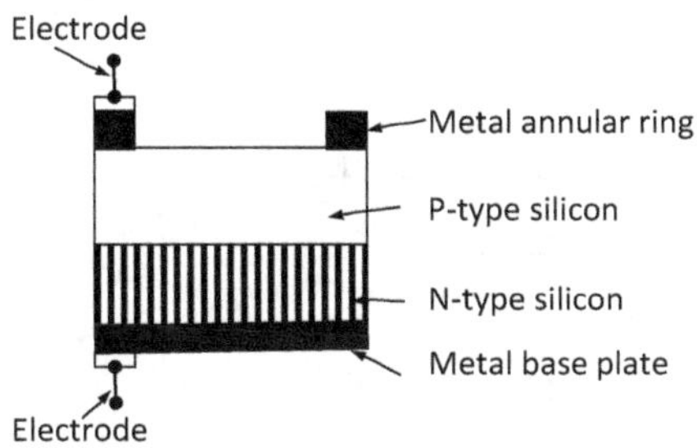

FIGURE 10.6 Silicon photovoltaic cell.

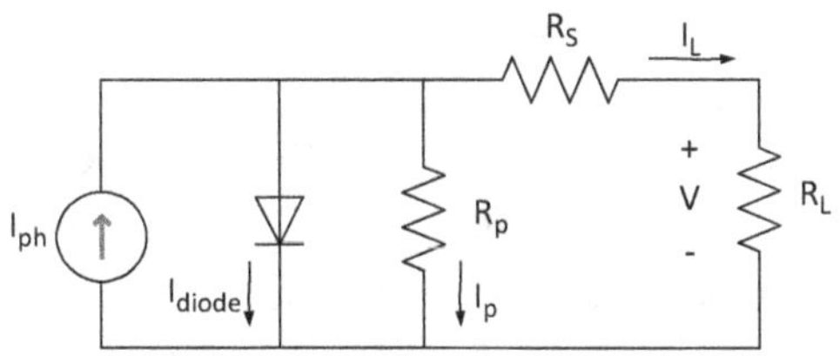

FIGURE 10.7 Simplified equivalent circuit of a silicon photovoltaic cell.

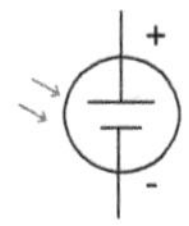

FIGURE 10.8 Electronic circuit symbol of a photovoltaic cell.

where

- V is the voltage across the diode.
- I_0 is the reverse saturation current of the diode.
- k is Boltzmann's constant.
- q is electronic charge.

The electronic circuit symbol of a photovoltaic cell is shown in Figure 10.8.

A photovoltaic cell can be used for light intensity measurements since its short-circuit current is proportional to the light intensity. For instrumentation applications, the cell is connected to a high-impedance amplifier, such as a noninverting amplifier, as shown in Figure 10.9. The op amp provides a buffer between the photovoltaic cell and any circuitry connected to it.

If V_{OC} is the open-circuit voltage of the photovoltaic cell, then the output voltage in Figure 10.9 is given as

$$V_O = V_{OC}\left(1 + \frac{R_2}{R_1}\right) \tag{10.4}$$

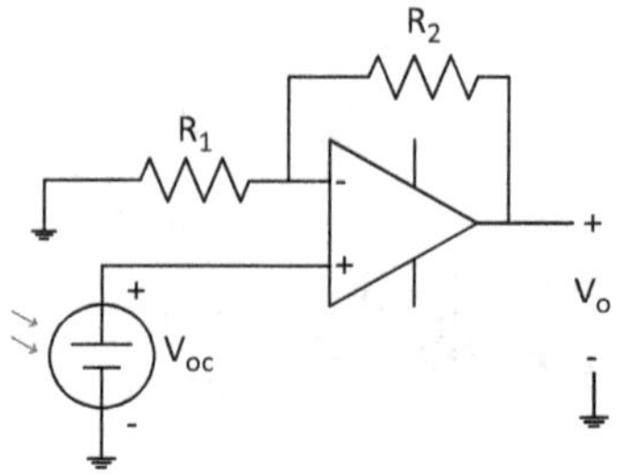

FIGURE 10.9 Photovoltaic cell with a buffer amplifier.

TABLE 10.2
Open-Circuit Voltage of a Photovoltaic Cell versus Light Intensity

Light Intensity (W/m^2)	Open-Circuit Voltage (V)
10	0.660
50	0.700
100	0.725
200	0.745
500	0.770
1000	0.790

Example 10.2: Intensity of Light on a Photovoltaic Cell

Figure 10.9 shows a photovoltaic cell connected to a noninverting amplifier. The data for light intensity and the open-circuit voltage are given in Table 10.2. If $R_1 = 1000\ \Omega$ and $R_2 = 6000\ \Omega$, plot the output voltage as a function of the light intensity.

Solution

Using Equation (10.4), the output voltage of Figure 10.9 is obtained with respect to light intensity, via the open-circuit voltage shown in Table 10.2.

MATLAB® CODE

```
% Solution to Example 10.2
% Output voltage versus light intensity of a photovoltaic cell
%
vs = 5; % source voltage
r1 = 1000; % resistance R1
r2 = 6000; % resistance R1
li = [10 40 100 200 500 1000]; % li is light intensities
voc = [0.66 0.7 0.725 0.745 0.77 0.79]; % open cct voltages
len = length(li);
```

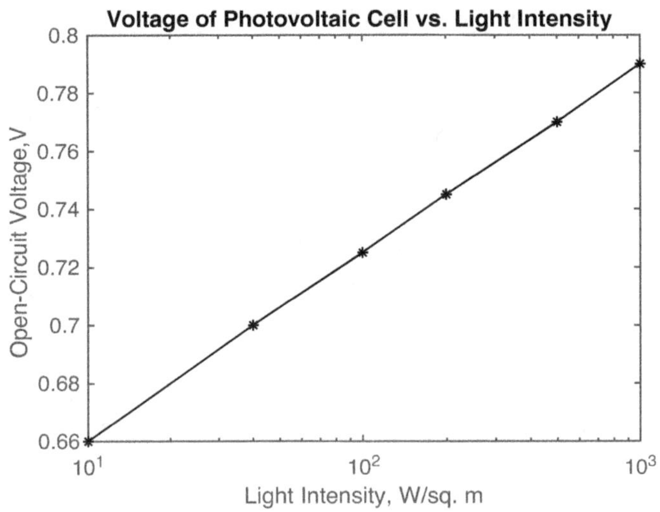

FIGURE 10.10 Open-circuit voltage of a photovoltaic cell versus light intensity.

```
% vo is the output voltage
for i=1:len
    vo(i) = voc(i)*(1+r2/r1);
    end
figure (1)
semilogx(li,voc,li,voc,'*')
title('Open-Circuit Voltage of Photovoltaic Cell vs Light
Intensity')
xlabel('Light Intensity, W/sq. m')
ylabel('Open-Circuit Voltage, V')
figure (2)
semilogx(li,vo,li,vo,'*')
title('Output Voltage of Op Amp versus Light Intensity')
xlabel('Light Intensity, W/sq. m')
ylabel('Output Voltage, V')
```

Figure 10.10 shows a plot of the open-circuit voltage of the photovoltaic cell versus light intensity, and Figure 10.11 shows a plot of output voltage versus light intensity.

10.4 PHOTO-EMISSIVE CELL AND PHOTOMULTIPLIERS

The photo-emissive cell has two electrodes, cathode and anode. The electrodes are housed inside a glass or metal envelope. The glass or metal envelope might have a vacuum or contain small traces of inert gas. The cathode is the photo-emissive surface and is normally coated with light-sensitive materials such as antimony, silver, cesium, or bismuth. The anode might be a thin rod or a loop of wire built using tungsten or platinum. The anode is connected to a positive terminal of a source, such that the electrons released from the cathode will be collected at the anode. Figure 10.12 shows a structure of a photo-emissive cell, and Figure 10.13 shows the electronic circuit symbol of the cell.

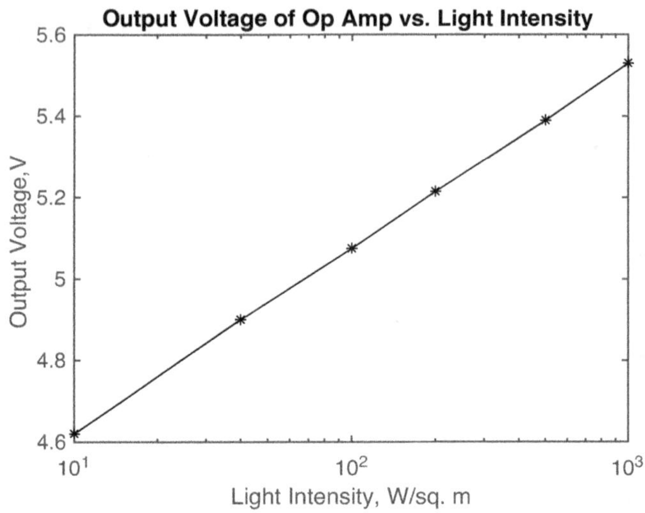

FIGURE 10.11 Output voltage versus light intensity.

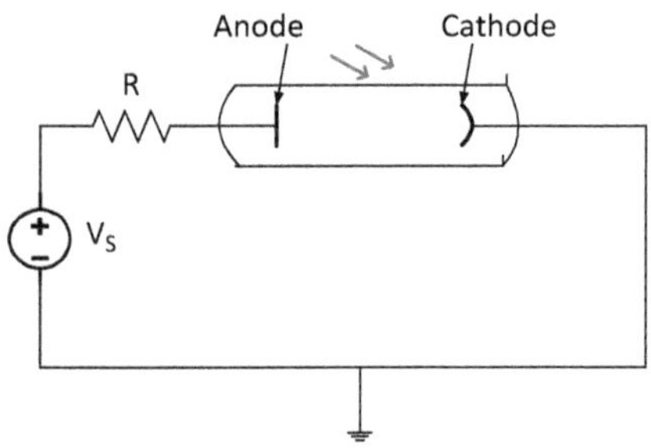

FIGURE 10.12 Photo-emissive cell with a voltage source and series resistor.

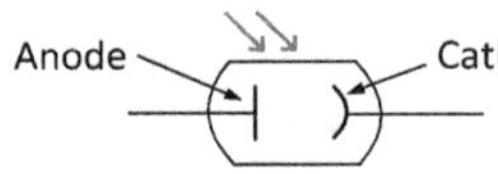

FIGURE 10.13 Electronic circuit symbol of a photo-emissive cell.

Photo-emissive cells operate based on the photoelectric effect phenomenon. When a photon strikes a sensitive surface of a photo-emissive device (normally cathode), an electron is released if its energy E ($= h\upsilon$) exceeds the work function of the sensitive surface of the device. The released electrons are collected by the positively charged anode. The resulting photocurrent is proportional to the number of photons (i.e., to the intensity of light).

A photo-emissive cell with a vacuum has a small output current I_O that is linearly related to the impinging light intensity. The current I_O is normally so small that it cannot be measured directly with an ammeter, so the current is converted to voltage by passing the current through a resistance $R_{O,}$ and the output voltage is obtained through Ohm's Law. A circuit for the voltage measurement is shown in Figure 10.14, and the output voltage $V_O = R_O I_O$.

A gas-filled photo-emissive cell is initially evacuated of air and then refilled with ionizing inert gas, such as argon. The photoelectrons emitted by the cathode create electrons and positive ions when electrons collide with the gas molecules. Because of this secondary emission, the gas-filled photo-emissive cells produce higher output current than the vacuum-type photo-emissive cells.

Current that flows through the photo-emissive devices when there is no light impinging on the photo-emissive surface is called *dark current*. For gas-filled tubes, the dark current is in the order of 10^{-7} to 10^{-8} A, whereas for vacuum-type devices, the dark current is smaller, from 10^{-8} to 10^{-9} A.

At low light intensities, the photo-emissive process is not sensitive. The photo-emissive cell is made more efficient and sensitive by using a photomultiplier tube (PMT). The latter contains a cathode and a number of positively charged anodes, called dynodes. Using a voltage divider network, the dynodes are maintained at successively higher voltages. A PMT is shown in Figure 10.15.

When light falls on the photo-emissive cathode, photoelectrons will be emitted. The electrons are accelerated by a voltage to the first dynode. When the accelerated electrons strike the first dynode, secondary electrons from the dynode surface are dislodged. These electrons are accelerated by a higher voltage and produce the same effect at the second dynode. This process continues through the various dynodes until the electrons are collected by the last dynode. Typically, 7–14 dynodes are used, and gains of 10^5 to 10^6 are common. The sensitivity of the photomultipliers is, therefore, very high. The response time of photomultipliers is also very high.

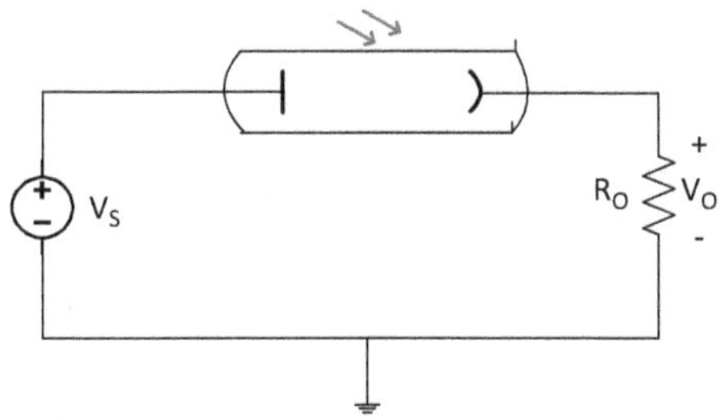

FIGURE 10.14 Measurement of current from a photo-emissive cell.

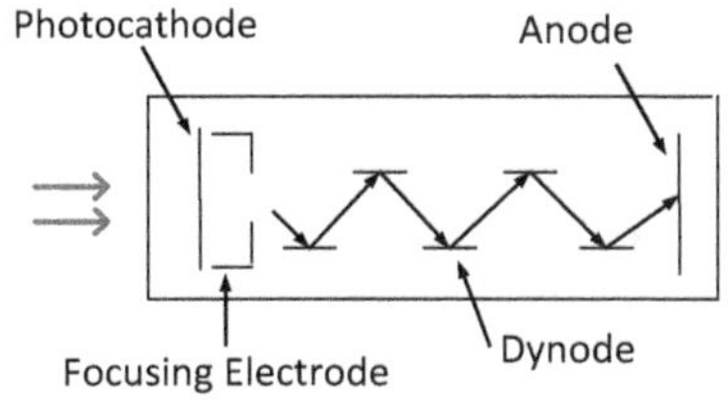

FIGURE 10.15 Simplified diagram of a photomultiplier tube.

TABLE 10.3
Current of a Photo-emissive Cell versus Light Flux

Light Flux (Lumens)	Anode Current (μA)
1.0	29
1.5	43
2.0	57
2.5	71
3.0	86
3.5	100
4.0	114
4.5	129

Example 10.3: Light Intensity on a Photo-Emissive Cell

Figure 10.14 shows a photo-emissive cell connected to a resistor R_L. The data for light flux and the current flowing through the photo-emissive cell are given in Table 10.3. If $R_L = 1.0$ KΩ, (i) determine the expression for the output voltage and (ii) plot the output voltage as a function of the light flux.

Solution

Using Ohm's Law, the output voltage is given as

$$V_O = (R_L)(i_O) \tag{10.5}$$

MATLAB® CODE

```
% Solution to Example 10.3
% Output voltage versus light flux of a photo-emissive cell
circuit
rl = 1000; % resistance RL
lf = [1  1.5  2.0  2.5  3.0  3.5  4.0  4.5]; % lf is light flux
(Lumens)
acur = [29e-6 43e-6 57e-6 71e-6 86e-6 100e-6 114e-6 129e-6];%
anode current
len = length(lf);
% vo is the output voltage
for i=1:len
    vo(i) = rl*acur(i);
    end
figure (1)
plot(lf,acur,lf,acur,'*')
title('Anode Current vs Light Flux')
xlabel('Light Flux, Lumens')
ylabel('Anode Current, A')
figure (2)
plot(lf,vo,lf,vo,'*')
```

```
title('Output Voltage versus Light Flux')
xlabel('Light Flux, Lumens')
ylabel('Output Voltage, V')
```

Figure 10.16 shows a plot of the anode current of the photo-emissive cell versus light flux, and Figure 10.17 shows a plot of output voltage versus light flux.

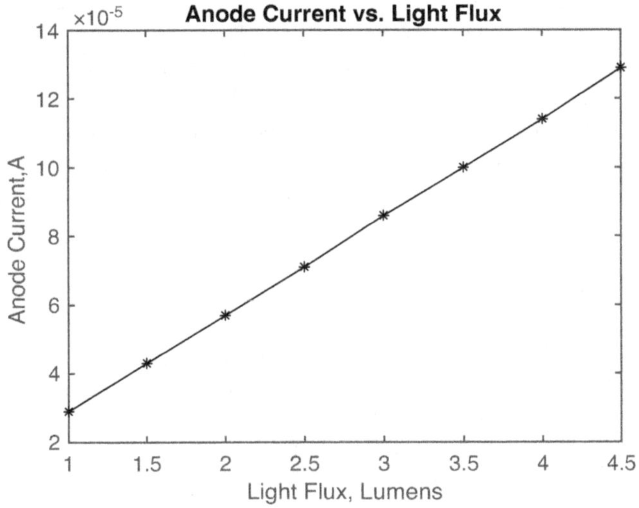

FIGURE 10.16 Anode current of a photo-emissive cell versus light flux.

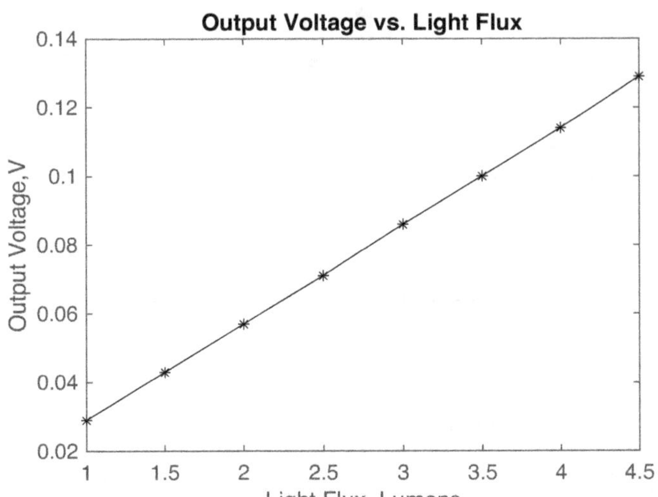

FIGURE 10.17 Output voltage of a photo-emissive cell versus light flux.

10.5 PHOTODIODES

A photodiode is similar to an ordinary pn junction diode, but it generates hole–electron pairs when light or photon is incident on it. The diode can be used to generate solar power, to detect ultraviolet or IR rays, and to perform light measurements. Photodiodes can be constructed using silicon, germanium, or gallium arsenide.

In a pn junction photodiode, light illuminates the N-type region, photons are absorbed into the region, and hole–electron pairs are formed that can move in the structure under the influence of a bias voltage. The pn junction photodiode can be made to operate in the photoconductive mode if the device is reversed-biased or in the photovoltaic mode if no bias is applied. When the diode is reversed-biased (photoconductive mode), a large electric field is created in the depletion region. The field drives the electrons onto the p-type terminal and the holes into the n-type terminal region. The flow of the carriers constitutes a flow of minority carriers, which is termed leakage current or reverse leakage current. The leakage current is normally proportional to the light intensity.

The electronic circuit symbol of a photodiode is shown in Figure 10.18. The equivalent circuit of the photodiode is shown in Figure 10.19. There is a pn junction that produces the current I_{diode}. R_p is parallel resistance due to the leakage current flowing through the p-n junction. R_s is a series resistance due to the combined resistance of contacts, metal grids, and p and n layers.

The current I_L, shown in Figure 10.19, can be written as follows:

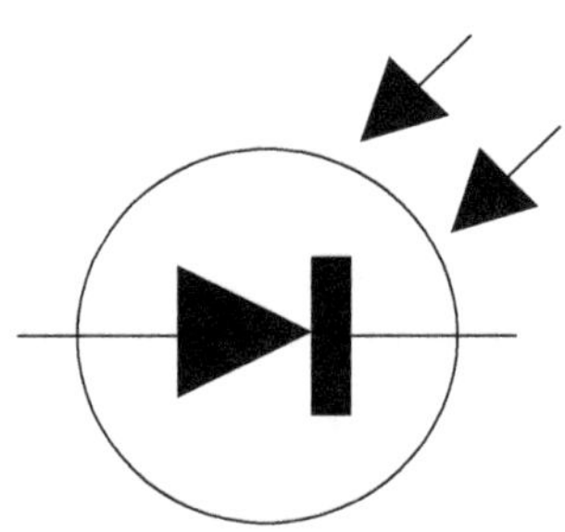

FIGURE 10.18 Electronic circuit symbol of a photodiode.

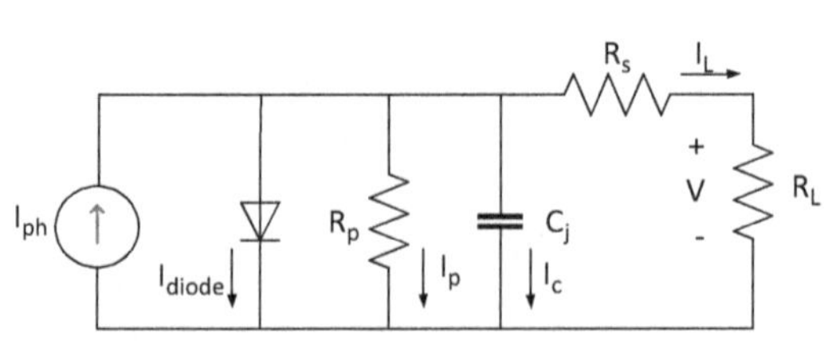

FIGURE 10.19 Equivalent circuit of a photodiode cell.

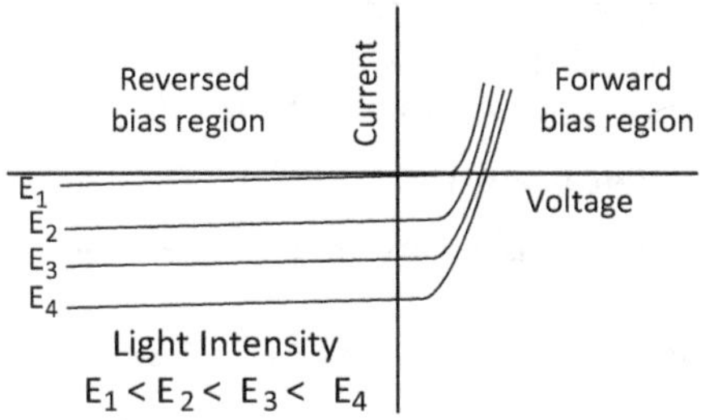

FIGURE 10.20 Current versus voltage characteristics of a photodiode.

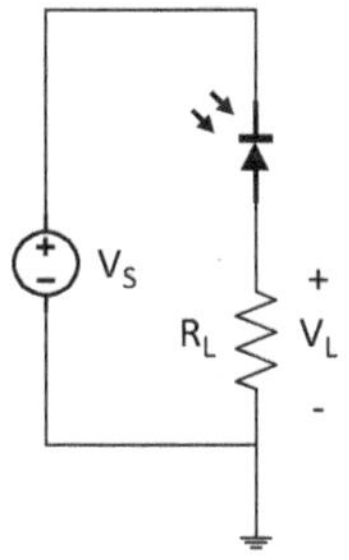

FIGURE 10.21 A photodiode circuit.

$$I_L = I_{ph} - I_{\text{diode}} - I_c - I_p = I_{ph} - I_0\left(e^{q(V+I_L R_s)/(nKT)} - 1\right) - I_c - \frac{V + I_L R_s}{R_p} \quad (10.6)$$

where

V is the voltage across the diode.
I_0 is the reverse saturation current of the diode.
I_c is the current through junction capacitance, $C_{j.}$
I_p is the current flowing through resistance, $R_{p.}$
k is Boltzmann's constant.
q is the electronic charge.

A typical current versus voltage characteristic of a photodiode is shown in Figure 10.20. To find the effect of light on the photodiode, Figure 10.21 can be used.

Example 10.4: Light Intensity on a Photodiode

Figure 10.21 shows a photodiode connected to a resistor R_L. The data for light intensity and current flowing through a reversed-biased photodiode are given in Table 10.4. If $R_L = 1.0$ KΩ and $V_S = 5$ V, (i) determine the expression for the output voltage and (ii) plot the output voltage as a function of the light intensity.

TABLE 10.4
Current of a Photodiode versus Light Intensity

Light Intensity (mW/cm²)	Current (μA)
2	−100
4	−200
6	−300
8	−400
10	−500

Solution

Using Ohm's Law, the output voltage is given as

$$V_o = (R_L)(i_{\text{diode}}) \tag{10.7}$$

MATLAB® CODE

```
% Solution to Example 10.4
% Output voltage versus light intensity of photodiode circuit
%
rl = 1000; % resistance RL
li = [2 4 6 8 10]; % li is light intensity in mW/sq. cm
pcur = [100e-6  200e-6  300e-6  400e-6  500e-6];% photodiode
current
len = length(li);
% vo is the output voltage
for i=1:len
    vo(i) = rl*pcur(i);
    end
figure (1)
plot(li,pcur,'k', li,pcur,'k*')
title('Photodiode Current versus Light Intensity')
xlabel('Light Intensity in mW/sq. cm')
ylabel('Photodiode Current, A')
figure (2)
plot(li,vo,'k', li,vo,'k*')
title('Output Voltage versus Light Intensity')
xlabel('Light Intensity in mW/sq. cm')
ylabel('Output Voltage, V')
```

Figure 10.22 shows a plot of the photodiode current versus light intensity, and Figure 10.23 shows a plot of output voltage versus light intensity.

10.6 PHOTOTRANSISTORS

A phototransistor is a combination of a photodiode and a transistor. A bipolar junction phototransistor is a transistor whose base current is supplied through electrons generated by a photodiode. When light illuminates the base of a bipolar junction transistor, photocurrent is generated. The base current from the photocurrent is then

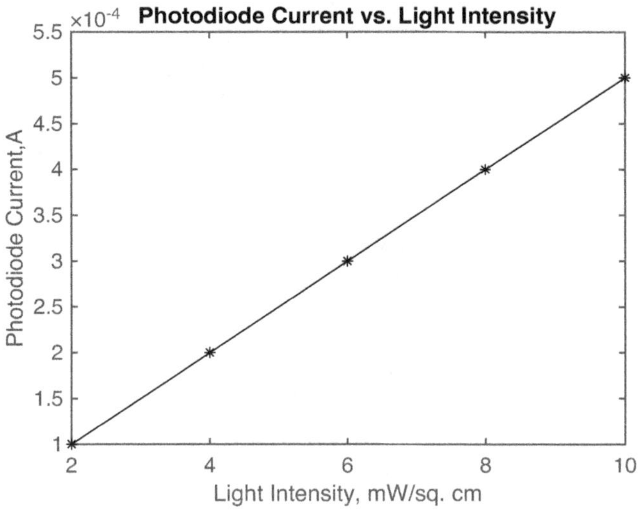

FIGURE 10.22 Photodiode current versus light intensity.

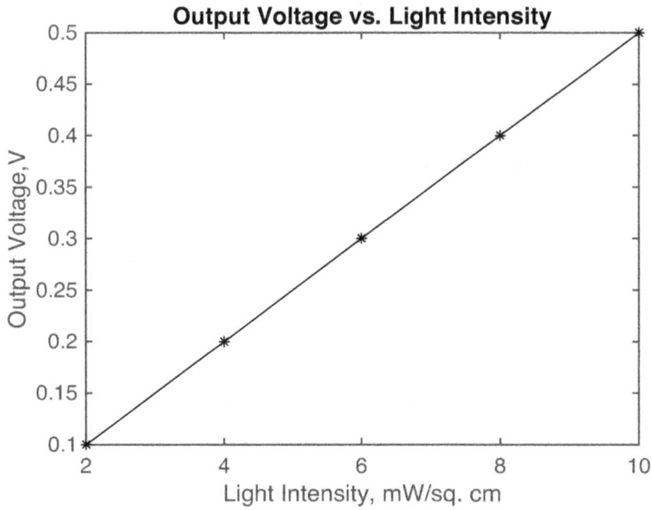

FIGURE 10.23 Output voltage versus light intensity.

amplified by the transistor to yield large collector current. The base region of a phototransistor is normally increased so as to provide an area where incident light can illuminate and generate hole–electron pairs. A side view of an NPN bipolar phototransistor is shown in Figure 10.24. The electronic circuit symbol and equivalent circuit of a phototransistor are shown in Figures 10.25 and 10.26, respectively. A typical current versus voltage characteristic of a phototransistor is shown in Figure 10.27.

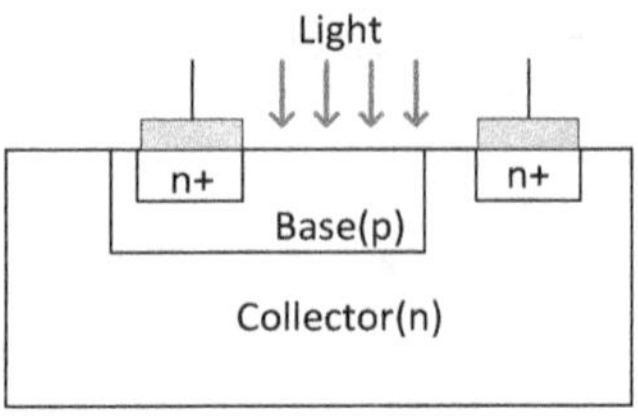

FIGURE 10.24 A side view of an NPN bipolar phototransistor.

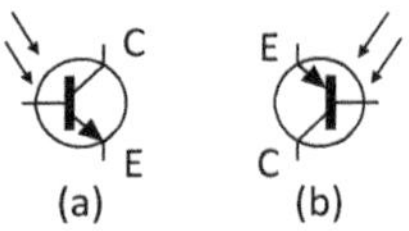

FIGURE 10.25 Electronic circuit symbol of (a) NPN and (b) PNP phototransistors.

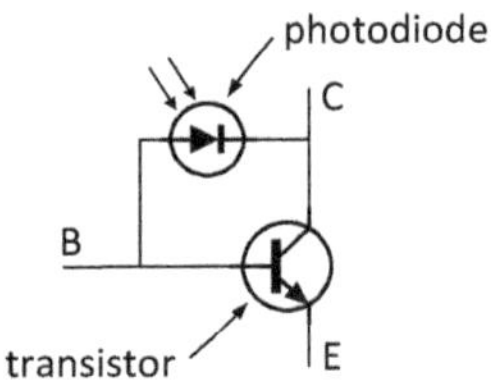

FIGURE 10.26 Equivalent circuit of an NPN phototransistor.

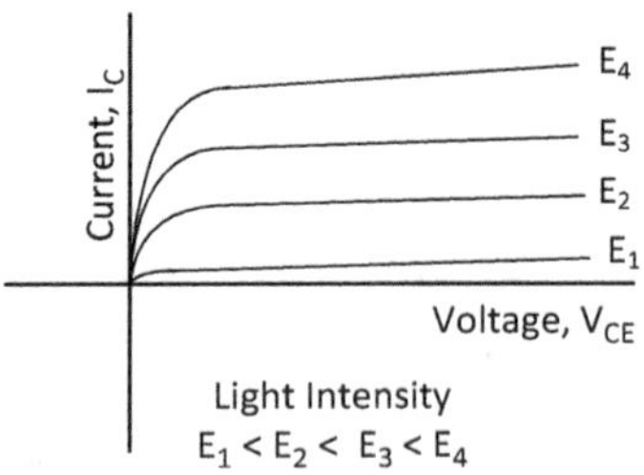

FIGURE 10.27 Current versus voltage characteristics of a phototransistor.

Figure 10.28 shows the electronic circuit symbol of a metal oxide semiconductor (MOS) phototransistor. In the case of the MOS phototransistor, the light illuminates the gate junction of the MOS phototransistor and changes the gate-to-source voltage.

Similar to a typical bipolar junction transistor, a phototransistor suffers from the temperature variation of collector current and leakage current. There is nonlinearity between the collector current and light intensity. This is due to the nonlinear

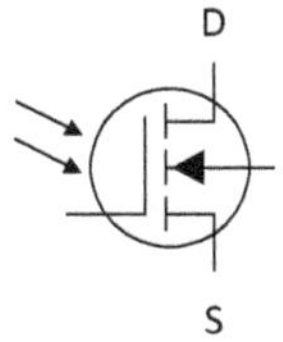

FIGURE 10.28 Electronic circuit symbol of an MOS phototransistor.

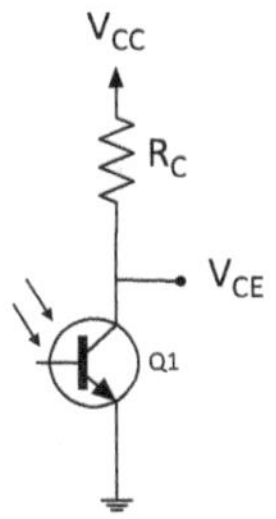

FIGURE 10.29 Common-emitter amplifier with a phototransistor.

relationship between the transistor β and the collector current. Figure 10.29 may be used to determine the effect of light intensity on the operation of the phototransistor.

Example 10.5: Light Intensity on a Phototransistor

Figure 10.29 shows a phototransistor connected to a resistor R_C and voltage source V_{CC}. The data for light intensity and current flowing through the phototransistor are given in Table 10.5. If $R_C = 2.5$ kΩ and $V_{CC} = 5$ V, (i) determine the expression for the output voltage and (ii) plot the output voltage as a function of the light intensity.

Solution

Using Ohm's Law, the output voltage is given as

TABLE 10.5
Current of a Phototransistor versus Light Intensity

Light Intensity (mW/cm²)	Collector Current (mA)
10	0.22
20	0.38
30	0.59
40	0.75
50	0.88

$$V_o = V_{CE} = V_{CC} - I_C R_C \tag{10.8}$$

MATLAB® CODE

```
% Solution to Example 10.5
% Output voltage versus light intensity of phototransistor circuit
%
rc = 2.5e3; % resistance of Rc,
vcc = 5.0; % source voltage
li = [10 20 30 40 50]; % li is light intensity in mW/sq. cm
ccur = [0.18e-3 0.35e-3 0.53e-3 0.70e-3 0.87e-3]; % collector
current
len = length(li);
% vo is the output voltage
for i=1:len
    vo(i) = vcc - rc*ccur(i);
    end
figure (1)
plot(li,ccur,'k', li,ccur,'k*')
title('Phototransistor Collector Current versus Light Intensity')
xlabel('Light Intensity in mW/sq. cm')
ylabel('Phototransistor Collector Current, A')
figure (2)
plot(li,vo,'k', li,vo,'k*')
title('Output Voltage versus Light Intensity')
xlabel('Light Intensity in mW/sq. cm')
ylabel('Output Voltage, V')
```

Figure 10.30 shows the plot of the output voltage versus light intensity of the phototransistor circuit shown in Figure 10.29.

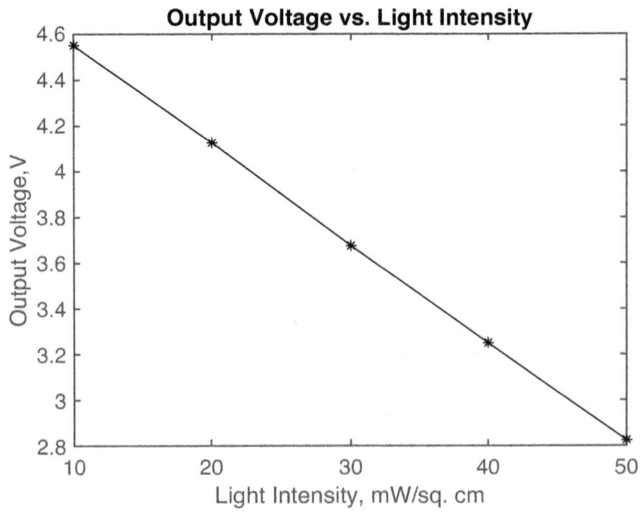

FIGURE 10.30 Output voltage versus light intensity of a phototransistor circuit.

10.7 THERMAL DETECTORS

Thermal detectors respond to IR, light, or ultraviolet radiation. When the detectors absorb electromagnetic radiation, they get heated and convert the electromagnetic radiation into thermal energy. In this section, we shall describe three thermal detectors: (i) bolometers, (ii) thermopiles, and (ii) pyroelectric detectors.

10.7.1 Bolometer

A bolometer consists of two matched elements that are temperature-sensitive. A bolometer can be constructed with metals (such a platinum, germanium, and nickel) or with semiconductor devices (thermistor). Usually, thermistors are used for bolometers. One of the elements is blackened with absorptive coating, and it is illuminated. The other is shielded from light illumination. Two matched elements are connected to a bridge circuit. This is shown in Figure 10.31. R_{T1} is thermistor with a heat-absorbing coating. R_{T2} is a thermistor shielded from radiation. With incident electromagnetic radiation, R_{T1} undergoes a change in resistivity, leading to imbalance of the bridge circuit. The element R_{T2} is needed to compensate for Joule heating (I^2R) of the elements and drift. A bolometer is used to measure IR radiation.

Example 10.6: Bridge Circuit with a Bolometer for Power Measurements

Figure 10.31 shows a bridge circuit with a bolometer. The radio-frequency (RF) power intensity versus resistance of the bolometer is given in Table 10.6. If $R_1 = R_2 = R_{T2} = 9\ \text{k}\Omega$ and $V_S = 5\ \text{V}$, (i) determine the expression for the output voltage and (ii) plot the output voltage as a function of the RF power intensity.

Solution

Using Ohm's Law and Kirchoff's voltage law, the output voltage of the bridge circuit can be shown as follows:

$$V_o = V_S\left(\frac{R_1}{R_1 + R_2} - \frac{R_{T1}}{R_{T1} + R_{T2}}\right) \tag{10.9}$$

Substituting the values of

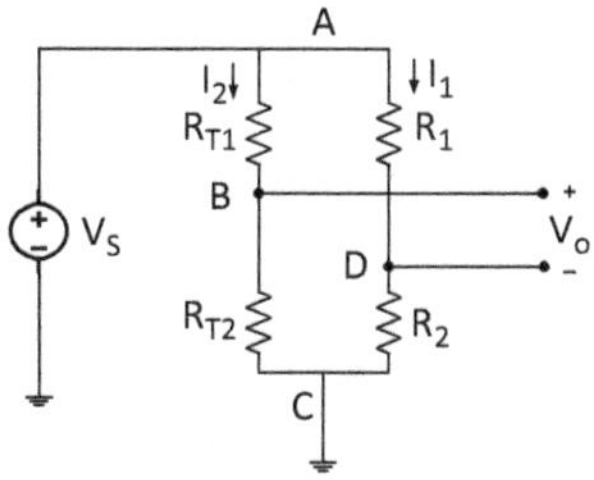

FIGURE 10.31 Bolometer with two matched thermistors R_{T1} and R_{T2}.

TABLE 10.6
RF Power Intensity versus Resistance of a Bolometer

RF Power Intensity (pW/μm²)	Bolometer Resistance R_{T1} (MΩ)
0.1	8.85
0.2	8.55
0.3	8.25
0.4	8.00
0.5	7.80
0.6	7.60
0.7	7.40
0.8	7.20
0.9	7.00
1.0	6.80

$R_1 = R_2 = R_{T2} = 9$ MΩ and $V_S = 5$ V, we have

$$V_o = 5\left(\frac{1}{2} - \frac{R_{T1}}{R_{T1} + 9\ \text{M}\Omega}\right)$$

MATLAB is used to plot the output voltage as the resistance R_{T1} is varied with RF power intensity.

MATLAB® CODE

```
% Solution to Example 10.6
% Output voltage versus RF Power on Bolometer
%
vs = 5.0; % source voltage
% rt1 is the resistance of the bolometer, RT1
rt1 = [8.85e6 8.55e6 8.25e6 8.0e6 7.8e6 7.6e6 7.4e6 7.2e6 7.0e6
6.8e6]; %
% rfpw is the RF Power
rfpw = [0.1e-12 0.2e-12 0.3e-12 0.4e-12 0.5e-12 0.6e-12 0.7e-12
0.8e-12 0.9e-12 1e-12];
len = length(rfpw);
% vo is the output voltage
for i=1:len
    vo(i) = vs/2  - vs*rt1(i)/(rt1(i)+9e6);
end
figure (1)
plot(rfpw,rt1,'k', rfpw,rt1,'k*')
title('Bolometer Resistance versus Power Intensity')
xlabel('Power Intensity in pW/sq. micrometer')
```

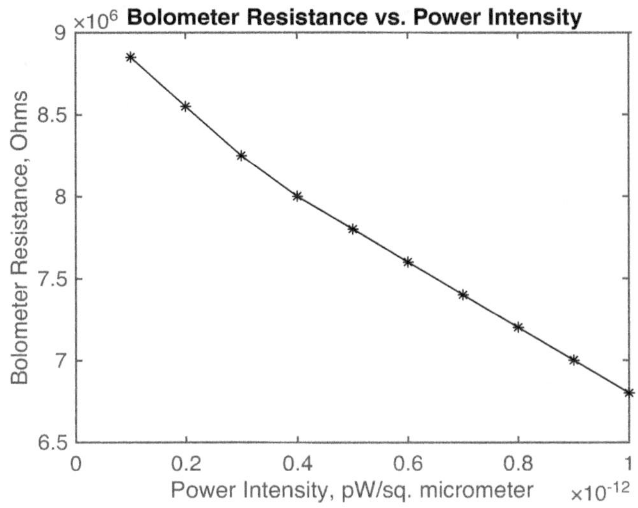

FIGURE 10.32 Bolometer resistance versus power intensity.

```
ylabel('Bolometer Resistance in Ohms')
figure (2)
plot(rfpw,vo,'k', rfpw,vo,'k*')
title('Output Voltage versus Power Intensity')
xlabel('Power Intensity in pW/sq. micrometer')
ylabel('Output Voltage, V')
```

Figure 10.32 shows a plot of the bolometer resistance versus power intensity, and Figure 10.33 shows the bridge output voltage versus the power intensity.

10.7.2 Thermopile

A thermopile converts thermal energy into electrical energy. It is composed of several thermocouples connected in series. Thermopiles generate a larger output voltage compared to a single thermocouple since several thermocouples are connected in series to obtain a thermopile. As mentioned in Section 6.6, a thermocouple is a sensor consisting of two different conductors that produce a voltage proportional to the temperature difference between the two ends of a pair of conductors. Figure 10.34 shows a thermocouple, and Figure 10.35 shows a thermopile with two thermocouples.

If the temperature at the IR absorber is T_X and the reference temperature is T_{ref}, then the voltage generated by the thermocouple is

$$V_0 = S(T_X - T_{\text{ref}}) \tag{10.10}$$

where S is the Seebeck coefficient.

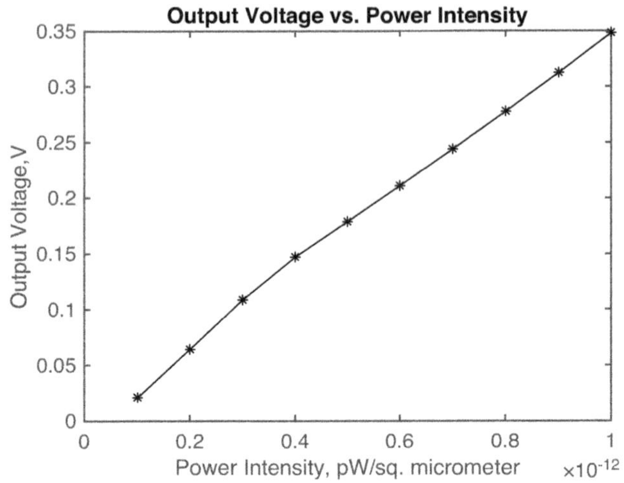

FIGURE 10.33 Bridge output voltage versus power intensity.

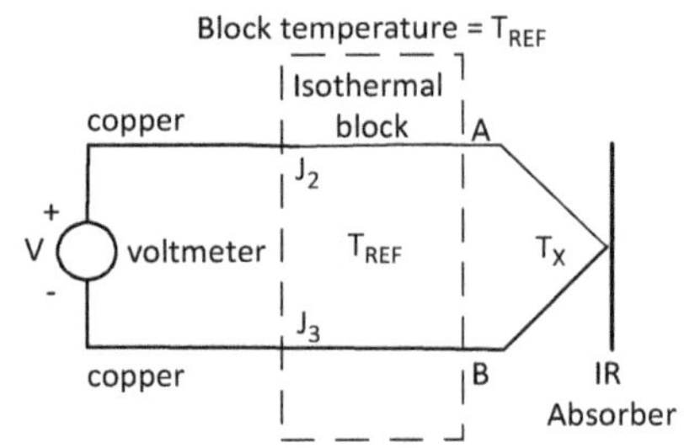

FIGURE 10.34 Thermocouple with an infrared absorber.

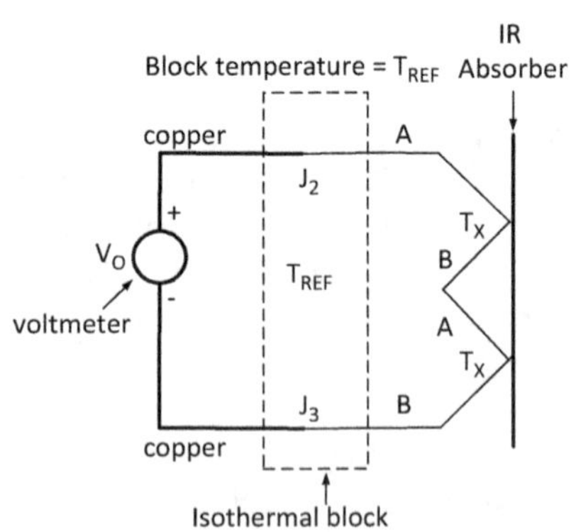

FIGURE 10.35 Thermopile with two thermocouples.

If there are N thermocouples connected in series for a particular thermopile, then the voltage generated by the thermopile is given as

$$V_0 = N.\,S\left(T_X - T_{\text{ref}}\right) \tag{10.11}$$

Example 10.7: Thermopile with Eight T-type Thermocouples

A thermopile is constructed by mounting eight T-type thermocouples connected together in series. If the reference junction is maintained at 0°C and each measuring junction is exposed to an IR radiation of 75°C, (i) determine the output of the circuit, and (ii) if the temperature of the junctions exposed to the IR radiation increased from 75°C to 150°C, plot the output voltage of the circuit with respect to temperature.

Solution

We use the NIST polynomial coefficients, shown in Table 6.8, to calculate the thermocouple voltage generated at temperatures 0°C and 75°C. We multiply the voltage generated by one thermocouple by 8 to get the output voltage of the thermopile consisting of eight thermocouples.

We use the MATLAB® code to do the calculations. The plot of the output voltage of the thermopile versus temperature is shown in Figure 10.36.

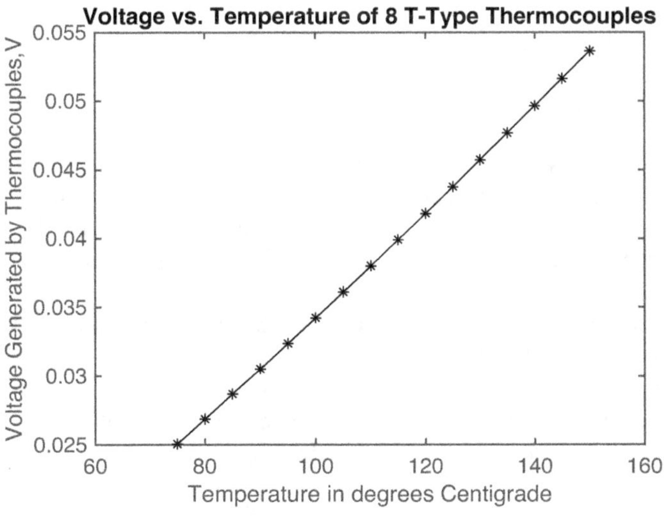

FIGURE 10.36 Voltage versus temperature of eight T-type thermocouples.

MATLAB® CODE

```
% Solution to Example 10.7
% From Table 6.8, all the
% polynomial coefficients for Type T thermocouple is c2
c2 = [-2.75112902e-17 4.54791353e-14 -3.0815759e-11 1.09968809e-8
...
    -2.18822568e-6 2.06182434e-4 3.32922279e-2 38.748106364 0];
% thermocouple voltages at 0 and 75 degrees centigrade are
calculated
b0 = polyval(c2,0)*1.0e-6
b75 = polyval(c2,75)*1.0e-6
% Thermocouple voltage for temperature difference between 75 and
0 Cent.
vb= b75 - b0;
vb8 = 8*vb; % voltage for the 8-T thermocouples
fprintf('8 T-Type thermocouple voltage using all poly coeff. is
%9.3e volts \n',vb8)
T = 75:5:150;
len = length(T);
for i =1:len
    v(i) = 8*polyval(c2, T(i))*1.0e-6;
end
plot(T,v,'k', T,v,'k*')
title('Voltage vs. Temperature of 8 T-Type Thermocouples')
xlabel('Temperature in degrees Centigrade')
ylabel('Voltage Generated by Thermocouples')
```

Answer

The eight T-Type thermocouple voltage using all poly coefficients is 2.506e-02 V.

10.7.3 Pyroelectric Detector

Pyroelectric detectors operate on the principle of pyroelectric effect, whereby a change in the surface charge occurs with a change in the temperature of the material. Whereas the pyroelectric effect creates a transient voltage when the temperature changes, in thermoelectric effect, a permanent voltage is created when there is a temperature gradient.

A simplified structure of a pyroelectric detector is shown in Figure 10.37. It consists of two electrodes. Sandwich between the electrodes is a pyroelectric material which may be terbium gallium garnet (TGG) or lead titanate zirconate (PZT). An absorbing layer (black layer) is applied to the upper electrode. When an IR radiation impinges of the black coating, the pyroelectric layer heats up and creates a surface

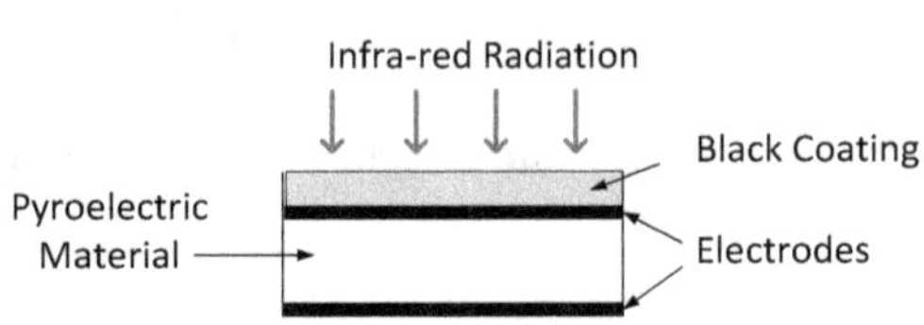

FIGURE 10.37 Simplified structure of a pyroelectric detector.

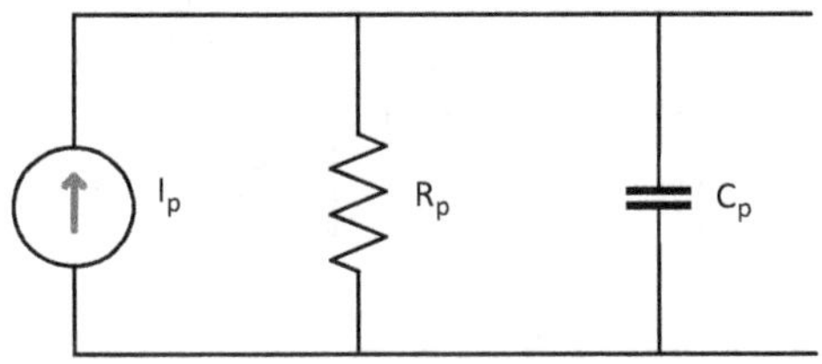

FIGURE 10.38 Equivalent circuit of a pyroelectric detector.

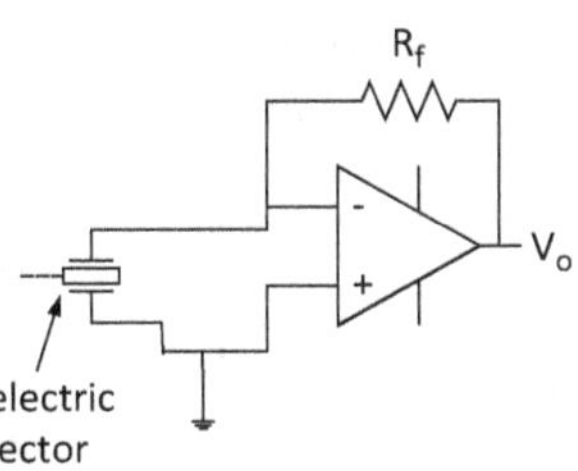

FIGURE 10.39 Current-to-voltage converter circuit for a pyroelectric detector.

charge and a transient voltage. An op amp circuit, connected to the detector, converts the surface charge to signal voltage.

Figure 10.38 shows the equivalent circuit of a pyroelectric detector. I_p is the current produced by the pyroelectric detector when radiation impinges on it. The pyroelectric detector is essentially capacitive, and C_p is the capacitance of the detector. R_p is the shunt resistance of the detector. Pyroelectric detectors respond to changes in the signal level, but not to DC. The time constant of pyroelectric sensors is in the picoseconds range.

A circuit that can be used to convert the pyroelectric signal to output voltage is a current-to-voltage converter shown in Figure 10.39.

PROBLEMS

Problem 10.1 For the inverting amplifier shown in Figure P10.1, (i) determine the expression for the output voltage; (ii) Table P10.1 shows the resistance of the photoconductive detector R_p as a function of light intensity. Assuming that $V_S = 5\,\text{V}$ and $R_2 = 1.0\ \text{k}\Omega$, Plot the output voltage as a function of the light intensity.

Problem 10.2 For the bridge circuit shown in Figure P10.2, (i) determine the expression for the output voltage; (ii) Table P10.2 shows the resistance of the photoconductive detector as a function of light intensity. Assuming that $V_S = 5\,\text{V}$ and $R_1 = R_2 = R_3 = 5.0\ \text{k}\Omega$, plot the output voltage as a function of the light intensity.

Problem 10.3 For the bridge circuit shown in Figure P10.3, (i) determine the expression for the output voltage; (ii) Table P10.3 shows the resistance of

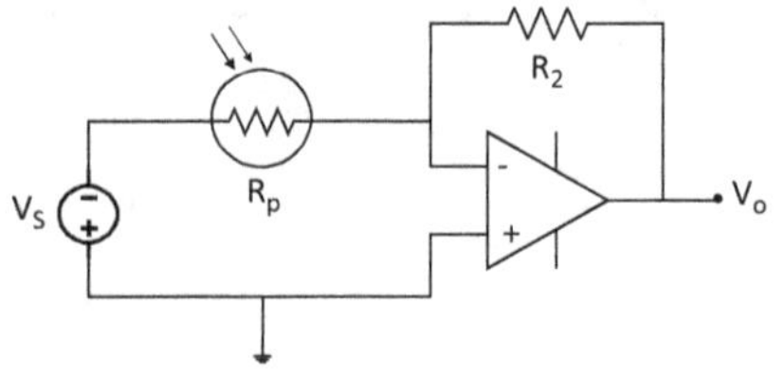

FIGURE P10.1 Inverting amplifier with a photoconductive detector.

TABLE P10.1
Resistance of a Photoconductive Sensor versus Light Intensity

Light Intensity (mW/cm²)	Resistance of a Photoconductive Sensor, R_p (Ω)
10	4000
30	2500
50	1500
70	1000
90	650
110	400

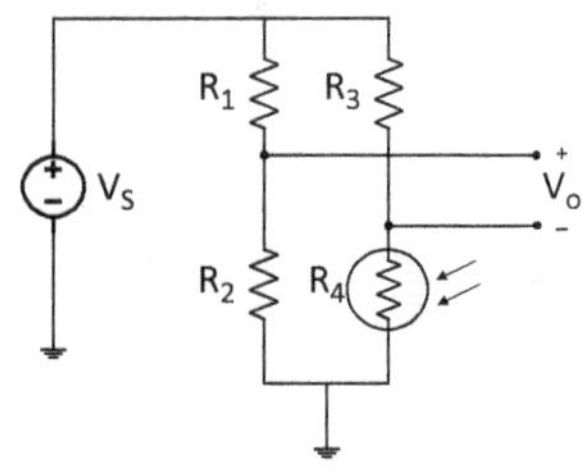

FIGURE P10.2 Wheatstone bridge circuit with a photoconductive sensor.

TABLE P10.2
Resistance of a Photoconductive Sensor versus Light Intensity

Light Intensity (mW/cm²)	Resistance of a Photoconductive Sensor, R_4 (kΩ)
10	11.6
20	8.5
30	6.3
40	4.8
50	3.5
60	2.4
70	1.6
80	1.0

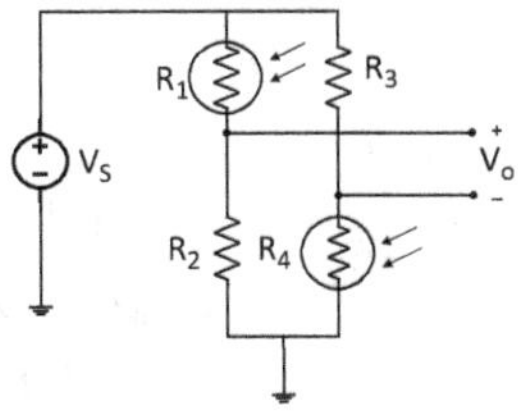

FIGURE P10.3 Wheatstone bridge circuit with a photoconductive sensor with light shown on two detectors.

TABLE P10.3
Resistance of a Photoconductive Sensor versus Light Intensity

Light Intensity (μW/cm²)	Resistance of Photoconductive Sensors, $R_1 = R_4$ (Ω)
0.1	800
1	250
10	60
100	10
1000	3
10000	0.2

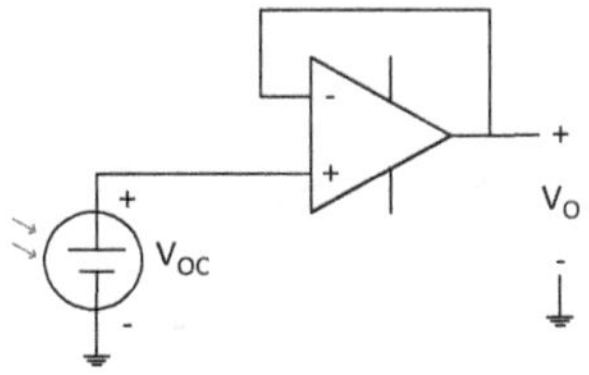

FIGURE P10.4 Photovoltaic cell connected to a buffer amplifier.

the photoconductive detector as a function of light intensity. Assuming that $V_S = 5\,\text{V}$ and $R_2 = R_3 = 250\ \Omega$, plot the output voltage as a function of the light intensity.

Problem 10.4 Data obtained from a light-dependent resistor are shown in Table P10.4. If the relation between resistance R and power P can be expressed as $R = a\left(P^b\right)$, find a and b.

Problem 10.5 Figure P10.4 shows a photovoltaic cell connected to a buffer amplifier. The data for light intensity and the open-circuit voltage, V_{OC}, are given in Table P10.5. (i) Determine the expression for the output voltage, and (ii) plot the output voltage as a function of the light intensity.

Problem 10.6 Figure P10.5 shows a photo-emissive cell connected to a resistor R_L. For an anode voltage of 50 V, the anode current versus light flux of the photo-emissive cell is given in Table P10.6. For $R_L = 1.5\ \text{k}\Omega$, (i) determine

TABLE P10.4
Power versus Resistance of a Light-Dependent Resistor

Power, P (mW)	Resistance of a Light-Dependent Resistor, R (Ω)
0.2	4.12e6
0.4	1.79e6
0.6	1.10e6
0.8	0.78e6
1.0	0.59e6
1.2	0.48e6
1.4	0.39e6
1.6	0.34e6
1.8	0.29e6
2.0	0.26e6

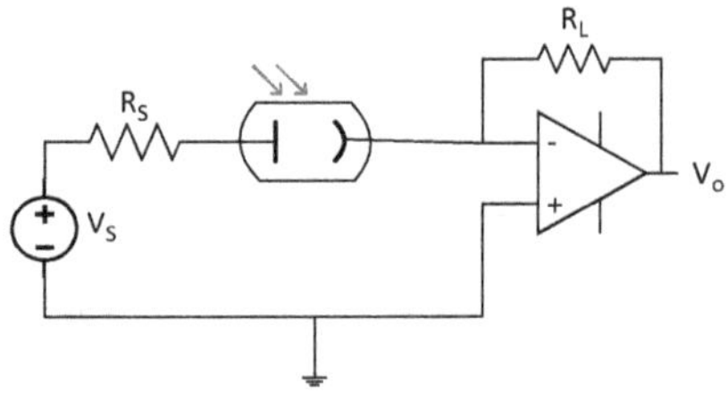

FIGURE P10.5 Photo-emissive cell connected to an operational amplifier circuit.

TABLE P10.5
Open-Circuit Voltage of a Photovoltaic Cell versus Light Intensity

Light Intensity (mW/m^2)	Open-Circuit Voltage, V_{OC} (V)
0.5	0.77
1.0	0.80
2.0	0.82
3.0	0.83
4.0	0.84
5.0	0.85
6.0	0.86
7.0	0.87
8.0	0.88

TABLE P10.6
Current of a Photo-Emissive Cell versus Light Flux

Light Flux (Lumens)	Anode Current (μA)
1.0	43.5
1.5	64.5
2.0	85.5
2.5	106.5
3.0	129.0
3.5	150.0
4.0	171.0
4.5	193.5

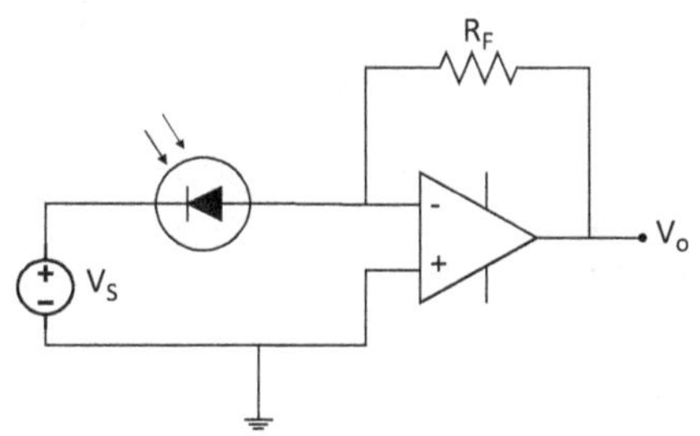

FIGURE P10.6 Circuit of a photodiode with an operational amplifier.

TABLE P10.7
Current of a Photo-Emissive Cell versus Light Intensity

Light Intensity (mW/sq. cm.)	Photodiode Current (μA)
0.5	1.4
1	3.0
2	6.0
5	15.0
10	30.0
20	60.0

the expression for the output voltage, and (ii) plot the output voltage as a function of the light intensity.

Problem 10.7 Figure P10.6 shows a photodiode circuit. The photodiode is reversed-biased, and for a reversed-biased voltage of up to 100 V, the relationship between the reversed-biased photocurrent with respect to light intensity is shown in Table P10.7. For $R_F = 2.5$ kΩ, (i) determine the expression for the output voltage V_O, and (ii) plot the output voltage as a function of the light intensity.

Problem 10.8 For the circuit shown in Figure P10.7, the relationship between the light intensity and the photodiode current is given in Table P10.8. (a) If the photodiode has the maximum light intensity of 20 mW/sq. cm and the operational amplifier saturates at ± 3.3 V, what will be the maximum value of the resistance R_F? (b) With the resistance value calculated in part (a), determine the output voltage when the light intensity changes from 1 to 20 mW/sq. cm.

Problem 10.9 For the phototransistor circuit shown in Figure P10.8, the illumination is varied such that the maximum light intensity on the transistor is 50 mW/sq. cm. If it is known that for each 5 mW/sq. cm. of light intensity, 10 μA of base current is produced, determine the collector-emitter voltage when the light intensity varies from 5 to 50 mW/sq. cm. Assume that β of the transistor is 100, $V_{CC} = 10$ V, and $R_C = 0.5$ kΩ, at the light intensity of 5 mW/sq. cm., the base current is 10 μA.

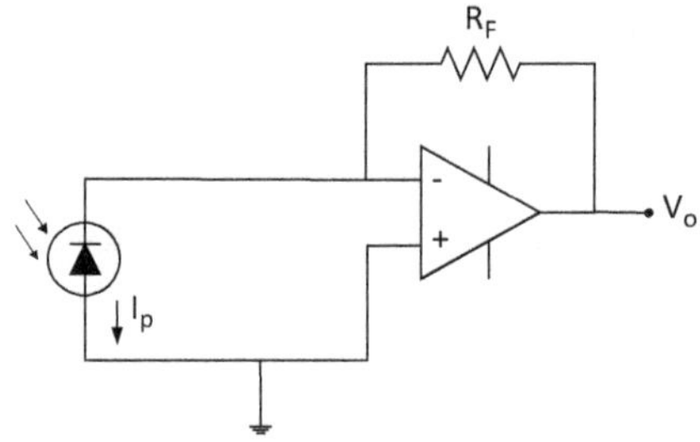

FIGURE P10.7 Photodiode circuit with an unknown feedback resistor.

TABLE P10.8
Current of a Photodiode versus Light Intensity

Light Intensity (mW/sq. cm.)	Photodiode Current, I_p (μA)
1	7.0
2	13.0
5	30.0
10	60.0
20	120.0

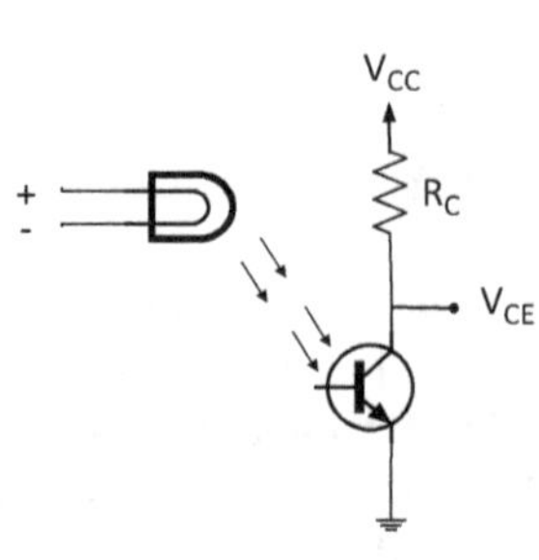

FIGURE P10.8 Phototransistor circuit.

Problem 10.10 A thermopile is constructed by mounting six J-type thermocouples connected together in series. If the reference junction is maintained at 0°C and each measuring junction is exposed to IR radiation of 85°C, (a) determine the output of the circuit, and (b) if the temperature of the junctions exposed to the IR radiation increased from 85°C to 120°C , plot the output voltage of the circuit with respect to temperature.

BIBLIOGRAPHY

Arunachalam, Rammohan and Chidambaram Kumar, "Performance Analysis of Photoresistor and Phototransistor for Automotive's Halogen and Xenon Bulbs Light Output," *IOP Conference Series: Materials Science and Engineering*, Vol. 263, p. 062056, 2017.

Attia, John O., *Electronics and Circuit Analysis Using MATLAB*, CRC Press, 2nd Edition, 2004.

Bhan, Rajinder Kumar, Raghvendra Sahai Saxena, Chhittar Ram Jalwania, and Shashi Kumar Lomash, "Uncooled Infrared Microbolometer Arrays and their Characterization Techniques," *Defence Science Journal*, Vol. 59, Issue 6, pp. 580–589, 2009.

Bolton, William, *Instrumentation and Control Systems*, 3rd Edition, Elsevier, 2021.

Calvi, Sabrina, Matteo Rapisarda, Antonio Valletta, Mattia Scagliotti, Stefania De Rosa, Luca Tortora, Paolo Branchini, Luigi Mariucci, "Highly Sensitive Organic Phototransistor for Flexible Optical Detector Arrays," *Organic Electronics*, Vol. 102, p. 106452, 2022.

Chakraborty, Abhisek, Giulia Lucarelli, Jie Xu, Zeynab Skafi, Sergio Castro-Hermosa, A.B. Kaveramma, R. Geetha Balakrishna, and Thomas M. Brown, "Photovoltaics for Indoor Energy Harvesting," *Nano Energy*, Vol. 128, p. 109932, 2024.

Chapman, Stephen J., *MATLAB Programming for Engineers*, 7th Edition, Cengage Learning, May 16, 2024.

Chen, Lin, Xingchao Wang, Qian Sen, Xiong Zhang, Qi Wu, Lishuang Ma, Pengxiang Zeng, and Chen Sun, "Optimizing the Design of Ultrafast Photomultiplier Tubes," *Optics Express*, Vol. 31, pp. 29975–29985, 2023.

Chen, Xiao, Haoyu Hu, Jiacheng Zhou, Yuan Li, Lei Wan, Zhen Cheng, Junwei Chen, Jun Xu, and Ru Zhou, "Indoor Photovoltaic Materials and Devices for Self-powered Internet of things applications," *Materials Today Energy*, Vol. 44, p. 101621, 2024.

de Haas, Johan and Pieter Dorenbos, "Methods for Accurate Measurement of the Response of Photomultiplier Tubes and Intensity of Light Pulses," *IEEE Transactions on Nuclear Science*, Vol. 58, pp. 1290–1296, 2011.

Dhahi, Th. S., Alaa Kamal Yousif Dafhalla, Omer Elsier Tayfour, Azath Mubarakali, Abdulrahman Saad Alqahtani, Amira Elsir Tayfour Ahmed, Mohamed Elshaikh Elobaid, Tijjani Adam, and Subash C.B. Gopinath, "Advances in Nano Sensors for Monitoring and Optimal Performance Enhancement in Photovoltaic Cells," *iScience*, Vol. 27, Issue 4, p. 109347, 2024.

Etter, Delores, *Engineering Problem Solving with MATLAB*, 3rd Edition, Pearson, 2021.

Fusetto, Samuele, Antonio Aprile, Piero Malcovati, and Edoardo Bonizzoni, "Readout IC Architectures and Strategies for Uncooled Micro-Bolometers Infrared Focal Plane Arrays: A Review," *Sensors*, Vol. 23, p. 2727, 2023.

Ida, Nathan, *Sensors, Actuators, and Their Interfaces: A Multidisciplinary Introduction*, 2nd Edition, Institution of Engineering and Technology, 2020.

Jaeger, Richard C. and Travis Blalock, *Microelectronic Circuit Design (ISE)*, 6th Edition, McGraw Hill, 2022.

Liang, Ting, Yihao Guan, Cheng Lei, Xuezhan Wu, and Yuehang Bai, Jijun Xiong and Lei Qi "Design and Fabrication of a Low-Cost Thermopile Infrared Detector," *Micromachines*, Vol. 12, p. 1134, 2021.

Liu, Xiao-Long, Yang Zhao, Wen-Jing Wang, Sheng-Xiang Ma, Xi-Jing Ning, Li Zhao, and Jun Zhuang, "Photovoltaic Self-Powered Gas Sensing: A Review," *IEEE Sensors Journal*, Vol. 21, Issue 5, pp. 5628–5644, 2021.

Mariya Aleksandrova, Chaitali Jagtap, Vishal Kadam, Sandesh Jadkar, Georgi Kolev, and Krassimir Denishevand Habib Pathan, "An Overview of Microelectronic Infrared Pyroelectric Detector," *Engineered Science*, Vol. 16, pp. 82–89, 2021.

Martín-Luna, Pablo, Daniel Esperante, José Vicente Casaña, Antonio Fernández Prieto, Nuria Fuster-Martínez, Iris García Rivas, Benito Gimeno, Damián Ginestar, Daniel González-Iglesias, José Luis Hueso, Hannah Andrea Leptin, Gabriela Llosá, Pablo Martinez-Reviriego, Jaime Riera, Pablo Vázquez Regueiro, and Fernando Hueso-González, "Effects of the Passive Voltage Divider in a Photomultiplier Tube: Analytical Model, Simulations and Experimental Validation," *Sensors and Actuators A: Physical*, Vol. 381, p. 116057, 2025.

Morris, Alan S. and Reza Langari, *Measurement and Instrumentation: Theory and Applications*, 3rd Edition, Academic Press, 2020.

Neumann, Norbert and Victor Banta, "Comparison of Pyroelectric and Thermopile Detectors," *AMA Conferences 2013, Proceedings IRS 2013*, pp. 139–143, 2013.

Northrop, Robert B., *Introduction to Instrumentation and Measurements*, 3rd Edition, CRC Press, 2014.

Obagade, T. A., "Comparative Analysis of Photoelectric Characteristics of Photosensors: A Case Study between Photoresistor, Photodiode, and Phototransistor," *IOSR Journal of Applied Physics*, Vol. 14, Issue 3, pp. 1–12, 2022.

Priyadarshini, Rashmi, Mrinal Sarvagya, Bonothu Naresh, and Ravi M.G. Kumar, "Design and Development of Readout Electronics for Thermopile Array Detector Using FPGA," *IEEE* 2024 *5th International Conference on Circuits, Control, Communication and Computing (I4C)*, pp. 143–147, 2024.

Sedra, Adel S. and Kenneth Smith, *Microelectronic Circuits*, 8th International Edition, Oxford University Press, 2020.

Sharma, Chandrika, Nidhi Sharma, Poornima Sharma and Veerendra Kumar, "A Review of BJT Based Phototransistor," *International Journal of Engineering Research & Technology (IJERT)*, Vol. 3, Issue 4, pp. 306–310, 2014.

Srivishnu, K.S., Manne Naga Rajesh, Seelam Prasanthkumar and Lingamallu Giribabu, "Photovoltaics for Indoor Applications: Progress, Challenges and Perspectives," *Solar Energy*, Vol. 264, p. 112057, 2023.

Valentine, Daniel T. and Brian D. Hahn, *Essential MATLAB for Engineers and Scientists*, 8th Edition, Academic Press, Sept. 26, 2022.

Wang, Yaoguang, Guofu Cao, Liangjian Wen, and Yifang Wang, "A New Optical Model for Photomultiplier Tubes," *The European Physical Journal C*, Vol. 82, p. 329, 2022.

Whatmore, Roger W. and Samuel J. Ward, "Pyroelectric Infrared Detectors and Materials—A Critical Perspective," *Journal of Applied Physics*, Vol. 133, p. 080902, 2023.

Wright, Anthony G., *The Photomultiplier Handbook*, Oxford University Press, 2017.

11 Sensors for Internet of Things

11.1 INTRODUCTION

Internet of Things (IoT) may be described as physical things with IP addresses that allow the physical things to send or receive data through the internet. It is projected that billion things will be connected to the internet, and they will continue to grow with time. There are several industries that will benefit from the growth of IoT applications. They include, among others, the automotive, retail, healthcare, and agriculture.

In industrial manufacturing plants, sensors can be installed in production lines to detect anomalies in production, reduce machine failures, and cut operating costs. In the automotive industry, data can be collected from sensors in various automotive systems to assist auto-manufacturers to improve their products. In the public sector, IoT sensors may collect data on government-owned utilities and traffic signals and deploy resources to alleviate issues with traffic congestion and utility outages. In agriculture, IoT sensors may be installed to monitor soil conditions. In addition, IoT sensors may be installed to monitor air quality in the environment.

11.2 TYPICAL BLOCK DIAGRAM OF IOT APPLICATIONS

A block diagram of a typical IoT application is shown in Figure 11.1. The sensor block will have sensing devices such as temperature, humidity, and flow sensors, which will sense the physical changes to be measured and convert them to electrical signals.

The controller controls the sensors and provides electrical power to them, if needed. In addition, the controllers will convert analog-to-digital signals. The processor will take the digital signal, remove unwanted data, and process the signal before some action is taken on the data. The controller and processor actions can be performed by a microprocessor.

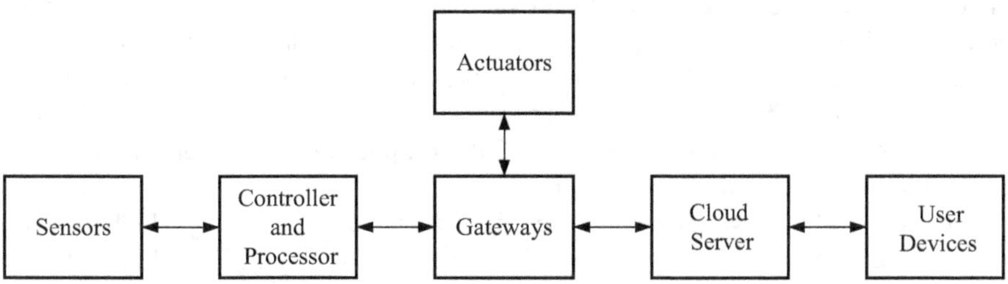

FIGURE 11.1 Block diagram of a typical IoT application.

DOI: 10.1201/9781003765554-11

The gateway routes the data processed by the microcontroller or other processors to the cloud. The gateway can be ESP8266 NodeMCU, an open-source low-cost Wi-Fi module. The ESP8266 can connect to a Wi-Fi network to access the internet, or it can create its own Wi-Fi wireless network to which other devices can connect to. The gateway may also be a local area network, wide-area network, or modem.

The processed data from the microcontroller may be connected to the server through the gateway. Some of the cloud servers may be Amazon Web Services (AWS), Google Server, or Microsoft Azure. The cloud server may have applications for proper utilization of the collected data. Some of the IoT applications may include the following: ThingSpeak, ThingBoard, and Blynk. The analysis of data in the cloud server may provide insights into data, allowing the user to perform some actions based on various information in the data.

The user device is the main device where the user may observe the status of events, read data from IoT sensors, and control events and processes. The user device may be a smartphone, laptop, or desktop computer.

The actuators may be motors, solenoids, relays, and electromagnets. The actuators may change an electrical signal or data into a mechanical motion.

11.3 SENSORS FOR IoT APPLICATIONS

This section will discuss some popular sensors for IoT applications. They will include the following sensors: (i) accelerometer, (ii) barometric pressure, (iii) force, (iv) gas, (v) Hall effect, (vi) light, (vii) magnetic field, (viii) PIR motion, (ix) rainfall, (x) capacitive soil moisture, (xi) sound, (xii) temperature, (xiii) touch, (xiv) ultrasonic, (xv) vibration, and (xvi) water flow.

11.3.1 Accelerometer Sensor

In general, capacitive accelerometer sensors have two parallel plates and a diaphragm. As the sensor experiences acceleration, the diaphragm moves and changes the distances between the parallel plates. This results in a change in the differential capacitance of the plate–diaphragm–plate assembly, which is proportional to the acceleration on the sensor.

The Analog Device ADXL377 is a three-axis capacitive accelerometer. The ADXL377 accelerometer is a micro-electro-mechanical system (MEM)-based sensor. It has a mass suspended by springs on a silicon substrate. As the accelerating force is applied to the sensor, the suspended mass moves relative to fixed plates. The change in displacement of the suspended mass from the fixed plates leads to a change in the capacitance of the sensor. The change in capacitance is converted to voltage by an electronic circuit in the module, and the output voltage is proportional to the acceleration experienced by the sensor.

The ADXL377 operates on single supply voltages of 1.8 to 3.6 V. The bandwidth can be selected by connecting capacitors with the XOUT, YOUT, and ZOUT pins. The bandwidth can range from 0.5 to 1300 Hz for the *X*-axis and the range of 0.5 to 1000 Hz for the *Z*-axis. In addition, the ADXL377 accelerometer can measure

TABLE 11.1
Pin Configuration of an ADXL377 Capacitive Accelerometer Sensor

Pin Number	Name	Function
1	VCC	Power supply to the module. The supply voltage is 3.3–5.0 V.
2	GND	Ground
3	XOUT	Analog output voltage for the *X*-axis.
4	YOUT	Analog output voltage for the *Y*-axis.
5	ZOUT	Analog output voltage for the *Z*-axis.
6	ST	Self-test pin

TABLE 11.2
Analog Devices Capacitive Accelerometers (with Voltage Source of 3.3 V)

Devices	Acceleration Measurement Range	Ratiometric Output
ADXL335	± 3G	0 V at −3G to 3.3 V at 3G and 1.65 V at 0G
ADXL326	± 16G	0 V at −16G to 3.3 V at 16G and 1.65 V at 0G
ADXL377	± 200G	0 V at −200G to 3.3 V at 200G and 1.65 V at 0G

acceleration up to ±200G in the *X*, *Y*, and *Z* axes. The ADXL377 capacitive accelerometer sensor has the pin configuration shown in Table 11.1.

Apart from the Analog Devices ADXL377 accelerometer, there are other capacitive accelerators produced by Analog Devices. The devices are shown in Table 11.2. The ADXL377 can be used for detection of the following: (i) concussion and head trauma, (ii) large force event, (iii) vibration, and (iv) motion.

11.3.2 Barometric Pressure Sensor

The BMP280 is a widely used sensor for measuring absolute barometric pressure. The sensor is based on the piezo-resistive MEMS technology. Data from the sensor can be transferred to a microcontroller through a I2C interface. Since pressure changes with altitude, the BMP280 barometric pressure sensor can also be used as an altimeter with an accuracy of ±1.0 m. The sensor can measure pressures from 300 to 1100 hPa with an absolute accuracy of ±1 hPa. [Note that: 1 hectopascal (hPa) equals 100 pascals.] Because of its small size ($2.0 \times 2.5 \times 0.93\,\text{mm}^3$), the BMP280 is suitable for mobile devices' applications. The BMP 280 can be used for the following: (i) floor detection, (ii) weather forecasting, and (iii) vertical velocity indicator. The pressure sensor has the pin configuration shown in Table 11.3.

Example 11.1 Atmospheric Pressure and Elevation above Sea Level

Table 11.4 shows the atmospheric pressure versus elevation above sea level of a barometric pressure sensor. Determine the atmospheric pressure at the following

TABLE 11.3
Pin Configuration of BMP280 Barometric Pressure Sensor

Pin Number	Name	Function
1	GND	Ground
2	CSB	Chip select
3	SDI	Serial data
4	SCK	Serial clock input
5	SDO	Serial data output
6	VDDIO	Digital interface supply
7	GND	Ground
8	VDD	Analog supply voltage

TABLE 11.4
Elevation versus Atmospheric Pressure of a Barometric Pressure Sensor

Elevation (m)	Atmospheric Pressure (kPa)
0	100
2000	78
4000	62
6000	48
8000	35
10000	25
12000	20
14000	15

elevations: (i) 3000 m and (ii) 10,000 m. In addition, determine the equation of best fit between elevation at sea level and atmospheric pressure, assuming a polynomial fit.

Solution

MATLAB® PROGRAM

```
% Example 11.1
% Elevation versus Atmospheric Pressure of Barometric Pressure
Sensor
elev = 0:2000:14000; % Elevation above sea level in meters
atm_pr = [100 78 62 48 35 25 20 15]; % Atmospheric pressure in
kPa
n = length(elev);
%
% coefficient
pfit = polyfit (elev, atm_pr, 2);
```

```
% Equation is of the form atm_pr = c*elev*elev + b*elev + a
a = pfit(3)
b = pfit(2)
c = pfit(1)
%
for i = 1:n
    atm_fit(i) = c*elev(i)*elev(i) + b*elev(i) + a;
end
% Plot R versus T and best fit  model
plot (elev, atm_fit, 'k', elev, atm_pr, 'k*')
xlabel ('Elevation above Sea Level in meters')
ylabel('Atmospheric Pressure in kPa')
title('Best Fit Model of Atmospheric Pressure versus Elevation')
```

Answers

Atmospheric pressure at 3000 m, pr3000 = 70.2098 kPa.
Atmospheric pressure at 10,000 m, pr10,000 = 25.8036 kPa.
a = 99.3750.
b = −0.0107.
c = 3.3780e-07.

The equation of best fit is as follows:

$$\text{atm\_pr} = 3.38e-07(\text{elev})^2 - 0.01(\text{elev}) + 99.38 \tag{11.1}$$

where

atm_pr is the atmospheric pressure in kPa.
elev is the elevation above sea level in meters.

Figure 11.2 is a plot of atmospheric pressure versus elevation.

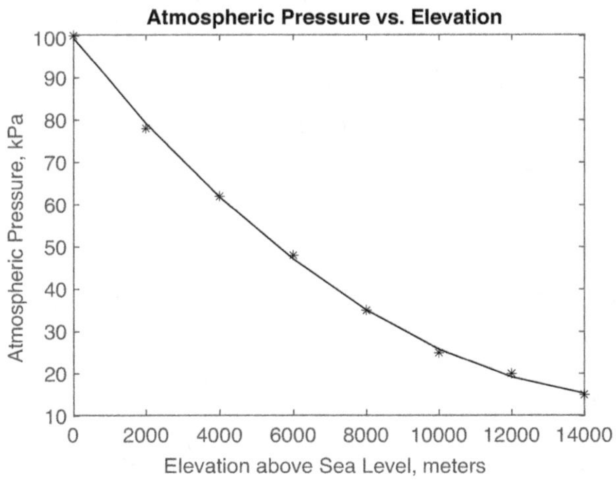

FIGURE 11.2 Plot of elevation above sea level versus atmospheric pressure of a barometric pressure sensor.

11.3.3 Force Sensor

Force sensitive resistors (FSRs) are two wire devices whose resistance decreases with increased force applied to the devices. The Interlink Electronics FSR402 is a force sensor with a diameter of 18.28 mm. It is made of two membranes, formed from a polymer thick film, that are separated by a thin airgap. The airgap is maintained by an adhesive around the edges. One of the membranes has two sets of electrically distinct interdigitated fingers, with each set connecting to one trace at the tail end. The second membrane is coated with FSR ink. When the membrane is pressed, the FSR ink bridges the two traces together and decreases the resistance. The more force is applied to the surface of the sensor, the resistance of the sensor decreases.

For the FSR 402, the resistance with no pressure is infinite (open circuit); it is about 100 kΩ with light pressure and about 200 Ω with maximum pressure. The range of force that can be measured is from 0 to 20 pounds (0 to 100 N). The force versus resistance characteristics of the FSRs can vary by 10% from one sensor to another. Thus, whereas the FSR sensors can detect weight, the FSR sensors cannot detect exact weight. However, they are good for detecting if an object has been pushed or squeezed.

Apart from FSR 402, there are other Interlink Electronic FSR sensors. They include FSR400, FSR404, FSR406, and FSR408. The most common Interlink FSRs are the FSR402 and FSR406. The FSR technologies have applications in the design of electronic drums and handheld gaming devices. Example 11.3 shows the resistance versus force of an FSR.

Example 11.2: Force versus Resistance of an FSR

Table 11.5 shows force versus resistance of an FSR. If the circuit, shown in Figure 11.3, is used to obtain the output voltage, plot the output voltage versus resistance of the FSR. Assume that RM = 5.0 kΩ and $V_{CC} = 5$ V

TABLE 11.5
Force versus Resistance of a Force Sensitive Resistor

Force (g)	Resistance, R_{FSR} (Ω)
40	60,000
50	35,000
60	27,000
70	18,000
80	11,000
90	8000
100	5000
200	90
300	63
400	50
500	40
600	37
700	34

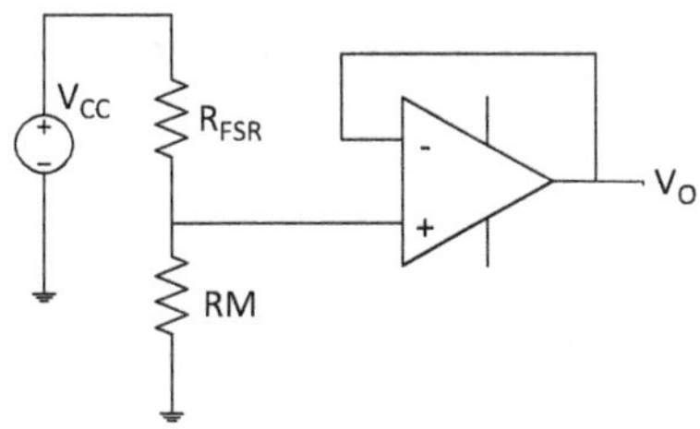

FIGURE 11.3 Operational amplifier circuit for obtaining the output voltage of a force-sensitive resistor.

Solution

Using Ohm's Law and noting that the op amplifier circuit is a non-inverting amplifier, the output voltage can be expressed as follows:

$$V_O = V_{CC} * \left(\frac{RM}{RM + R_{FSR}} \right) \tag{11.2}$$

MATLAB® CODE

```
% Solution to Example 11.2.
% Output voltage versus resistance of force sensitive resistance
%
rm = 5000; % resistance RM
vcc = 5; % source voltage
f = [40 50 60 70 80 90 100 200 300 400 500 600 700]; % force data
rfs = [60e3 35e3 30e3 20e3 13e3 9e3 5e3 90 63 50 40 37 34]; %
force sens res
len = length(rfs);
% vo is the output voltage
for i=1:len
    vo(i) = vcc*rm/(rm+rfs(i));
    end
figure (1)
plot(f,rfs,f,rfs,'*')
title('Force vs. Resistance of Force Sensitive Resistor')
xlabel('Force in grams')
ylabel('Resistance in Ohms')
figure (2)
plot(rfs,vo,rfs,vo,'*')
title('Resistance of FSR versus Output Voltage')
xlabel('Resistance in Ohms')
ylabel('Output Voltage in Volts')
```

Figure 11.4 shows the plot of the force versus resistance of the FSR, and Figure 11.5 shows the plot of the resistance of the FSR versus output voltage of the operational amplifier circuit shown in Figure 11.3.

11.3.4 Gas Sensors

Gas sensors can be used to detect various gases. The most common types of gas sensors are the MQ series. Table 11.6 shows the MQ series sensors and the corresponding

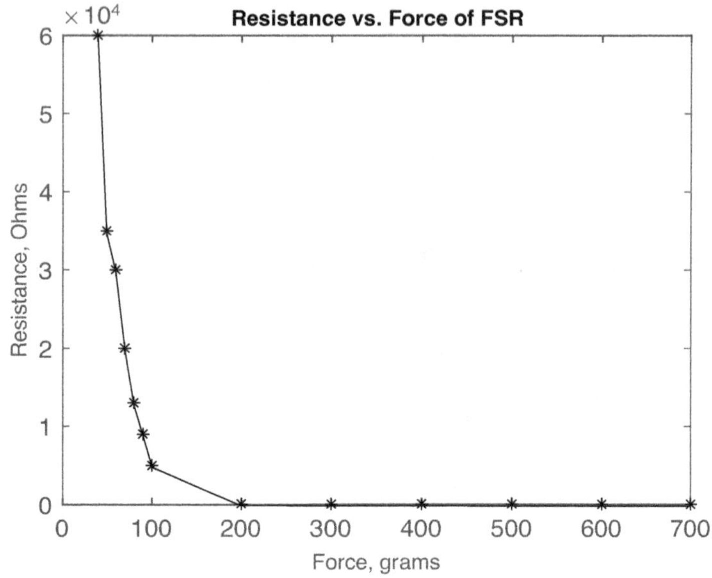

FIGURE 11.4 Force versus resistance of a force-sensitive resistor.

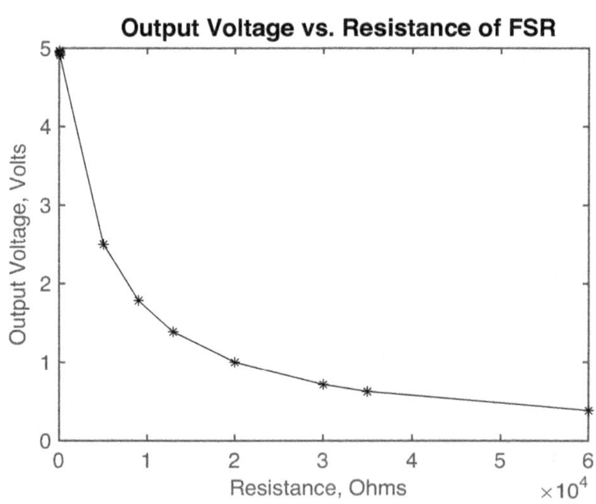

FIGURE 11.5 Resistance of a force-sensitive resistor versus output voltage.

gases that they can detect. The MQ-series sensors usually have a built-in heater. Thus, one has to wait some time to get stable readings. Only the MQ-2 sensor will be discussed in this book.

The MQ-2 is sensitive to LPG, propane, and hydrogen. The MQ-2 is a metal oxide semiconductor sensor. It has a micro-Al_2O_3 ceramic tube, a layer of tin oxide (SnO_2),

TABLE 11.6
Some MQ Series Gas Sensors

MQ Series Number	Gas That Can Be Detected
MQ-2	Smoke sensor
MQ-3	Alcohol sensor
MQ-4	Methane sensor
MQ-5	LPG gas sensor
MQ-6	Isobutane propane sensor
MQ-7	Carbon monoxide sensor
MQ-8	Hydrogen sensor
MQ-9	Carbon monoxide sensor
MQ-135	Air quality sensor

TABLE 11.7
Pin Configuration of MQ-2 Gas Sensor

Pin Number	Name	Description
1	VCC	Power supply to the sensor (5.0V DC)
2	GND	Ground
3	D0	Digital output. It is LOW when water is detected and HIGH otherwise.
4	A0	Analog output voltage varies from 0 to 5V based on the gas intensity.
5	H1	Pins 5 and 6 are used to provide a heating element to the sensor.
6	H2	Pins 5 and 6 are used to provide a heating element to the sensor.

and nickel–chromium alloy, which is used as a heater coil. The MQ-2 smoke sensor has SnO_2, which has low conductivity in a smoke-free environment. The SnO_2 is sensitive to gas. When the gas concentration increases, the device conductivity increases, and the device converts conductivity to a corresponding output signal. The operating voltage of the sensor is 5 V, and the analog output voltage is from 0 to 5 V depending on the gas concentration level. The digital output is either 0 or 5 V. The digital output can be used to detect the presence or absence of a gas. The analog output can be used to measure the gas concentration in ppm. The preheat duration of the sensor is about 20 seconds. Table 11.7 shows the pin configuration of MQ-2 Gas Sensor.

11.3.5 Hall Effect Sensor

The Hall effect sensor is based on the phenomenon whereby when a magnetic field is applied perpendicular to the direction of the current flow, it causes a voltage to be induced at right angles to both the current and magnetic field.

In the Hall effect sensor, the sensor has a constant current supplied to it, and a magnetic field perpendicular to the current induces a Hall voltage across the conductor. The voltage measured is proportional to the magnetic field strength.

TABLE 11.8
Pin Configuration of A1104 Hall Effect Sensor with LH Package

Pin Number	Name	Function
1	VCC	Connects power supply to the chip
2	VOUT	Output from the sensor
3	GND	Ground

The Allegro A1104 Hall effect sensor has a silicon chip with a voltage regulator, Hall voltage generator, amplifier, Schmitt trigger, and NMOS output transistor. The supply voltage of the A1104 Hall effect sensor ranges from 3.8 to 24 V. A typical output voltage is about 215 mV with a maximum voltage of 400 mV. The sensor has continuous-time operation.

The sensor has a variety of applications including (i) position sensing, (ii) current sensing, and (iii) speed detection. With regard to position sensing, by placing a magnet on a moving part and the Hall effect sensor on a stationary part, the relative distance of the moving and stationary part can be determined from the change in the sensor's output voltage. For current sensing, when a current passes through a wire, it creates a magnetic field. This is measured by the Hall effect sensor. This is a non-invasive current sensing method. For speed detection, a magnet may be mounted on a rotating object, and the number of revolutions per second can be counted by using the Hall sensor. The A1104 Hall effect sensor has the pin configuration shown in Table 11.8.

11.3.6 Light Sensor

A BH1750 Light sensor can be used to measure ambient light. It has an integrated circuit with the following: (i) A/D converter for converting analog illuminance into digital data; (ii) oscillator, which is used to clock the internal logic of the device; and (iii) photodiode, whose response is close to that of the human eye. Since it is a 16-bit ambient light sensor, it can measure ambient light with a minimum of 1 Lux and a maximum of 65535 Luxes. Table 11.9 shows the pin configuration of the BH1750 ambient light sensor.

The sensor uses the I2C communication protocol to create an interface with microcontrollers. The digital data are transmitted to a microcontroller through I2C pins SCL and SDA. SCL is used to provide clock pulse, and SDA is used to transfer the digital luminance value. The sensor supports continuous measurement or one-time measurement modes. In a continuous mode, ambient light is continuously measured. However, in the one-time mode, the sensor measures the ambient light value once and powers down. The spectral response of the sensor is similar to that of the human eye, and it is less affected by infrared radiation. The BH1750 ambient light sensor can be used to adjust brightness in a mobile phone and turn on/off automobile headlights based on ambient light conditions.

TABLE 11.9
Pin Configuration of a BH1750 Ambient Light Sensor

Pin Number	Name	Description
1	VCC	Power supply to the sensor is typically 3V. It can range from 2.4 to 3.6V.
2	GND	Ground
3	SCL	Serial Clock Line used to provide a clock for the I2C communication.
4	SDA	Serial Data Address is used to transfer data from the sensor to the microcontroller through I2C communication.
5	ADDR	A Device Address pin is used to select when more than two modules are connected.

TABLE 11.10
Pin Configuration of HMC5883L Magnetic Field Sensor

Pin Number	Name	Function
1	GND	Ground
2	VDD	Supply voltage (2.16–3.6V)
3	VDDIO	Supply voltage for the IO interface
4	DRDY	Data ready interrupt
5	SCL	I2C clock.
6	SDA	I2C data

11.3.7 Magnetic Field Sensor

The Compass module 3-axis HMC5883L magnetic field sensor is designed to measure low-field magnetic fields. The HMC5883L sensor has the following: (i) HMC 118X series magnetic-resistive sensors; (ii) ASIC containing amplifiers, strap drivers, offset cancellation circuit, and a 12-bit ADC. With applied voltage supply, the sensor converts an incident magnetic field into a differential voltage output. The sensor has low-voltage operation, and it is designed to measure both the direction and magnitude of the Earth's magnetic field from milli-gauss to 8 Gauss.

The sensor has two different power supplies. One of the supply voltages, VDD, is the power supply for internal circuitry of the sensor. The second voltage supply, VDDIO, is the power supply for the input–output interface. The voltages VDD and VDDIO can have the same value. But, VDDIO can be lower than VDD to allow compatibility of the sensor with other devices in the board.

The sensor has a 7-bit serial address and supports I2C communication protocols. The device support standard-mode (100 KHz) and fast-mode (400 KHz) operations. For these two modes of operation, and external pull-up resistors are required.

The HMC5883L magnetic field sensor can be used for navigation systems. It is found in mobile phones and personal navigation devices. The magnetic field sensor has the pin configuration shown in Table 11.10.

TABLE 11.11
Pin Configuration of the HC-SR501 PIR Motion Sensor

Pin Number	Name	Description
1	VCC	Power supply to the sensor (5.0 V DC)
2	D0	Digital output. It is HIGH when motion is detected and LOW otherwise.
3	GND	Ground

11.3.8 PIR Motion Sensor

An HC-SR501 passive infrared (PIR) motion sensor can detect the presence of human beings or animals. The sensors can be used in applications where light intensity is used to switch on or off a device. It can sense distances of up to 7 m within 120°. The sensing distance may be adjusted for a distance between 3 and 7 m. In addition, it can be set for repeat triggers or a single trigger. Furthermore, the sensor can operate between the temperatures of –20°C and +80°C. Table 11.11 shows the pin configuration of HC-SR501 PIR Motion Sensor.

The typical value of VCC is 5 V but can work from 4.5 to 12 V. A digital pulse of 3.3 V will be triggered when motion is detected and a digital low (0 V) when there is no motion.

11.3.9 Rainfall Sensor

An FC-37 Raindrop sensor can be used to sense rain. It is also known as YL-83. The sensor has (i) an electronic control board and (ii) a rain collector board that collects water drops. The electronic control board has a built-in potentiometer for sensitivity adjustments of the digital output. It also has a power LED that lights up when the sensor is turned on. In addition, it has a digital output LED which lights up when there is a digital output. Its working voltage is 5 V. Furthermore, it has both digital and analog outputs. There is a digital pin for the digital output and an analog pin for the analog output.

The rain collector board has been nickel-coated in the form of lines on the board. When a rain drop is present, the resistance of the collector board decreases because water is a conductor and the presence of water connects the nickel lines in parallel, reducing the voltage drop across the resistance. The more intense the rainfall, the lower the resistance and lower the output voltage. Figure 11.6 is a simplified diagram of a rainfall collector board of an FC-37 Raindrop sensor. Table 11.12 shows the pin configuration of an FC-37 Raindrop sensor.

11.3.10 Capacitive Soil Moisture Sensor

Capacitive soil moisture sensors normally have two parallel plates that use soil as the dielectric material. The advantage of using a capacitive soil sensor is that it has the ability to minimize the influence of ionic activities in the soil, compared to resistive soil moisture sensors.

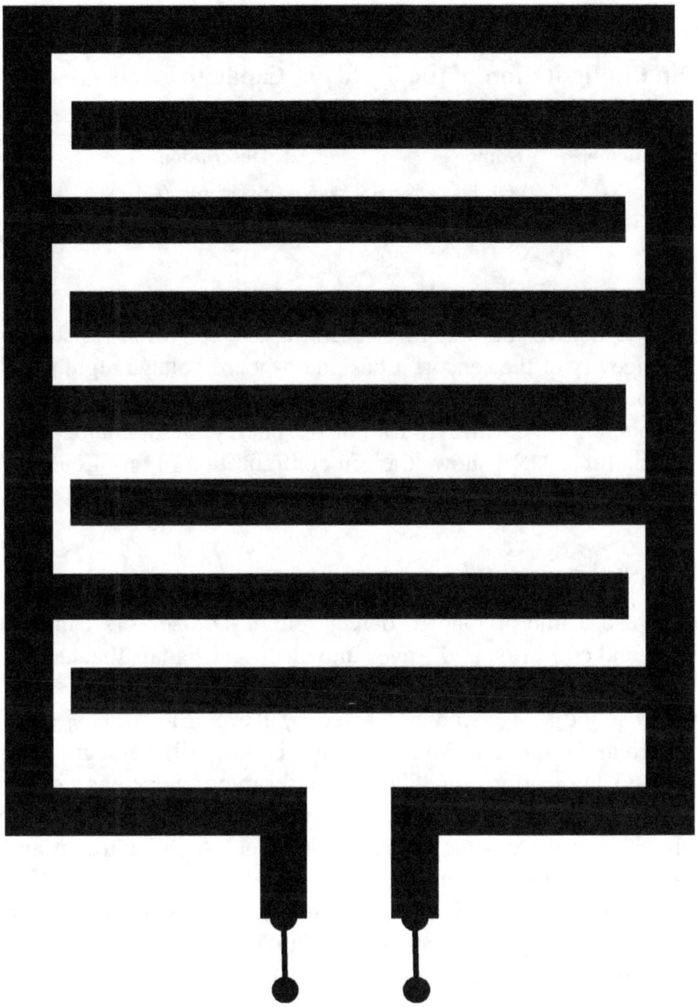

FIGURE 11.6 Simplified diagram of a rain collector board.

TABLE 11.12
Pin Configuration of an FC-37 Raindrop Sensor

Pin Number	Name	Description
1	VCC	Power supply to the sensor (3.0–5.0 V DC)
2	GND	Ground
3	D0	Digital output. It is LOW when water is detected and HIGH otherwise.
4	A0	Analog output. It varies with the amount of water detected.

TABLE 11.13
Pin Configuration of the SEN0193 Capacitive Soil Moisture Sensor

Pin Number	Name	Description
1	VOUT	Analog output voltage from 0 to 3 V
2	VCC	Power supply (3.0–5.5 V DC)
3	GND	Ground

SEN0193, a capacitive soil moisture sensor, uses a corrosion-resistant material to increase the longevity of the sensor. It has an on-board voltage regulator that allows the sensor to operate from a 3.3 to 5.5-V power source and output DC voltages from 0 to 3 V. For compatibility with Arduino or Raspberry Pi, an analog-to-digital converter is needed. Table 11.13 shows the pin configuration of an SEN0193 capacitive soil moisture sensor.

11.3.11 Sound Sensor

Sound sensors use a microphone to detect sound. The sensors can also measure sound intensity and convert sound waves into electrical signals that can be amplified or processed or connected to a microcontroller.

The KY-038 sound sensor, manufactured by Keyes Electronics, consists of the following: (i) condenser microphone, (ii) two LEDs, (iii) comparator operational amplifier, and (iv) a potentiometer. The condenser microphone has a diaphragm that vibrates in response to sound waves that impinge on the microphone. The movement of the diaphragm and the change in spacing between the diaphragm and the other plate of the condenser microphone produces an electrical signal that corresponds to the sound waves picked up by the microphone. Of the two LEDs of the KY-038 sound sensor, one LED lights up when the module is connected to a power supply. The other LED turns on when an incoming audio signal exceeds a threshold value set for an application.

The sound sensor can be powered by a 3.3 or 5.0-V DC source. The LM393 op amp comparator converts an incoming analog signal from the microphone into a digital signal. There is a 10-kΩ potentiometer that is used to set a reference voltage for the op amp. In addition, the potentiometer is used to generate a reference voltage for the analog output of the module. The

The KY-038 sound sensor module has the pin configuration shown in Table 11.14. The sound sensor can be used to detect sounds for security systems and can also be used in speech recognition systems where sound is converted into text.

11.3.12 Temperature Sensors

DHT11 and DHT22 are temperature and humidity sensors, respectively. The sensors have a capacitor for humidity and a thermistor for temperature measurements. They have analog-to-digital converters that provide digital output signals for temperature

TABLE 11.14
Pin Configuration of the KY-038 Sound Sensor

Pin Number	Name	Function
1	VCC	Power supply to the module. The supply voltage is 3.3–5.0 V.
2	DOUT	Digital output LOW indicates no sound detected by the sensor, and HIGH shows that sound has been detected.
3	AOUT	Analog output that gives an analog reading directly from the microphone.
4	GND	Ground

TABLE 11.15
Comparison between DHT11 and DHT22 Temperature Sensors

	DHT11	DHT22
Cost	Ultralow cost	Low cost
Power	3–5 V power and I/O	3–5 V power and I/O
Current	2.5 mA maximum during conversion	2.5 mA maximum during conversion
Humidity range	Good for 20%–80% humidity with 5% accuracy	Good for 0%–100% humidity with 2% to 5% accuracy
Temperature range	Good for 0°C to 50°C temperature readings with ±2 °C accuracy	Good for −40°C to 80°C temperature readings with ±0.5°C accuracy
Sampling rate	1 Hz	0.5 Hz

TABLE 11.16
Pin Configuration of the DHT11 Temperature Sensor

Pin Number	Name	Description
1	VCC	Power supply (3.0–5.5 V DC)
2	DATA	Digital data output
3	NC	Not connected
3	GND	Ground

and humidity measurements. Table 11.15 shows the comparison between DHT11 and DHT22. It can be seen from Table 11.15 that DHT22 has more accurate readings than DHT11. In addition, DHT22 has a larger range for humidity and temperature measurements. Table 11.16 shows the pin configuration of the DHT11 temperature sensor.

It is recommended that (i) a pull-up resistor (typically 4.7 to 10 kΩ) be connected between the DATA pin and VCC. In addition, a cable length of less than 20 m should be used to prevent signal degradation.

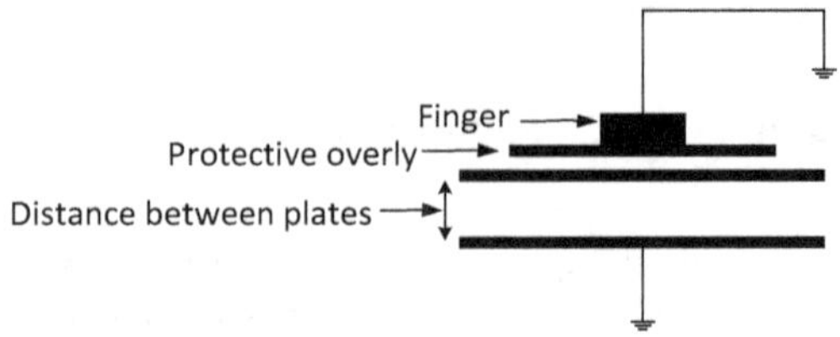

FIGURE 11.7 Simplified structure of capacitive switch sensor TTP223.

11.3.13 Touch Sensor

Sunrom Electronics TTP223 is a capacitive switch sensor. It is also called a touch button or touch switch. The sensor consists of two capacitances, as shown in Figure 11.7. The capacitance C_S is formed by two plates with a dielectric between them. The switch sensor uses the human body as part of the circuit. The finger capacitance, C_f, is a capacitance induced when a finger touches a protective overlay of the sensor grid layer. The total capacitance, C_T,is given as follows:

$$C_T = C_s + C_f \tag{11.3}$$

where

C_S is the capacitance between the touch pad and the ground.
C_f is the capacitance induced by the finger.

The touch sensor module has an oscillator whose frequency of oscillation depends upon the capacitance of the touch sensor module. When the sensor pad is touched, there is a change in capacitance that leads to a change in the oscillator frequency. A frequency detector circuit tracks the oscillator frequency, and when the frequency shift is above a threshold value, the circuit triggers an event, showing that the sensor pad has been touched.

The touch sensor module accepts a power supply voltage of 3.3 to 5.5 V DC. The output can be configured to be active HIGH or LOW. In addition, the output can be configured to toggle between ON/OFF or be momentarily ON when the pad is touched. The touch area of the TTP223 capacitive touch switch module is about $11 \times 10.5\,\text{mm}^2$. The sensor module has an LED that lights up when the module is connected to a power supply source. The TTP223 touch sensor module has the pin configuration shown in Table 11.17.

The TTP223 touch sensor can be used to replace direct button key, keypads, and mechanical switches. In addition, the TTP223 touch sensor is used in mobile phones, iPods, and small home appliances.

TABLE 11.17
Pin Configuration of the TTP223 Touch Switch Sensor

Pin Number	Name	Function
1	SIG	The output voltage can be configured for Active HIGH or Active LOW. In addition, the output voltage can Toggle or be Momentarily ON.
2	VCC	Power supply to the module. The supply voltage is 3.3–5.0 V.
3	GND	Ground

11.3.14 Ultrasonic Sensor

The HC-SR04 ultrasonic sensor has an ultrasonic transmitter, receiver, and a control circuit. The sensor works by sending a sound wave from the ultrasonic transmitter. The waves bounce off from an object and return to an ultrasonic receiver. When the transmitter is activated, it transmits a burst of eight pulses at 40 kHz. The pulse pattern makes the transmitter signal unique and allows the receiver to discriminate the transmitter pattern from other ultrasonic background noise signals. The sensor can measure the object at distances that range from 2 to 400 cm (1 in to 13 ft).

For the transmitter to generate the ultrasonic pulses, the TRIG pin is set HIGH for 10 μs. The transmitter sends out eight cycles of ultrasonic burst, the ECHO pin is set as HIGH when the ultrasonic burst is sent, and it starts detecting for reflected pulses from an object. If there is no object or reflected pulse, the ECHO pin will time out after 38 ms and go back to the LOW state. If a reflected pulse is received, the ECHO pin will time out sooner than 38 ms. Based on the amount of time the ECHO pin was set as HIGH, one can determine the distance the sound traveled.

If Δt is the time for the ultrasonic signal to be transmitted and received at the ECHO input, then the distance d from the transmitter to the reflecting object is given as

$$d = (\Delta t)(V_{\text{sound}})/2 \qquad (11.4)$$

where

V_{sound} is the velocity of sound in air (340 m/s)

The ultrasonic sensor can be used for: (i) detecting objects, (ii) measuring the distance to a nearest object, and (iii) avoiding objects in robotic projects. The HC-SR04 ultrasonic sensor module has the pin configuration shown in Table 11.18.

11.3.15 Vibration Sensor

The SW-420 Vibration sensor module is a nondirectional vibration sensor. The module consists of a vibration sensor, LM 393 voltage comparator, a potentiometer, LEDs, current limiting resistor, and capacitors for biasing and noise filtering. The SW-420

TABLE 11.18
Pin Configuration of the HC-SR04 Ultrasonic Sensor

Pin Number	Name	Function
1	VCC	Power supply to the module. The working voltage is 3.3–5.0 V.
2	TRIG	Trigger pulse that prompts the transmitter to send ultrasonic pulses.
3	ECHO	Echo pulse output.
4	GND	Ground

TABLE 11.19
Pin Configuration of the SW-420 Vibration Sensor

Pin Number	Name	Function
1	VCC	Connects power supply to the module
2	GND	Ground
3	DO	Digital output. It passes the digital output of the built-in LM393 comparator

Vibration sensor detects the magnitude of a vibration. When there is a vibration, the sensor closes or opens an electrical contact. The triggered switch can be a relay or a semiconductor device. By default, the vibration sensor is in conduction state when there is no vibration (i.e., the switch is closed). When vibration is detected, the contacts open and the resistance increases. The increased resistance causes a pulse to be generated. The generated pulse is passed to the LM393 voltage comparator, whose output is connected to the digital output pin of the module.

The 10-kΩ potentiometer is used to adjust the sensitivity of the sensor. There are two in-built LEDs. There is a power LED to indicate that power has been supplied to the module, and there is a status LED that indicates the status of the digital output. The operating voltage of the module is from 3.3 to 5.0 V. The vibration sensor can be used to detect earthquakes and moving objects. The SW-420 Vibration sensor module has the pin configuration shown in Table 11.19.

11.3.16 Water Flow Sensor

YS-S201 is a water flow measurement sensor. The sensor has a plastic valve body, rotor, and Hall effect sensor. The sensor is connected between water inlet and outlet valves. When water flows through the sensor, the flowing water forces the rotor, which has a magnet attached to it to turn. The water flow rate interferes with the magnetic flux, which is sensed by a Hall effect sensor. The Hall effect sensor generates output pulses proportional to the magnetic flux with every revolution of the rotor. For every rotation of the water rotor, the Hall effect sensor produces a pulse signal at the output pin that is proportional to the water flow rate.

TABLE 11.20
Pin Configuration of the YS-S201 Water Flow Sensor

Pin Number	Name	Function
1	VCC (red wire)	Supply voltage from 4.5 to 24 V
2	GND (black wire)	Ground
3	Output (yellow wire)	Output voltage of the sensor

The flow sensor can measure the water flow rate from 1 to 30 L/min. The normal working voltage is from 4.5 to 24 V. The water temperature should be less than 120°C, and the water pressure should be less than 2.0 MPa. The sensor can be used to monitor water usage and to regulate the flow of water in water heaters and vending machines. The flow sensor has the pin configuration shown in Table 11.20.

Example 11.3: Flow Rate versus Frequency of the Water Flow Sensor

Table 11.21 shows the flow rate versus pulse frequency of the YS-S201Water Flow Sensor. Determine the line of best fit between the flow rate and the frequency of the pulses.

Solution

MATLAB® CODE

```
% Solution to Example 11.3
% Water Flow SensorT
frate = 120:120:720; % flow rate in liters per hour
freq = [16 32.5 49.3 65.5 82 92.2]; % frequency in Hz
% D = -30e-3:10e-3:30e-3;
% V = [-5.2e-3 -4.7e-3 -2.9e-3 0 1.8e-3 4.8e-3 5.3e-3];
len = length(frate);
%
% Linear equation is y = mx + b
% coefficient
vfit = polyfit (frate, freq, 1);
b = vfit(2);
m = vfit(1);
%
vfit = m*frate + b;
% Plot frate versus freq and best fit  model
plot (frate, vfit, 'k', frate, freq, 'k*')
xlabel ('Flow Rate in Liters/hour')
ylabel('Frequency in Hz')
title('Frequency of Water Flow Sensor versus Flow Rate')
% slope of equation
m
% constant term of equation
b
```

TABLE 11.21
Flow Rate versus Frequency of Pulses from the YS-S201 Water Flow Sensor

Flow Rate (L/h)	Frequency (Hz)
120	16.0
240	32.5
360	49.3
480	65.5
600	82.0
720	92.2

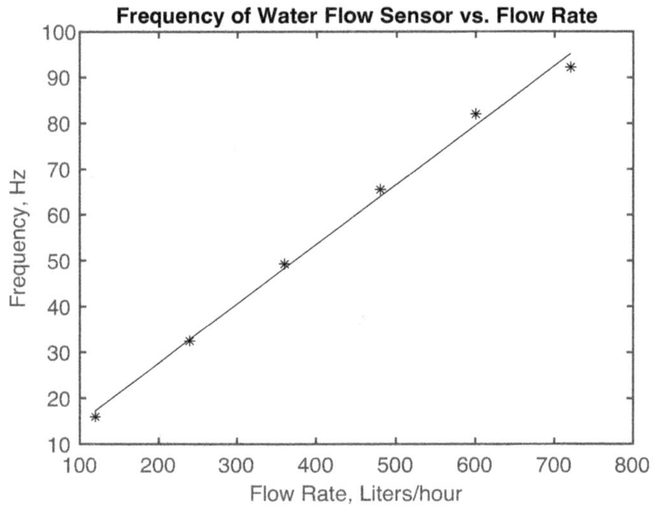

FIGURE 11.8 Flow rate versus frequency of the water flow sensor.

Answers

$m = 0.1299$; $b = 1.6800$.

The equation of the line is given as

$$\text{freq} = 0.13(\text{frate}) + 1.68 \tag{11.5}$$

where

freq is the frequency in Hz.
frate is the flow rate in L/hr.

Figure 11.8 shows the line of best fit.

PROBLEMS

Problem 11.1 Table P11.1 shows the flow rate versus pulse frequency of a water flow sensor. Determine the line of best fit between the flow rate versus the frequency of the pulses.

Problem 11.2 Table P11.2 shows the atmospheric pressure versus elevation above sea level of a barometric pressure sensor.

Determine the atmospheric pressure at the following elevations: (i) 5000 ft and (ii) 2100 ft. In addition, determine the equation of best fit between elevation at sea level versus atmospheric pressure.

TABLE P11.1
Flow Rate versus Frequency of Pulses from a Water Flow Sensor

Flow Rate (L/h)	Frequency (Hz)
100	19.81
200	37.73
300	55.93
400	73.74
500	91.81
600	109.69
700	127.75
800	145.71

TABLE P11.2
Elevation versus Atmospheric Pressure of a Barometric Pressure Sensor

Elevation (ft)	Atmospheric Pressure (psi)
0	14.8
2000	13.6
4000	12.6
6000	11.7
8000	10.8
10,000	10.2
12,000	9.3
14,000	8.7
16,000	8.1
18,000	7.2
20,000	6.9
22,000	6.1
24,000	5.6
26,000	5.1
28,000	4.9
30,000	4.3

Problem 11.3 At atmospheric pressure, the air densities at different temperatures are shown in Table P11.3.

Find the equation of best fit.

Problem 11.4 Table P11.4 shows force applied to an FSR sensor versus output voltage of the circuit shown in Figure P11.1. If the RM = 10.0 kΩ and $V_{CC} = 5$ V, determine the resistances of the FSR for various forces applied to the sensor.

Problem 11.5 Table P11.5 shows force versus resistance of an FSR. If the circuit, shown in Figure P11.1, is used to obtain the output voltage, plot the output voltage versus resistance of the FSR. Assume that RM = 2.0 kΩ and $V_{CC} = 5$ V.

TABLE P11.3
Temperature versus Air Density at Atmospheric Pressure

Temperature (°C)	Air Density (kg/m³)
0	1.10
100	0.95
200	0.75
300	0.60
400	0.56
500	0.44
600	0.40
700	0.37

TABLE P11.4
Force versus Output Voltage of the Operational Amplifier Circuit

Force (g)	Output Voltage (V)
100	1.7
200	2.2
300	2.5
400	2.7
500	2.9
600	3.1
700	3.2
800	3.3
900	3.4
1000	3.5

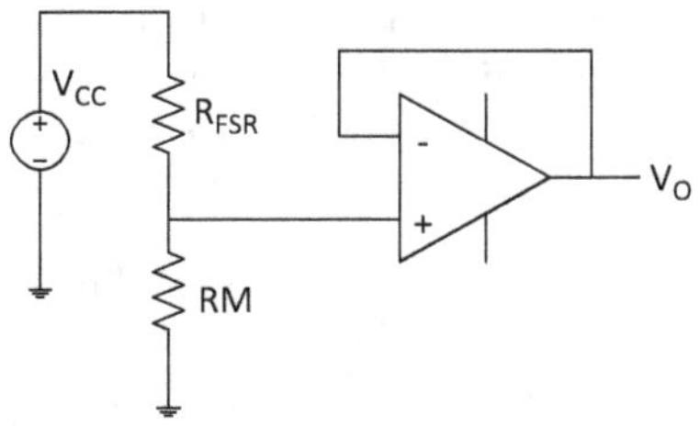

FIGURE P11.1 A circuit for obtaining the output voltage of a force-sensitive resistor.

TABLE P11.5
Force versus Resistance of a Force-Sensitive Resistor

Force (g)	Resistance (Ω)
20	30,000
50	10,000
100	6000
500	2000
1000	1100
5000	400
7000	300
10,000	150

Problem 11.6 Table P11.5 shows force versus resistance of an FSR. If the relationship between force and resistance is modeled by a power relation $R = a(\text{Force})^n$, find a and n.

Use
$\text{Log}(R) = \log(a) + n\log(\text{Force})$

BIBLIOGRAPHY

Ada, Lady, "DHT11, DHT22 and AM2302 Sensors," Sensors/Weather, January 22, 2025. https://learn.adafruit.com/dht/overview. (Accessed September 5, 2025).

Ajiboye, Aye, Opadiji, Jayeola Femi, Abdulrahman Yusuf and Olusogo Popoola, "Analytical Determination of Load Resistance Value for MQ-series Gas Sensors: MQ-6 as Case Study. *TELKOMNIKA Telecommunication Computing Electronics and Control*, Vol. 19, Issue 2, pp. 575–582, 2021.

Allegro Microsystems, A1101, A1102, A1103, A1104, A1105, A1106 Hall Effect Switches Datasheets. https://www.allegromicro.com/~/media/Files/Datasheets/A110x-Datasheet.ashx. (Accessed September 5, 2025).

Analog Devices, ADXL377 Accelerometer Data Sheet. https://www.analog.com/media/en/technical-documentation/data-sheets/ADXL377.pdf. (Accessed September 5, 2025).

Arunachalam, Rammohan and Chidambaram Kumar, "Performance Analysis of Photoresistor and Phototransistor for Automotive's Halogen and Xenon Bulbs Light Output," *IOP Conference Series: Materials Science and Engineering*, Vol. 263, p. 062056. 2017.

Attia, John O., *Electronics and Circuit Analysis Using MATLAB*, CRC Press, 2nd Edition, 2004.

Bolton, William, *Instrumentation and Control Systems*, 3rd Edition, Elsevier, 2021.

Bosch BMP 280 Digital Pressure Sensor, Data Sheet, Document Number BST-BMP280-DS001-26, October 2021. https://www.bosch-sensortec.com/media/boschsensortec/downloads/datasheets/bst-bmp280-ds001.pdf (Accessed September 10, 2025).

Chakraborty, Abhisek, Giulia Lucarelli, Jie Xu, Zeynab Skafi, Sergio Castro-Hermosa, Appaneravanda B. Kaveramma, R. Geetha Balakrishna, and Thomas Malcolm Brown, "Photovoltaics for Indoor Energy Harvesting," *Nano Energy*, Vol. 128, p. 109932, 2024.

Chapman, Stephen J., *MATLAB Programming for Engineers*, 7th Edition, Cengage Learning, May 16, 2024.

Das, Debashis, "How a KY-038 Sound Sensor works and how to Interface it with ESP32?" Circuit Digest, November 25, 2022. https://circuitdigest.com/microcontroller-projects/interface-ky038-sound-sensor-with-esp32. (Accessed September 10, 2025).

Dfrobot, "Capacitive Soil Moisture Sensor SKU:SEN0193. https://wiki.dfrobot.com/Capacitive_Soil_Moisture_Sensor_SKU_SEN0193. (Accessed September 10, 2025).

Eleobo, "How MQ-2 Gas Smoke Sensor Works: A Complete Guide to Understanding the MQ2 Gas Sensor," https://eleobo.com/how-mq-2-gas-smoke-sensor-works/. (Accessed September 10, 2025).

Fawad, "Getting Started with SW-420 Vibration Sensor using Arduino: A Step-by-Step Guide," Programming Digest, February 15, 2024. https://programmingdigest.com/getting-started-with-sw-420-vibration-sensor-using-arduino-a-step-by-step-guide/. (Accessed September 10, 2025).

Handson Technology, HC-SR04 Ultrasonic Sensor Module User Guide. https://www.handsontec.com/pdf_files/hc-sr04-User-Guide.pdf. (Accessed September 15, 2025).

Handson Technology, HC-SR501 Passive Infrared (PIR) Motion Sensor User Guide. https://handsontec.com/dataspecs/SR501%20Motion%20Sensor.pdf. (Accessed September 15, 2025).

Ida, Nathan, *Sensors, Actuators, and Their Interfaces: A Multidisciplinary Introduction*, 2nd Edition, Institution of Engineering and Technology, 2020.

Interlink Electronics, Force Sensing Resistor FSR 402, Product Datasheet. https://www.interlinkelectronics.com/fsr-402. (Accessed September 15, 2025).

Jaeger, Richard C. and Travis Blalock, *Microelectronic Circuit Design (ISE)*, 6th Edition, McGraw Hill, 2022.

Khanal, Kapalik, Grishma Ojha, Sandeep Chataut and Umesh Ghimire, "IoT-Based Real-Time Soil Health Monitoring System for Precision Agriculture," *International Research Journal of Engineering and Technology*, Vol. 11, Issue 7, pp. 470–478, 2024.

Li, Yujie, Huimin Lu, Hyoungseop Kim and Seiichi Serikawa, "Touch Switch Sensor for Cognitive Body Sensor Networks," *Journal of Computer Communications*, Vol. 146, pp. 32–38, 2019.

Mamtaz Alam, "Rain Detection using Raindrop Sensor Module & Arduino," February 2, 2025. https://how2electronics.com/rain-detection-using-raindrop-sensor-module-arduino/. (Accessed September 15, 2025).

Morris, Alan S. and Reza Langari, *Measurement and Instrumentation: Theory and Applications*, 3rd Edition, Academic Press, 2020.

Northrop, Robert B., *Introduction to Instrumentation and Measurements*, 3rd Edition, CRC Press, 2014.

Obagade, Thompson Adewumi, "Comparative Analysis of Photoelectric Characteristics of Photosensors: A Case Study between Photoresistor, Photodiode, and Phototransistor," *IOSR Journal of Applied Physics*, Vol. 14, Issue 3, pp. 1–12, 2022.

Panda, Sagarika, Savita Mehlawat, Neeraj Dhariwal, Ashwani Kumar, Amit Sanger, "Comprehensive Review on Gas Sensors: Unveiling Recent Developments and Addressing Challenges," *Materials Science and Engineering: B*, Vol. 308, Article 117616, 2024.

Parallax, Compass Module 3-Axis HMC5883L Magnetic Field Sensor. https://learn.parallax.com/kickstarts/compass-module-3-axis-hmc5883l/. (Accessed September 15, 2025).

Radi, Murtiningrum, Fajar S. Muzdrikah, M. Shohibun Nuha and Fathi A. Rizqi, "Calibration of Capacitive Soil Moisture Sensor (SKU: SEN0193)," *2018 4th International Conference on Science and Technology (ICST)*, Yogyakarta, Indonesia, pp. 1–6, 2018.

Ramachandran, Rajesh, "The Analysis of Different Types of IoT Sensors and Security Trend as Quantum chip for Smart City Management," *IOSR Journal of Business and Management*, Vol. 20, Issue 1, pp. 55–60, 2018.

Sedra, Adel S. and Kenneth Smith, *Microelectronic Circuits*, 8th International Edition, Oxford University Press, 2020.

Sharma, Anukriti, Sharad Sharma, and Dushyant Gupta, "A Review of Sensors and Their Application in Internet of Things (IOT)," *International Journal of Computer Applications*, Vol. 174, Issue 24, pp. 27–34, 2021.

Sharma, Chandrika, Nidhi Sharma, Poornima Sharma and Veerendra Kumar, "A Review of BJT Based Phototransistor," *International Journal of Engineering Research & Technology (IJERT)*, Vol. 3, Issue 4, pp. 306–310, 2014.

Siepert, Bryan, "Adafruit BH1750 Ambient Light Sensor," January 21, 2025. https://learn.adafruit.com/adafruit-bh1750-ambient-light-sensor#. (Accessed September 15, 2025).

Swagatam, "How to Use TTP223 Capacitive Touch Switch Module," Homemade Circuits Projects, February 1, 2024. https://www.homemade-circuits.com/ttp223-capacitive-touch-module-explained/. (Accessed September 15, 2025).

Thu, Si, Nilar Aung and Ko Maung, "A Calibration Technique for Water Flow Sensor YF-S201," *International Journal of Trend in Research and Development*, Vol. 5, Issue 5, pp. 152–154, 2018.

Valentine, Daniel T. and Hahn, Brian D., *Essential MATLAB for Engineers and Scientists*, 8th Edition, Academic Press, Sept. 26, 2022.

Index

www.ingramcontent.com/pod-product-compliance
Lightning Source LLC
Chambersburg PA
CBHW070626230726
48635CB00003B/36

* 9 7 8 1 0 4 1 2 8 6 1 6 5 *